Hydrogels

The demand for advanced energy devices such as high-performance batteries, supercapacitors, fuel cells, electrolyzers, and flexible/wearable devices is increasing rapidly. To meet such demand, high-performance and stable materials that could be used as active materials in these devices are much needed. This book focuses on the use of hydrogels in such emerging applications.

The main objective of this book is to provide current, state-of-the-art development in hydrogel-based materials, their applications in energy, and their future challenges. This book covers the entire spectrum of hydrogels for their applications in a range of energy devices in terms of materials, various synthetic approaches, architectural aspects, design and technology of energy devices, and challenges. This book covers the fundamentals of hydrogels, various composites of hydrogels, design concepts, different technologies, and applications in the diverse energy area. All chapters are written by experts in these areas around the world, making this a suitable textbook for students and providing new guidelines to researchers and industries working in these areas.

This book includes topics such as various approaches to synthesizing hydrogels, their characterizations, and emerging applications in the energy area. Fundamentals of energy devices, working principles, and their challenges are also covered. This book will provide new directions to scientists, researchers, and students to better understand hydrogel-based materials and their emerging applications in energy.

Series in Materials Science and Engineering

The series publishes cutting edge monographs and foundational textbooks for interdisciplinary materials science and engineering. It is aimed at undergraduate and graduate level students, as well as practicing scientists and engineers. Its purpose is to address the connections between properties, structure, synthesis, processing, characterization, and performance of materials.

Flame Retardant Polymeric Materials, A Handbook
Xin Wang and Yuan Hu

2D Materials for Infrared and Terahertz Detectors
Antoni Rogalski

Fundamentals of Fibre Reinforced Composite Materials
A. R Bunsell. S. Joannes, A. Thionnet

Fundamentals of Low Dimensional Magnets
Ram K. Gupta, Sanjay R Mishra, Tuan Anh Nguyen, Eds.

Emerging Applications of Low Dimensional Magnets
Ram K. Gupta, Sanjay R Mishra, Tuan Anh Nguyen, Eds.

Handbook of Silicon Carbide Materials and Devices
Zhe Chuan Feng, Ed.

Bioelectronics: Materials, Technologies, and Emerging Applications
Ram K. Gupta and Anuj Kumar, Eds.

Advances in 3D Bioprinting
Roger J. Narayan, Ed.

Hydrogels: Fundamentals to Advanced Energy Applications
Anuj Kumar and Ram K. Gupta, Eds.

Series Preface
The series publishes cutting edge monographs and foundational textbooks for interdisciplinary materials science and engineering.

Its purpose is to address the connections between properties, structure, synthesis, processing, characterization, and performance of materials. The subject matter of individual volumes spans fundamental theory, computational modeling, and experimental methods used for design, modeling, and practical applications. The series encompasses thin films, surfaces, and interfaces, and the full spectrum of material types, including biomaterials, energy materials, metals, semiconductors, optoelectronic materials, ceramics, magnetic materials, superconductors, nanomaterials, composites, and polymers.

It is aimed at undergraduate and graduate level students, as well as practicing scientists and engineers.

Proposals for new volumes in the series may be directed to Carolina Antunes, Commissioning Editor at CRC Press, Taylor & Francis Group (Carolina.Antunes@tandf.co.uk).

Hydrogels
Fundamentals to Advanced Energy Applications

Edited by
Anuj Kumar and Ram K. Gupta

CRC Press is an imprint of the
Taylor & Francis Group, an **informa** business

Cover Image: Shutterstock Image ID 2152184203

First edition published 2024
by CRC Press
6000 Broken Sound Parkway NW, Suite 300, Boca Raton, FL 33487-2742

and by CRC Press
4 Park Square, Milton Park, Abingdon, Oxon, OX14 4RN
CRC Press is an imprint of Taylor & Francis Group, LLC

© 2024 selection and editorial matter, Ram K. Gupta and Anuj Kumar; individual chapters, the contributors

Reasonable efforts have been made to publish reliable data and information, but the author and publisher cannot assume responsibility for the validity of all materials or the consequences of their use. The authors and publishers have attempted to trace the copyright holders of all material reproduced in this publication and apologize to copyright holders if permission to publish in this form has not been obtained. If any copyright material has not been acknowledged please write and let us know so we may rectify in any future reprint.

Except as permitted under U.S. Copyright Law, no part of this book may be reprinted, reproduced, transmitted, or utilized in any form by any electronic, mechanical, or other means, now known or hereafter invented, including photocopying, microfilming, and recording, or in any information storage or retrieval system, without written permission from the publishers.

For permission to photocopy or use material electronically from this work, access www.copyright.com or contact the Copyright Clearance Center, Inc. (CCC), 222 Rosewood Drive, Danvers, MA 01923, 978-750-8400. For works that are not available on CCC please contact mpkbookspermissions@tandf.co.uk

Trademark notice: Product or corporate names may be trademarks or registered trademarks and are used only for identification and explanation without intent to infringe.

ISBN: 9781032385129 (hbk)
ISBN: 9781032398235 (pbk)
ISBN: 9781003351566 (ebk)

DOI: 10.1201/9781003351566

Typeset in Palatino
by Deanta Global Publishing Services, Chennai, India

Contents

Preface vii
Editor Biographies ix
List of Contributors xi

Part I Fundamentals

1. Hydrogels: An Introduction to Advanced Energy Applications 3
   Jasvinder Kaur, Dipak Kumar Das, Anuj Kumar, and Ram K. Gupta

2. Hydrogels: Synthesis and Recent Advancements in
   Electrochemical Energy Storage 19
   *Dawson Blanco, Teddy Mageto, Felipe M. de Souza,
   Anuj Kumar, and Ram K. Gupta*

3. Approaches for Hydrogel Synthesis 47
   Ramji Kalidoss and Radhakrishnan Kothalam

4. Architecture of Hydrogels 67
   Tamara D. Erceg and Nevena R. Vukić

5. Characteristics of Hydrogels 83
   *Jagadeeshwar Kodavaty, Suresh Kumar Yatirajula,
   Abhishek Kumar Gupta, and Rejeswara Reddy Erva*

Part II Hydrogels for Energy Applications

6. Hydrogels and Their Emerging Applications 103
   *Atta Rasool, Muhammad A. u. R. Qureshi, Atif Islam, and
   Shumaila Fayyaz*

7. Hydrogels for 3D Frameworks 127
   *Ismail Can Karaoglu, Esra Yalcin, Ayesha Gulzar, Isilay Goktan, and
   Seda Kizilel*

8. 3D Printing of Hydrogels 147
   *Janet de los Angeles Chinellato Díaz, Santiago P. Fernandez Bordín,
   Ramses S. Meleán, and Marcelo R. Romero*

9 Hydrogels for CO_2 Reduction 169
   Jesús A. Claudio-Rizo, Martín Caldera-Villalobos,
   Denis A. Cabrera-Munguía, and María I. León-Campos

10 Recent Advancements in Hydrogels for Electrocatalytic
   Activities 189
   Allen Davis, Teddy Mageto, Felipe M. de Souza, Anuj Kumar, and Ram
   K. Gupta

11 Hydrogels for Metal-Ion Batteries 213
   Swathi Pandurangan, Dhavalkumar N Joshi, Arun Prasath Ramaswamy,
   and Vinod Kumar

Part III Emerging Applications, Challenges, and Future Perspectives

12 Hydrogels for Metal-Air Batteries 235
   Debasrita Bharatiya, Biswajit Parhi, Sachit Kumar Das,
   and Sarat K Swain

13 Hydrogels for Flexible/Wearable Batteries 257
   Runwei Mo

14 Hydrogels for Flexible/Wearable Supercapacitors 273
   Jintao Guo, Xiaodong Qin, Wenkun Jiang, and Yinghui Han

15 Hydrogels for Wearable Electronics 293
   Zahra Karimzadeh, Mansour Mahmoudpour, Abolghasem Jouyban, and
   Elaheh Rahimpour

16 Hydrogels for the Removal of Water Content from
   Liquid Fuels 311
   Letícia Arthus, Patrícia Bogalhos Lucente Fregolente,
   Maria Regina Wolf Maciel, and Leonardo Vasconcelos Fregolente

17 Current Challenges and Perspectives of Hydrogels 335
   Dimpee Sarmah, Ashok Bora, and Niranjan Karak

Index 353

Preface

With the rapid development of technology and the advancement of lifestyle, the demand for advanced energy devices such as high-performance batteries, supercapacitors, fuel cells, electrolyzers, and flexible/wearable devices is increasing rapidly. To meet such demand, high-performance and stable materials which could be used as active materials in these devices are much needed. Many nanostructured materials such as 0D, 1D, 2D, and 3D materials are being used. Hydrogel-based materials are emerging as attractive materials for energy applications due to their smart and tailorable physiochemical properties. The properties, thus the applications, can be tuned by using different synthetic approaches, blending them with other materials, using approaches to architect their structures, and many more.

The main objective of this book is to provide current, state-of-the-art development in hydrogel-based materials, their applications in energy, and their future challenges. This book covers the basics of hydrogels and their applications in a range of energy devices in terms of materials, synthetic approaches, architectural aspects, design and technology of energy devices, and challenges.

Editor Biographies

Dr. Ram K. Gupta is an Associate Professor at Pittsburg State University. Dr. Gupta's research focuses on nanomagnetism, nanomaterials, green energy production and storage using conducting polymers and composites, sensors, electrocatalysts for fuel cells, optoelectronics and photovoltaics devices, organic-inorganic hetero-junctions for sensors, bio-based polymers, biocompatible nanofibers for tissue regeneration, scaffold and antibacterial applications, and biodegradable metallic implants. Dr. Gupta has published over 250 peer-reviewed articles, made over 350 national/international/ regional presentations, chaired many sessions at national/international meetings, and edited/wrote several books/chapters for the American Chemical Society, the Royal Society of Chemistry, CRC Press, Elsevier, Springer, and Wiley. He has received several million dollars for research and educational activities from external agencies. He is a serving associate editor, guest editor, and editorial board member for various journals.

Dr. Anuj Kumar is an Assistant Professor at GLA University, Mathura, India. His research focuses on molecular as well M-N-C electrocatalysts for H_2, O_2, and CO_2 involving electrocatalysis, nanomaterials, nanocomposites, fuel cells, water electrolyzers, nano-sensors, bio-inorganic chemistry, and macrocyclic chemistry. He has published more than 45 articles in reputed peer-reviewed international journals. He has also contributed more than 12 book chapters to Elsevier, Springer, CRC Press, and the Bentham Science book series. For outstanding contribution to his research field, he was awarded the "Best Young Scientist Award 2021" from the Tamil Nadu Association of Intellectuals and Faculty (TAIF) and GRBS Educational Charitable Trust, India, and the "Young Researcher Award 2020" by Central Education Growth and Research (CEGR), India. He is a serving section editor, guest editor, and editorial board member for various journals.

Contributors

Letícia Arthus
University of Campinas
School of Chemical Engineering
Department of Processes and
  Products Design
Campinas, São Paulo Brazil

Debasrita Bharatiya
Department of Chemistry
Veer Surendra Sai University of
  Technology
Odisha, India

Dawson Blanco
Department of Chemistry
Pittsburg State University
Pittsburg, Kansas, USA
National Institute for Materials
  Advancement
Pittsburg State University
Pittsburg, Kansas, USA

Patrícia Bogalhos Lucente Fregolente
Salesian University Center of São
  Paulo
Campinas, São Paulo, Brazil

Ashok Bora
Advanced Polymer and
  Nanomaterial Laboratory
Department of Chemical Sciences
India

Denis A. Cabrera-Munguía
Facultad de Ciencias Químicas
Universidad Autónoma de Coahuila
Unidad Saltillo
Coahuila, México

Martín Caldera-Villalobos
Facultad de Ciencias Químicas
Universidad Autónoma de Coahuila
Unidad Saltillo
Coahuila, México

Janet de los Angeles Chinellato Díaz
Universidad Nacional de Córdoba
Facultad de Ciencias Químicas
Departamento de Química Orgánica
Consejo Nacional de Investigaciones
  Científicas y Técnicas (CONICET)
Instituto de Investigación y
  Desarrollo en Ingeniería de
  Procesos y Química Aplicada
  (IPQA)
Argentina

Jesús A. Claudio-Rizo
Facultad de Ciencias Químicas
Universidad Autónoma de Coahuila
Unidad Saltillo
Coahuila, México

Sachit Kumar Das
Department of Chemistry
Veer Surendra Sai University of
  Technology
Odisha, India

Allen Davis
Department of Chemistry
Pittsburg State University
National Institute for Materials
  Advancement
Pittsburg State University
Pittsburg, Kansas, USA

Felipe M. de Souza
National Institute for Materials
 Advancement
Pittsburg State University
Pittsburg, Kansas, USA

Tamara D. Erceg
Department of Materials
 Engineering
Faculty of Technology Novi Sad
University of Novi Sad
Novi Sad, Serbia

Rejeswara Reddy Erva
Department of Biotechnology
National Institute of Technology
 Andhra Pradesh
Andhra Pradesh, India

Shumaila Fayyaz
Department of Chemistry
University of Lahore
Lahore, Pakistan

Santiago P. Fernandez Bordín
Universidad Nacional de Córdoba
Facultad de Ciencias Químicas
Departamento de Química Biológica
 Ranwell Caputto
Consejo Nacional de Investigaciones
 Científicas y Técnicas (CONICET)
Centro de Investigaciones en
 Química Biológica de Córdoba
 (CIQUIBIC)
Córdoba, Argentina

Isilay Goktan
Chemical and Biological
 Engineering
Koc University
Istanbul, Turkey

Ayesha Gulzar
Biomedical Science and Engineering
Koc University
Istanbul, Turkey

Ram K. Gupta
Department of Chemistry
Pittsburg State University
Pittsburg, Kansas, USA
and
National Institute for Materials
 Advancement
Pittsburg State University
Pittsburg, Kansas, USA

Jintao Guo
China Energy Longyuan
 Environmental Protection Co. Ltd.
China

Yinghui Han
College College of Resources and
 Environment, University of
 Chinese Academy of Sciences,
China

Atif Islam
Institute of Polymer and Textile
 Engineering
University of the Punjab
Lahore, Pakistan

Wenkun Jiang
Department of Mathematics and
 Physics
North China Electric Power
 University
Beijing, China

Dhavalkumar N Joshi
Department of Mechanical
 Engineering
Gyeongsang National University
Jinju, South Korea

Abolghasem Jouyban
Pharmaceutical Analysis Research
 Center
Tabriz University of Medical
 Sciences
Tabriz, Iran

Contributors

Ramji Kalidoss
Department of Biomedical
  Engineering
SRM Institute of Science and
  Technology, Ramapuram
Chennai, India

Niranjan Karak
Advanced Polymer and
  Nanomaterial Laboratory
Department of Chemical Sciences
India

Ismail Can Karaoglu
Chemical and Biological
  Engineering
Koc University
Istanbul, Turkey

Zahra Karimzadeh
Department of Medicinal Chemistry
Faculty of Pharmacy
Tabriz University of Medical
  Sciences
Tabriz, Iran

Jasvinder Kaur
Department of Chemistry
School of Sciences
IFTM University
Uttar Pradesh, India

Seda Kizilel
Chemical and Biological
  Engineering
Koc University
Istanbul, Turkey
and
Biomedical Science and Engineering
Koc University
Istanbul, Turkey

Jagadeeshwar Kodavaty
Department of Chemical
  Engineering
University of Petroleum and Energy
  Studies
Uttarakhand, India

Radhakrishnan Kothalam
Department of Science and
  Humanities
Sri Venkateswaraa College of
  Technology
Vadakkal, India

Anuj Kumar
Department of Chemistry
GLA University
Uttar Pradesh, India

Vinod Kumar
Department of Physics
The University of the West Indies
St. Augustine, Trinidad and Tobago

Dipak Kumar Das
Department of Chemistry
GLA University
Uttar Pradesh, India

Abhishek Kumar Gupta
Department of Chemical
  Engineering
Pandit Deendayal Energy University
Gujarat, India

Suresh Kumar Yatirajula
Department of Chemical
  Engineering
Indian Institute of Technology (ISM)
  Dhanbad
Jharkhand, India

María I. León-Campos
Facultad de Ciencias Químicas
Universidad Autónoma de Coahuila
Unidad Saltillo
Coahuila, México

Teddy Mageto
National Institute for Materials
 Advancement
Pittsburg State University
Pittsburg, Kansas, USA
Department of Physics
Pittsburg State University
Pittsburg, Kansas, USA

Mansour Mahmoudpour
Miandoab School of Nursing
Urmia University of Medical
 Sciences
Urmia, Iran

Ramses S. Meleán
Universidad Nacional de Córdoba
Facultad de Ciencias Químicas
Departamento de Química Orgánica
Consejo Nacional de Investigaciones
 Científicas y Técnicas (CONICET)
Instituto de Investigación y Desarrollo
 en Ingeniería de Procesos y
 Química Aplicada (IPQA)
Argentina

Runwei Mo
School of Mechanical and Power
 Engineering
East China University of Science
 and Technology
Shanghai, China

Xiaodong Qin
Department of Mathematics and
 Physics
North China Electric Power
 University
Beijing, China

Muhammad A. u. R. Qureshi
Department of Chemistry
Allama Iqbal Open University
Islamabad, Pakistan

Biswajit Parhi
Department of Chemistry
Veer Surendra Sai University of
 Technology
Odisha, India

Swathi Pandurangan
Laboratory for Energy Materials
 and Sustainability
Department of Green Energy
 Technologies
Pondicherry University
Pondicherry, India

Elaheh Rahimpour
Pharmaceutical Analysis Research
 Center
Tabriz University of Medical
 Sciences
Tabriz, Iran

Arun Prasath Ramaswamy
Laboratory for Energy Materials
 and Sustainability
Department of Green Energy
 Technologies
Pondicherry University
Pondicherry, India

Atta Rasool
School of Chemistry
University of the Punjab
Lahore, Pakistan

Maria Regina Wolf Maciel
University of Campinas
School of Chemical Engineering
Department of Processes and
 Products Design
Campinas, São Paulo, Brazil

Marcelo R. Romero
Universidad Nacional de Córdoba
Facultad de Ciencias Químicas
Departamento de Química Orgánica
Consejo Nacional de Investigaciones
  Científicas y Técnicas (CONICET)
Instituto de Investigación y
  Desarrollo en Ingeniería de
  Procesos y Química Aplicada
  (IPQA)
Argentina

Dimpee Sarmah
Advanced Polymer and
  Nanomaterial Laboratory
Department of Chemical Sciences
India

Sarat K Swain
Department of Chemistry
Veer Surendra Sai University of
  Technology
Odisha, India

Leonardo Vasconcelos Fregolente
University of Campinas
School of Chemical Engineering
Department of Processes and
  Products Design
Campinas, São Paolo, Brazil

Nevena R. Vukić
Department of Physics and
  Materials
Faculty of Technical Sciences Čačak
University of Kragujevac
Kragujevac, Serbia

Esra Yalcin
Biomedical Science and Engineering
Koc University
Istanbul, Turkey

Part I

Fundamentals

1

Hydrogels: An Introduction to Advanced Energy Applications

Jasvinder Kaur, Dipak Kumar Das, Anuj Kumar, and Ram K. Gupta

CONTENTS

1.1 Introduction 3
1.2 Nanostructured Hydrogels and Their Chemistry 5
1.3 Energy Applications of Hydrogels 6
    1.3.1 Hydrogels for Energy Storage 6
        1.3.1.1 Rechargeable Batteries 7
        1.3.1.2 Supercapacitors 10
    1.3.2 Hydrogels for Energy Conversion 11
1.4 Conclusions 13
References 14

1.1 Introduction

The scarcity of energy is one of many challenges in the current century [1]. The rapid growth of infrastructure, industries, and finance largely depends on the availability of energy resources globally. To fulfill energy demand, uncontrolled use of non-conventional, non-renewable fossil fuels is not only inviting uncertainty in the availability of energy resources for future generations, but at the same time, an alarming level of environmental pollution has been attained due to unlimited carbon emissions, creating a threat to the inhabitants of the world [2]. Although other traditional and renewable energy resources such as tidal, wind, and solar energy are potential substitutes for fossil fuels, they are uncertain and dependent on environmental issues such as climate and weather, as well as geographical features [3]. Hence, immediate innovation is required so that these renewable, eco-friendly resources can be efficiently utilized. Another renewable resource is the use of chemical energy via redox reactions, with the main task of extending it to man-made materials with the simultaneous control of e-transfer between reacting components as well as ensuring conversion of renewable energy to an electrical form and static storage [4]. This possibility has created tremendous interest in energy transformation electrochemically and also in storage technologies, which can interconvert electrical and chemical energy efficiently. The transport of ions and electrons or the charge-discharge process can realize energy storage that can ultimately be released in electrochemical conversion and storage devices. Compared with bulk material, improved diffusion kinetics and an enhanced number of active sites are observed in nanomaterials, which have been utilized in the form of energy converters and reserve wires to be employed for advanced energy devices [5].

Because these nanostructures have tailorable physical and chemical functionalities, nanostructured hydrogels made of crosslinked polymer physically or chemically created attraction are among a large number of nanomaterials [6]. These hydrogels possess polymeric properties like ionic conductivity, stretchability, flexibility, and electrochemical properties and create monolithic structures with additional properties such as a high surface area, a continued skeleton, and manageable crystallization (Figure 1.1) [7].

It has been reported, for example, that nanostructured conductive hydrogels are being used as a binder in Li-ion batteries (LIBs) to significantly improve electrochemical activity as well as cycling stability; an ionic conductive framework prepared from nanostructured hydrogel can be successfully used as a solid electrolyte in LIBs and new active electrodes or catalysts can be obtained by carbonizing nanostructured hydrogel for electrochemical energy conversion. Herein, a crucial review of the recent developments of

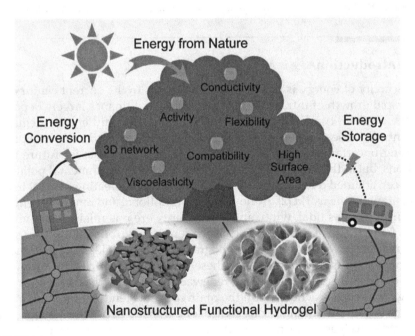

FIGURE 1.1
Improved energy storage and conversion technologies enabled by nanostructured hydrogel characteristics. (Adapted with permission [7], Copyright 2018, Wiley-VCH.)

nanostructured functional hydrogels has been made, from design to functionalization protocol. Representative nanostructured hydrogels, which proved successful in addressing prevailing challenges, have been focused on instead of reviewing the existing literature.

1.2 Nanostructured Hydrogels and Their Chemistry

Hydrogels, which are polymer networks that have been significantly swelled with water, are primary materials. Hydrogels, or hydrophilic gels, are polymer chain networks that can also be found in colloidal gel form, with water serving as the dispersion medium. Over the last few decades, scientists have defined hydrogels in several ways. The most widespread definition of a hydrogel is that it is a polymeric network that is inflated with water and forms through the simple reaction of one or more monomers. It has also been described as a polymeric material that can absorb and hold large amounts of water without dissolving in liquid water. Hydrogels have garnered a lot of interest over the past 20 years because of the extraordinary possibilities they show in a variety of applications. Due to the high proportion of water that they contain, they also have a degree of flexibility that is comparable to that of natural tissue. Hydrophilic functional groups connected to the polymeric backbone allow hydrogels to absorb water, while crosslinks between network chains provide the hydrogels' resistance to dissolution.

Several "traditional" chemical routes can be used to synthesize hydrogels. It can proceed with a single step, as in the case of polymerization and parallel crosslinking of multifunctional monomers, or it can proceed with multiple steps, beginning with the synthesis of polymer molecules with reactive groups and ending with their crosslinking, perhaps also by reacting polymers with suitable crosslinking agents. Polymer engineers are capable of adjusting qualities like biodegradability, mechanical strength, and chemical and biological reactions to stimuli [8] by designing and synthesizing polymer networks with molecular-scale control over structures, such as crosslinking density. Depending on a variety of criteria, hydrogels can be classified based on the following: origin (i): hydrogels are classified as natural or synthetic [9]; polymer (ii): homopolymer hydrogels (a polymer network derived from a single species of monomer) and copolymer hydrogels (a hydrogel containing more than two monomer species, each of which contains at least one hydrophilic component); chemical properties (iii): non-crystalline, semi-crystalline, and crystalline; appearance (iv): matrix, film, or micro

Hydrogels are made by constructing a three-dimensional (3D) nanostructured web of polymer chains through the process of crosslinking polymers or monomers. This strategy is influenced by the use of perspective in organic gels. Three-dimensional hydrogel is a polymer network consisting

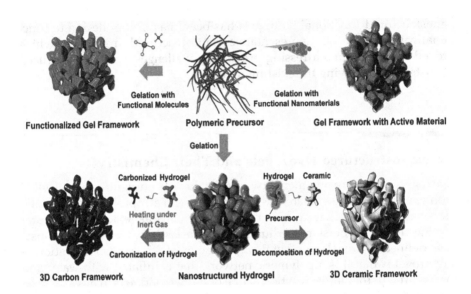

FIGURE 1.2
Representation of nanostructured hydrogel's functionalization, prepared by polymeric precursor, and molecules. (Adapted with permission [7], Copyright 2018, Wiley-VCH.)

of crosslinking through both irreversible and reversible interactions [10]. The concentration, kind of polymer chains used, and degree of crosslinking can all be used to manipulate the resulting molecular structure. Energy-related gadgets are also helpful for 3D nanostructures. Heteroatom doping [11], grafting copolymerization [12], chemical change [13], and the gathering of 1D and/or 2D nanostructures [14] are all examples of synthetic procedures that go into the construction of this nanostructure. Many different kinds of applications rely on the 3D nanostructures that are the result of the frequent combination of 1D and 3D nanostructures (Figure 1.2) [7]. Functional group titration allows for the initiation of desired nanostructures and physical and chemical characteristics in hydrogel frameworks. High-performance energy storage and conversion devices have also been designed using 3D carbon or ceramic frameworks formed from nanostructured hydrogels.

1.3 Energy Applications of Hydrogels

1.3.1 Hydrogels for Energy Storage

Two electrodes and an electrolyte are the main structural features of electrochemical energy storage devices. The device's efficiency is determined by how well electrons or ions are delivered, as well as how quickly reactions can occur on the electrodes. Cycling stability, as well as mechanical integrity, are two other aspects that play a critical role in achieving robust and energy-efficient devices. In this regard, flexibility, functionality, and a hierarchical and interconnected structure may offer added advantages to a nanostructured hydrogel. Various applications of hydrogel have been highlighted in this section. Hydrogen serves as a precursor for an electrode, a functional binder electrode composite, ceramic filler or functional components in the electrolyte for rechargeable batteries, and conductive electrodes in supercapacitors (SCs).

1.3.1.1 Rechargeable Batteries

The ultra-high theoretical capacity of 1675 mAh/g, five times more than the commonly used cathode for LIBs [15, 16] of rechargeable Li-S batteries, created interest among researchers in this field. Unfortunately, the insulating nature (electronic and ionic) of Li_2S/Li_2S_2 and the dissolution of intermediate polysulfides restrict their practical applications because of poor sulfur utilization and energy density [17–19]. Uniformity and distribution of Li_2S particles are uncertain and difficult to achieve, along with nullifying the shuttle effect due to the dissolution of polysulfides. Electrode materials requiring ionic and electronic conduction benefit greatly from the nanostructured polymer hydrogel's ability to manage volume change during charging and discharging operations. To improve rate performance through the complete use of active material, a 3D nanostructured hydrogel may provide first ion diffusion and electron conduction in the form of an electrode.

Although conducting polymers can improve Li-SBs, the weak reaction of polysulfide with the polymer results in polysulfide dissolution and thus lower columbic efficiency [20, 21]. These issues were addressed by fabricating $Ppy-MnO_2$ co-axial nanotubes to encapsulate sulfur, where PPy acts as a conduction framework and MnO_2 restricts the dissolution of polysulfide via chemisorption (Figure 1.3a) [22]. $Ppy-MnO_2$ co-axial nanotubes resulted in outstanding trapping ability for polysulfide tubular PPy and displayed significant flexibility and electron conductivity to enhance rate capabilities as well as cycling stability (Figure 1.3b). Carbon frameworks co-doped with P- and N-atoms were reported to have been prepared from phytic acid-doped polyaniline (P-PANI) hydrogels (Figure 1.3c) [23], where Li_2S NPs were accommodated by N and P-C frameworks free from binders (Figure 1.3d). Three-dimensional nanostructures and N/P doping display a stable capacity of 700 mAh/g with 99.4% columbic efficiency due to synergistic effects (Figure 1.3e). It has been conceived that the porous nanostructure of the N, P, and C networks provides a continuous electron transfer pathway along with hierarchical pores for Li ions. Nanoparticle connectivity promotes more electrolyte uptake, altering volume accommodation during charging and discharging and resulting in rapid ion diffusion and cycling stability. P-doped carbon, on the other hand, reduces the dissolution of polysulfide, restricting

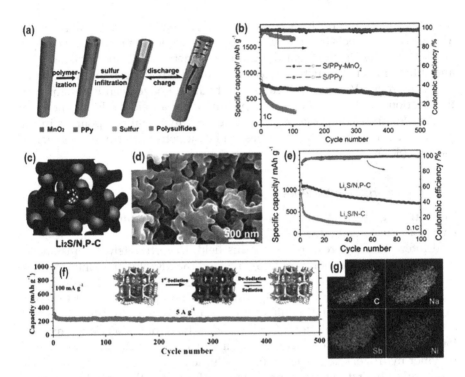

FIGURE 1.3
(a) Preparation and working mechanism for S/PPy–MnO$_2$ material. (b) Comparison between the performance of S/PPy–MnO$_2$ and S/PPy materials. (Adapted with permission [22], Copyright 2016, American Chemical Society.) (c) Illustration and (d) SEM picture for the Li$_2$S/N, P–C hybrid. (e) Stability cycling for Li$_2$S/N, P–C and Li$_2$S/N–C materials. (Adopted with permission [23], Copyright 2017, Wiley-VCH.) (f) Cycling stability, and (g) TEM picture for rGO@Sb–Ni. (Adapted with permission [26]. Copyright 2018, American Chemical Society.)

the shuttle effect because of the strong S-P interaction on the carbon framework. Because of the low cost and abundance of sodium, sodium-ion batteries (SIBs) are the next generation of batteries after LIBs. SIBs, like LIBs, require high-capacity anode materials, such as alloying type anodes; antimony is an alloying type anode, but drastic volume changes during alloying and de-alloying limit the amount of Sb that can be used in SIBs, which must withstand rapid capacity fading and 100% columbic efficiency [24, 25].

To address this issue and ultimately obtain rGO@Sb-Ni composite, Yu et al. synthesized a cyanogel-enabled approach to produce Sb-Ni-C turnery anodes [26]. The results of an electrochemical cycling performance study (Figure 1.3f) reveal ultra-long cycle stability of up to 500 cycles at 5 A/g and a high reversible capacity of 210 mAh/g. This composite was found to accommodate volume change during cycling and controlled the accumulation and pulverizing of the turnery anode (Figure 1.3g). It displayed a surprisingly high trapping capacity for undesired products, flexibility, and conductivity.

An electrode can be prepared from hydrogels as a precursor [27, 28], and this approach applies to a wide range of rechargeable batteries.

The active electrode is generally added with conductive additives and binders to obtain charging and discharging processes of LIBs following electronic as well as ionic conduction mechanisms. Polyvinylidene fluoride (PVDF) is used as a binder to fabricate such composite electrodes. The binder holds the active material as well as additives, in which the binder ensures mechanical integrity and the additive offers electronic conductivity between the active particle and the current collector [29]. Agglomeration of particles for extended diffusion length and sufficient inactive space created by aggregation prevent the full utilization of active material as well as lowering rate capacity and specific capacity [30]. In addition, conductive additives or binders, being inactive here, donate nothing to energy storage, resulting in limited energy density in general along with enhanced packaging costs. In this context, the nanostructured functional hydrogel has the potential to serve as the composite electrode because of its high electronic conductivity as well as mechanical strength [31–34]. In this context, Yu et al. reported having used a 3D gel framework as a multifunctional binder for active material and Fe_3O_4 NPs (Figure 1.4a) [34]. Crosslinking polypyrrole with CuPc Ts (C-PPy), capable of performing electronic conduction quickly, was used to fabricate highly conductive polymers. Because of its porous and continuous framework, the polymer gel framework offers mechanical balances to the electrode as well as 3D continuous structural support in the unformal dispersion of active particles, as well as the initiation of Li-ion transportation (Figure 1.4b). The

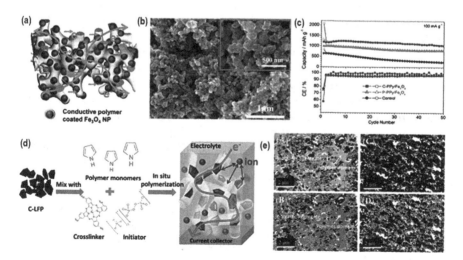

FIGURE 1.4
(a) Preparation strategy. (b) SEM pictures. (c) cycling stability/columbic efficiency for the C-PPy/Fe_3O_4 material. (Adapted with permission [34], Copyright 2017, Wiley-VCH.) (d) Synthesis and (e) TEM picture for the C-LFP/C-PPy electrode. (Adapted with permission [37], Copyright 2017, American Chemical Society.)

composite electrode (C-PPy/Fe_3O_4) displays a higher discharging specific capacity compared with Fe_3O_4 with a traditional PVDF binder (Figure 1.4c). There is an extension of the idea of a nanostructured functional hydrogel in terms of LIB cathodes [35–37]. A 3D inorganic gel framework was reported via in-situ polymerization of conductive polymer gel on $LiFePO_4$ (LEP) particles [37], which was found to be better rate capable and cycling performing with the initiation of electronic and ionic transport and mechanical integrity (Figure 1.4d). Agglomeration of LEP particles in composite (left) occurred with homogeneous distribution, as shown by the TEM image (Figure 1.4e).

1.3.1.2 Supercapacitors

SCs, a potential and efficient energy storage system having a rapid charge/discharge rate and a long cycle life, are capable of connecting rechargeable batteries as well as electrolytic capacitors. While traditional capacitors have dielectric material to store energy, SCs follow two different mechanisms: electric double-layer capacitance (EDLC) and electrochemical pseudocapacitance. In EDLC, electrostatic charges accumulate at the electrode-electrolyte interface, necessitating a large surface area for electrolyte ions as well as the electrical conductivity of the electrode [38]. The EDLC, therefore, may be improved in terms of performance by using a carbon-based electrode with a large surface area like carbon nanotubes (CNTs) [39], activated carbon [40], or graphene [41]. On the contrary, an electrochemical pseudo-capacitor uses a rapid and reversible faradic way for energy storage [33]. As electrode materials, polyaniline (PANI) [42], polypyrrole (PPy) [43], and their derivatives are used for SCs because of higher conductivity resulting from their large pi-conjugated structure. Although many conductive, flexible, and lightweight polymers can be used, the bulk of them have a rapid decaying capacitance, yielding poor cycling life. On the other hand, 3D nanostructured hydrogels do have pores with a continuous polymeric network to facilitate electron transport and a small diffusion path for electrolyte ions, as well as the ability to manage volume change during charge and discharge operations. Yu et al. [44] demonstrated the development of a 3D hierarchical nanostructured PPy hydrogel via an interfacial polymerization process. Excellent performance, even under bending conditions (Figure 1.5b), was observed due to the spongy PPy network, including exceptional rate performance and cycling stability (Figure 1.5c) with approximately 90% capacitance retention. In a similar study, Yu et al. demonstrated self-standing flexible conducting hydrogels for SCs through the application of 3D nanostructured conducting PANI/graphene-hydrogel (Figure 1.5d) [45]. Severe agglomeration of the structure during reduction was successfully controlled by strong intermolecular interactions between PANI and graphene. The flexible SCs demonstrate high energy density compared with earlier fibrous SCs. Performance even under the deformation of all-gel-state SCs (Figure 1.5e-f), shown by electrochemical analysis,

An Introduction to Advanced Energy Applications

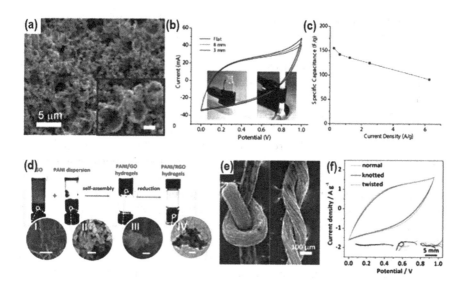

FIGURE 1.5
(a) SEM pictures of the 3D-PPy nanostructured hydrogel. (b) CV curves were recorded with the prepared materials under various experimental conditions. (c) Specific capacitance vs current densities. (Adapted with permission [46], Copyright 2014, Royal Society of Chemistry.) (d) Graphical pictures for the self-assembled PANI/GO hybrid hydrogels along with their SEM/TEM pictures. (e) The mechanical properties of PANI/GO. (f) CV curves in straight, curved, and twisted configurations of a fully hydrogel-state formed cell. (Adapted with permission [45], Copyright 2018, Wiley-VCH.)

indicates the possibility of using such a structure in flexible SCs in multiple geometries for wearable articles.

1.3.2 Hydrogels for Energy Conversion

Fuel cells and water electrolysis are considered to be the most promising alternatives to conventional fossil fuels, as electrochemical energy in the redox reaction of abundant H_2 and O_2 can be used with environmentally friendly by-products. Electrode reactions like oxygen reduction reaction (ORR), oxygen evolution reaction (OER), or hydrogen evolution reaction (HER) are mainly responsible for the performance of these innovations [11, 30]. H_2 obtained from HER appears to be eco-friendly fuel, while ORR and OER are thermodynamically reversible processes for charging and discharging [47]. Three types of catalysts are available for such electrochemical redox reactions: expensive metals, cheap metals, and catalysts free from metal; so far, Pt and its alloys are extensively used as the efficient electrodes of choice for ORR, OER, and HER [47]. Due to the high cost of Pt, out of the many compounds investigated, non-precious metal compounds and nitrogen-containing organic compounds were found to be promising for this kind of reaction.

Hetero-atomic doped carbon materials free from metals, as well as conductive polymers, have also been evaluated to assess their potential as efficient ORR, OER, and HER, which can be monitored cost-effectively [47]. Moreover, some specially designed nanostructured gels may find utility as catalysts without undergoing calcination. A nanostructured hydrogel may undergo carbonization after calcination, retaining hetero-atomic components in the polymeric network in the form of the dopant; thus, the distribution of the dopant can be monitored by using such a polymeric structure. Utilizing such concepts and protocols, new electrocatalysts were fabricated for ORR, OER, and HER because pure carbon material shows poor catalytic performance [48]. For instance, 3D porous carbon-based non-specious metal electrocatalysts were fabricated following a homogenous hydrogel-based bottom-up strategy [49, 50]. Jiang et al. reported making N, P-doped hierarchal porous carbon forms, then co-pyrolyzing the carbon form precursor with a poly(vinyl alcohol), polystyrene hydrogel composite as an in-situ template. The prepared carbon framework has monitorable N and P, a large surface area, and an interconnected macro-meso-porous structure with high electrocatalytic performance for ORR in acidic, basic, as well as neutral conditions [50]. Other than this, these materials can act as bi-functional catalysts for ORR and OER [48, 51] (Figure 1.6a). Hydrogel made of PANI that is crosslinked by phytic

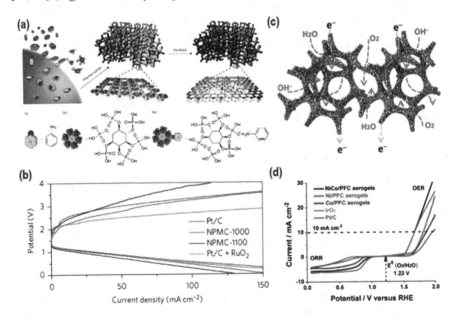

FIGURE 1.6
(a) Synthesis strategy for the nanocarbon foams using aniline, phytic acid, and complex. (b) Charge/discharge curves in Zn–air battery for the NPMC-1000, NPMC-1100, or traditional catalysts materials. (Adapted with permission [48], Copyright 2015, Nature Publishing Group.) (c) Electrocatalytic mechanism on prepared NiCo/PFC catalyst. (d) ORR and OER polarization curves. (Adapted with permission [51], Copyright 2016, American Chemical Society.)

acid as N and P doping sources may undergo simple pyrolysis to make such material. These bi-functional catalysts are initiated by molecular level regulation of hetero-atoms to act as air electrodes for rechargeable Zn-air batteries (Figure 1.6b). A new $K_2Ni(CN)_4/K_3Co(CN)_{6\text{-chitosan}}$ hybrid hydrogel was utilized to synthesize Ni/Co NPs anchored in porous fibrous carbon (PFc) form with good dispersion (Figure 1.6c) [51]. Carbon catalysts outperformed Pt/C and IrO_2 catalysts in terms of stability and performance for ORR and OER electrocatalytically (Figure 1.6d).

1.4 Conclusions

Hydrogels and their derivatives have significantly contributed to sustainable energy conversion and storage innovation. The principal principle involved in the functionalization of hydrogel-based materials is the complete exploitation of their intrinsic features, like their large surface area, tailorable chemical composition, and better ionic and/or electronic conductivity than bulk moieties, to fabricate unique nanostructured hydrogels with innovative functionalities for fabricated hydrogel-derived materials. As far as the future is concerned, nanostructured functional hydrogels and their derivatives must ensure complete creative utilization of their distinct and special features compared with traditional nanomaterials to address the limitations and challenges aimed at next-generation energy conversion and storage. The drawbacks and limitations of traditional nanomaterials like complicated production procedures, expensive nature, long process times, and scalability problems are economically addressed by facile synthesis and simple functionalization of polymer networks of nanostructured hydrogel, where 3D nanostructured carbon or ceramic frameworks can easily be produced, extending the range of functional materials and devices. In addition, good elasticity, self-healing ability, and thermoresponsive properties of nanostructured hydrogels make them materials of choice in new functional applications like smart electrolytes or flexible devices.

Moreover, the fact that the structure and components of the nanostructured hydrogel are chargeable readily and rationally make them an ideal and powerful platform to undergo extensive investigation of the basic mechanisms involved in many physicochemical processes, aiming at their suitability for the rational design of advanced energy technologies. For practical applications in energy devices, several aspects still need to be developed further. Although hydrogel is being used as a framework or precursor for the inorganic framework for energy storage devices, many disadvantages, like low-volumetric energy density as offered by non-porous structure and capacity fading during cycling or even intrinsic deuteriation of the structural material, remain to be resolved. The scalability of the hydrogel-derived inorganic framework is another issue to be taken care of. The hydrogel can be prepared on a large scale, however, when it is being used as a precursor, it is susceptible to shrinkage, resulting in cracks or getting de-shaped during drying or heat treatment. A hybrid gel framework (double interpenetrating network), and a novel polymer matrix, including optimization of the production method, are needed to make hydrogel, which has prospects for practical applications in energy storage/conversion devices. By implementing all of the strategies, the nanostructured functional hydrogel has a strong chance of overcoming all limitations and improving the limit of energy technologies that traditional nanomaterials cannot resolve or achieve.

References

1. Chu S, Majumdar A. Opportunities and challenges for a sustainable energy future. *Nature*. 2012;488(7411):294–303.
2. Manthiram A, Fu Y, Chung S-H, Zu C, Su Y-S. Rechargeable lithium–sulfur batteries. *Chemical Reviews*. 2014;114(23):11751–87.
3. Raza A, Lu J-Y, Alzaim S, Li H, Zhang T. Novel receiver-enhanced solar vapor generation: Review and perspectives. *Energies*. 2018;11(1):253.
4. Liu C, Li F, Ma LP, Cheng HM. Advanced materials for energy storage. *Advanced Materials*. 2010;22(8):E28–E62.
5. Shi Y, Zhou X, Yu G. Material and structural design of novel binder systems for high-energy, high-power lithium-ion batteries. *Accounts of Chemical Research*. 2017;50(11):2642–52.
6. Shi Y, Peng L, Ding Y, Zhao Y, Yu G. Nanostructured conductive polymers for advanced energy storage. *Chemical Society Reviews*. 2015;44(19):6684–96.
7. Zhao F, Bae J, Zhou X, Guo Y, Yu G. Nanostructured functional hydrogels as an emerging platform for advanced energy technologies. *Advanced Materials*. 2018;30(48):1801796.
8. Raccichini R, Varzi A, Passerini S, Scrosati B. The role of graphene for electrochemical energy storage. *Nature Materials*. 2015;14(3):271–9.
9. Xia W, Mahmood A, Zou R, Xu Q. Metal–organic frameworks and their derived nanostructures for electrochemical energy storage and conversion. *Energy & Environmental Science*. 2015;8(7):1837–66.
10. Döring A, Birnbaum W, Kuckling D. Responsive hydrogels–structurally and dimensionally optimized smart frameworks for applications in catalysis, micro-system technology and material science. *Chemical Society Reviews*. 2013;42(17):7391–420.
11. He Y, Han X, Du Y, Zhang B, Xu P. Heteroatom-doped carbon nanostructures derived from conjugated polymers for energy applications. *Polymers*. 2016;8(10):366.
12. Nasef MM, Gürsel SA, Karabelli D, Güven O. Radiation-grafted materials for energy conversion and energy storage applications. *Progress in Polymer Science*. 2016;63:1–41.

13. Pérez-Madrigal MM, Estrany F, Armelin E, Díaz DD, Alemán C. Towards sustainable solid-state supercapacitors: Electroactive conducting polymers combined with biohydrogels. *Journal of Materials Chemistry A*. 2016;4(5):1792–805.
14. Rong Q, Lei W, Chen L, Yin Y, Zhou J, Liu M. Anti-freezing, conductive self-healing organohydrogels with stable strain-sensitivity at subzero temperatures. *Angewandte Chemie International Edition*. 2017;56(45):14159–63.
15. Seh ZW, Sun Y, Zhang Q, Cui Y. Designing high-energy lithium–sulfur batteries. *Chemical Society Reviews*. 2016;45(20):5605–34.
16. Manthiram A, Chung SH, Zu C. Lithium–sulfur batteries: Progress and prospects. *Advanced Materials*. 2015;27(12):1980–2006.
17. Ji X, Nazar LF. Advances in Li–S batteries. *Journal of Materials Chemistry*. 2010;20(44):9821–6.
18. Zheng G, Zhang Q, Cha JJ, Yang Y, Li W, Seh ZW, et al. Amphiphilic surface modification of hollow carbon nanofibers for improved cycle life of lithium sulfur batteries. *Nano Letters*. 2013;13(3):1265–70.
19. Zhang J, Huang H, Bae J, Chung SH, Zhang W, Manthiram A, et al. Nanostructured host materials for trapping sulfur in rechargeable Li–S batteries: Structure design and interfacial chemistry. *Small Methods*. 2018;2(1):1700279.
20. Xiao L, Cao Y, Xiao J, Schwenzer B, Engelhard MH, Saraf LV, Nie Z, Exarhos GJ, Liu J. A soft approach to encapsulate sulfur: Polyaniline nanotubes for lithium-sulfur batteries with long cycle life. *Advanced Materials*. 2012;24:1176–81.
21. Yang Y, Yu G, Cha JJ, Wu H, Vosgueritchian M, Yao Y, et al. Improving the performance of lithium–sulfur batteries by conductive polymer coating. *ACS Nano*. 2011;5(11):9187–93.
22. Zhang J, Shi Y, Ding Y, Zhang W, Yu G. In situ reactive synthesis of polypyrrole-MnO_2 coaxial nanotubes as sulfur hosts for high-performance lithium–sulfur battery. *Nano Letters*. 2016;16(11):7276–81.
23. Zhang J, Shi Y, Ding Y, Peng L, Zhang W, Yu G. A conductive molecular framework derived Li_2S/N, P-codoped carbon cathode for advanced lithium–sulfur batteries. *Advanced Energy Materials*. 2017;7(14):1602876.
24. Allan PK, Griffin JM, Darwiche A, Borkiewicz OJ, Wiaderek KM, Chapman KW, et al. Tracking sodium-antimonide phase transformations in sodium-ion anodes: Insights from operando pair distribution function analysis and solid-state NMR spectroscopy. *Journal of the American Chemical Society*. 2016;138(7):2352–65.
25. He M, Kravchyk K, Walter M, Kovalenko MV. Monodisperse antimony nanocrystals for high-rate Li-ion and Na-ion battery anodes: Nano versus bulk. *Nano Letters*. 2014;14(3):1255–62.
26. Wu P, Zhang A, Peng L, Zhao F, Tang Y, Zhou Y, et al. Cyanogel-enabled homogeneous Sb–Ni–C ternary framework electrodes for enhanced sodium storage. *ACS Nano*. 2018;12(1):759–67.
27. Hasegawa G, Ishihara Y, Kanamori K, Miyazaki K, Yamada Y, Nakanishi K, et al. Facile preparation of monolithic LiFePO4/carbon composites with well-defined macropores for a lithium-ion battery. *Chemistry of Materials*. 2011;23(23):5208–16.
28. Xu J-J, Wang Z-L, Xu D, Meng F-Z, Zhang X-B. 3D ordered macroporous $LaFeO_3$ as efficient electrocatalyst for Li–O_2 batteries with enhanced rate capability and cyclic performance. *Energy & Environmental Science*. 2014;7(7):2213–9.
29. Markevich E, Salitra G, Aurbach D. Influence of the PVdF binder on the stability of $LiCoO_2$ electrodes. *Electrochemistry Communications*. 2005;7(12):1298–304.

30. Bock DC, Kirshenbaum KC, Wang J, Zhang W, Wang F, Wang J, et al. 2D cross sectional analysis and associated electrochemistry of composite electrodes containing dispersed agglomerates of nanocrystalline magnetite, Fe_3O_4. ACS Applied Materials & Interfaces. 2015;7(24):13457–66.
31. Xie J, Gu P, Zhang Q. Nanostructured conjugated polymers: Nanostructured conjugated polymers: Toward high performance organic electrodes for rechargeable batteries. ACS Energy Letters. 2017;2:1985–96.
32. Liu G, Xun S, Vukmirovic N, Song X, Olalde-Velasco P, Zheng Battaglia HVS, Wang L, Yang W. Polymers with tailored electronic structure for high capacity lithium battery electrodes. Advanced Materials. 2011;23:4679.
33. Kovalenko I, Zdyrko B, Magasinski A, Hertzberg B, Milicev Z, Burtovyy R, et al. A major constituent of brown algae for use in high-capacity Li-ion batteries. Science. 2011;334(6052):75–9.
34. Shi Y, Zhang J, Bruck AM, Zhang Y, Li J, Stach EA, et al. A tunable 3D nanostructured conductive gel framework electrode for high-performance lithium ion batteries. Advanced Materials. 2017;29(22):1603922.
35. Lai C-H, Ashby DS, Lin TC, Lau J, Dawson A, Tolbert SH, et al. Application of poly (3-hexylthiophene-2, 5-diyl) as a protective coating for high rate cathode materials. Chemistry of Materials. 2018;30(8):2589–99.
36. Ling M, Qiu J, Li S, Yan C, Kiefel MJ, Liu G, et al. Multifunctional SA-PProDOT binder for lithium ion batteries. Nano Letters. 2015;15(7):4440–7.
37. Kwon YH, Minnici K, Huie MM, Takeuchi KJ, Takeuchi ES, Marschilok AC, et al. Electron/ion transport enhancer in high capacity Li-ion battery anodes. Chemistry of Materials. 2016;28(18):6689–97.
38. Zhang LL, Zhao X. Carbon-based materials as supercapacitor electrodes. Chemical Society Reviews. 2009;38(9):2520–31.
39. Park AKKW, Moon J, Bae DJ, Lim SC, Lee YS, Lee YH. Electrochemical properties of high-power supercapacitors using single-walled carbon nanotube electrodes. Advanced Functional Materials. 2001;11(5):387.
40. Li B, Zhang C, Dai F, Xiao Q, Yang L, Cai M, et al. Nitrogen-doped activated carbon for high energy hybrid supercapacitor. Energy & Environmental Science. 2016;9:102.
41. Peng L, Peng X, Liu B, Wu C, Xie Y, Yu G. Ultrathin two-dimensional MnO_2/graphene hybrid nanostructures for high-performance, flexible planar supercapacitors. Nano Letters. 2013;13(5):2151–7.
42. Chen W, Rakhi R, Alshareef HN. Capacitance enhancement of polyaniline coated curved-graphene supercapacitors in a redox-active electrolyte. Nanoscale. 2013;5(10):4134–8.
43. Sharma RK, Rastogi A, Desu S. Manganese oxide embedded polypyrrole nanocomposites for electrochemical supercapacitor. Electrochimica Acta. 2008;53(26):7690–5.
44. Li J, Illeperuma WR, Suo Z, Vlassak JJ. Hybrid hydrogels with extremely high stiffness and toughness. ACS Macro Letters. 2014;3(6):520–3.
45. Li P, Jin Z, Peng L, Zhao F, Xiao D, Jin Y, et al. Stretchable all-gel-state fiber-shaped supercapacitors enabled by macromolecularly interconnected 3D graphene/nanostructured conductive polymer hydrogels. Advanced Materials. 2018;30(18):1800124.

46. Shi Y, Pan L, Liu B, Wang Y, Cui Y, Bao Z, et al. Nanostructured conductive polypyrrole hydrogels as high-performance, flexible supercapacitor electrodes. *Journal of Materials Chemistry A*. 2014;2(17):6086–91.
47. Liu X, Dai L. Carbon-based metal-free catalysts. *Nature Reviews Materials*. 2016;1(11):1–12.
48. Zhang J, Zhao Z, Xia Z, Dai L. A metal-free bifunctional electrocatalyst for oxygen reduction and oxygen evolution reactions. *Nature Nanotechnology*. 2015;10(5):444–52.
49. You B, Kang F, Yin P, Zhang Q. Hydrogel-derived heteroatom-doped porous carbon networks for supercapacitor and electrocatalytic oxygen reduction. *Carbon*. 2016;103:9–15.
50. Jiang H, Zhu Y, Feng Q, Su Y, Yang X, Li C. Nitrogen and phosphorus dual-doped hierarchical porous carbon foams as efficient metal-free electrocatalysts for oxygen reduction reactions. *Chemistry–A European Journal*. 2014;20(11):3106–12.
51. Fu G, Chen Y, Cui Z, Li Y, Zhou W, Xin S, et al. Novel hydrogel-derived bifunctional oxygen electrocatalyst for rechargeable air cathodes. *Nano Letters*. 2016;16(10):6516–22.

2

Hydrogels: Synthesis and Recent Advancements in Electrochemical Energy Storage

Dawson Blanco, Teddy Mageto, Felipe M. de Souza,
Anuj Kumar, and Ram K. Gupta

CONTENTS

2.1 Introduction 19
2.2 Synthesis of Hydrogels 22
2.3 Hydrogels for Supercapacitors 25
    2.3.1 Hydrogel-based Electric Double-Layer Capacitors 26
    2.3.2 Hydrogel-based Redox-Active Pseudocapacitors 27
    2.3.3 Hydrogel-based Hybrid Supercapacitors 28
2.4 Hydrogels for Batteries 31
    2.4.1 Metal-Ion Batteries 31
    2.4.2 Metal-Air Batteries 32
    2.4.3 Metal-Sulfur Batteries 34
2.5 Hydrogels for Fuel Cells 36
2.6 Hydrogels for Flexible Devices 39
2.7 Conclusion and Perspectives 43
References 44

2.1 Introduction

The proper storage and distribution of energy is a core subject for nations. The success of properly providing energy for society can indirectly impact important human well-being factors such as life expectancy, happiness, and access to resources, among others. Finding efficient, continuous, and eco-friendly ways to harvest energy, aside from the equilibrated dissemination of it, is a recurrent concern. There has been the introduction of green energy generation processes such as solar, wind, tide, hydroelectric, and biomass, which can serve as sustainable technologies. However, even though these ways to acquire energy are eco-friendly and renewable, their relatively lower output and intermittence create some hindrances to their larger application and reliability [1]. Despite this challenge, the United States Energy Information Administration (EIA) has predicted that from 2020 to 2050 there will be a general increase in the consumption of non-renewable sources such as petroleum and natural gas, as well as a more drastic increase in the use of renewable sources, including wind, solar, and hydroelectric [2].

Based on this scenario, there have been considerable efforts from the scientific community to develop novel materials that can be suitable for energy storage and generation applications. There has been the development of energy storage devices such as supercapacitors, batteries, fuel cells, and hydrolyzers, among many others. However, for the development of these high-end devices, there is a plethora of electroactive materials that can be used such as transition metal oxides, dichalcogenides, and bi or trimetallic, along with their composites with conducting polymers or carbon-based nanomaterials that can further enhance their electrochemical performance. The suitability of these materials for energy storage devices is based on some of their inherent properties, which are appreciable electron conductivity as well as surface area. In this sense, it is highly desired that electroactive materials can conduct electrons to facilitate electron transfer processes during energy storage or electrocatalytic steps. On top of that, they should present a relatively higher surface area to allow the permeation or diffusion of the electrolyte, which can greatly influence their capacitance. One of the classes of materials that can present these properties is hydrogels, which are crosslinked polymer networks that, due to their highly porous structure, can adsorb large amounts of liquid or gases. Their network structure can be formed through gelation, which is the phenomenon of polymer chains in solution crosslinking to form a single large macrostructure that contains all polymeric units. This type of structure can prevent the free flow of certain species yet present a high surface area.

Hydrogels can be classified based on the materials used in their synthesis, crosslinking method, stimuli response, and ionic charge [3]. One of the first synthesized hydrogels was reported by O. Wichterle, and D. Lím [4], who pointed out the mechanical similarities between polyvinyl alcohol (PVA) hydrogel and living tissue, showing their advantage over hard plastics in biotechnological applications. Alongside that, hydrogel is becoming more relevant since, in 2019, the market for hydrogel products was around $22.1 billion, and is anticipating a 6.7% growth to $31.4 billion by 2027 [5]. This growth can be explained by their wide range of applications within the fields of biology, medicine energy storage, and catalysis due to their relatively low cost and biocompatibility. For instance, agarose is a naturally sourced heteropolysaccharide that, after being formed into a gel, is used in the field of biology as a substrate for bacterial and fungal cultures [6], and modern soft contact lenses are hydrogels constructed from the monomer poly(2-hydroxyethyl methacrylate-co-methacrylic acid) (poly(HEMA-co-MAA)) [7]. Other medical applications include self-healing artificial skin bandages, wound packing with antibacterial properties, and biosensors [8]. The latter requires long-term compatibility with human tissue, electron conductivity, appreciable mechanical properties, and the ability to retain these properties under stress. Hydrogels constructed with naturally sourced biomolecules such as DNA, peptides, and polysaccharides (e.g., agarose, chitosan, cellulose) or constructed with otherwise biocompatible materials, such as polyethylene glycol (PEG), polycaprolactone, and polylactide, show almost no cytotoxicity [9].

Recently, hydrogels have garnered attention for their applications in electrochemical devices where their highly tunable properties facilitate their use as an alternative for many of their components. Hydrogels excel when used as an electrode because of their high surface area and greatly interconnected network, encouraging ion transport. When combined with heteroatom dopants, redox activity can be enhanced [10]. Another important aspect of hydrogels is their suitability as both binder and conductive material for rechargeable batteries which can improve electrochemical stability as well as conductivity. It can heighten the proportion of active to inactive materials in the battery [11]. This all serves the goal of increasing the battery's capacitance. The hydrogel can also substitute the typical liquid or solid-state electrolyte of a battery or supercapacitor. Hence, due to the great versatility and applicability of hydrogels, these materials can be classified based on five aspects: the source, preparation approach, crosslinking, ionic charge, and stimuli response. Each aspect can be further divided as shown in the complete scheme provided in Figure 2.1.

A gel electrolyte provides high ion permeability and a reasonable degree of chemical and mechanical stability while avoiding undesired reactions with the electrode material, which is a problem encountered with liquid electrolytes. Hydrogels can also be created to have tunable interfacial adhesion [12], providing an alternative to solid-state electrolytes. Mechanical stability, alongside self-healing properties, is of the greatest importance to flexible devices: those which are likely to experience bending or other stress. An issue that remains is that aqueous hydrogel electrolytes are constrained to an operating voltage window of roughly 1.2 V. Comparatively, organic electrolytes can reach nearly 4.0 V but are environmentally hazardous [13]. Hydrogels, being a polyelectrolyte network, can also serve as the electrolyte in fuel cells, typically as a proton or anion exchange membrane. In the following sections, the synthesis options for hydrogels are discussed followed by a deeper exploration of electrochemical devices and how hydrogels can be engineered to suit each.

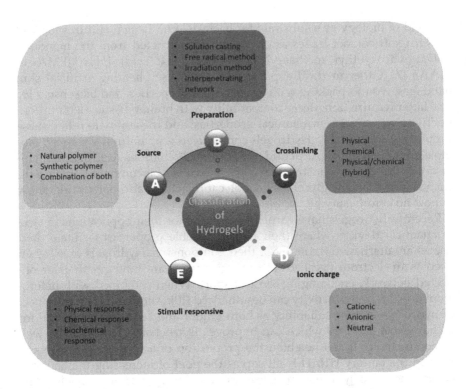

FIGURE 2.1
Classifications of hydrogels by source, preparation, crosslinking, ionic charge, and stimuli responsiveness. (Adapted from reference [12]. Copyright 2020 by the authors. Licensee MDPI, Basel, Switzerland. This article is an open-access article distributed under the terms and conditions of the Creative Commons Attribution (CC BY) license.)

2.2 Synthesis of Hydrogels

Through logical engineering of the gelation process, hydrogels can be created to possess unique characteristics to better suit targeted applications. The choice of monomer or polymer building blocks, method, and degree of crosslinking are key to modulating the characteristics of the resulting hydrogel and creating a highly specialized material. Importantly, the monomers used need not necessarily be a traditional covalently bonded chain. It is crucial that polymers suitable for hydrogel present reactive groups to promote crosslinking and/or attractive intermolecular interactions that can aid in the formation of a networked structure.

The polymer chains within the gel are the main contributors to its characteristics, so the choice of monomer is an influential variable in controlling the resulting hydrogel's properties. In this sense, the functionality and chemical characteristics of the polymeric chain strongly influence the hydrogel's properties. For example, the water retention of the gel can be controlled by the presence of hydrophilic groups such as –OH, –NH$_2$, –COOH, and –SO$_3$H. Or the presence of dandling groups can induce a physical entanglement in the structure. Hydrogels can be classified as natural, synthetic, or hybrid based on the origin of their constituents. Natural hydrogels are mostly created from polysaccharides and proteins, resulting in a highly biocompatible yet relatively cheap-to-produce material. One shortcoming of natural hydrogels is that they tend to be mechanically weak. Synthetic hydrogels, in contrast, are created from non-naturally sourced materials. Their synthesis is more reproducible due to the predictability of their components.

A hydrogel with different structures can be obtained based on the monomers utilized. Based on that, homopolymerization involves only one type of monomer, while copolymerization involves two or more monomers. Copolymerization allows the combination of multiple properties and their modulation based on the ratio between each of the constituent monomers in the final chain. The careful selection of monomers with certain chemical groups and functionality allows one to control the degree of crosslinking, which defines some of the gel's properties. In this sense, either one monomer can be used to obtain a homopolymer or different monomers can be introduced in the same system to obtain copolymers. Homopolymers lead to the formation of a single network type of structure whereas the use of a copolymer can lead to the formation of either a semi-interpenetrating network (IPN) or a fully IPN. Within this line, the schematic of polymers and types of networks are provided in Figure 2.2.

Being that crosslinking is the process in which the gel is rendered insoluble, it follows that the mechanical effects of swelling and elasticity are controlled by the method used, be it the degree to which the gel is crosslinked, the type of crosslinking bond, or the use of another entanglement method such as forming an IPN. Generally, the more crosslinked a structure is, the stronger and more inelastic it tends to become. The crosslinking of hydrogels can be classified into two main types: chemical and physical. Chemically crosslinked gels are more stable than physically crosslinked ones but usually require an added crosslinking agent, or in the case of radical polymerization, an initiator [14]. However, some of these initiators can be cytotoxic, which restricts the hydrogel's ability to be used for some biomedical applications. Crosslinking agents and initiators can be avoided by using fast and reliable click reactions, such as 1,3-dipolar addition and Diels–Alder. Other covalent crosslinking methods include aldol condensation, imine formation, esterification, and Michael addition.

Physically crosslinked gels are mostly based on intermolecular interactions rather than covalent bonds but usually do not require a crosslinking agent. The lower strength allows the bond to dynamically break and reform, giving the gel self-healing properties and enhanced recyclability. Self-healing properties are explored more in-depth in the final section of this chapter relating to flexible devices. Types of physical crosslinking methods include

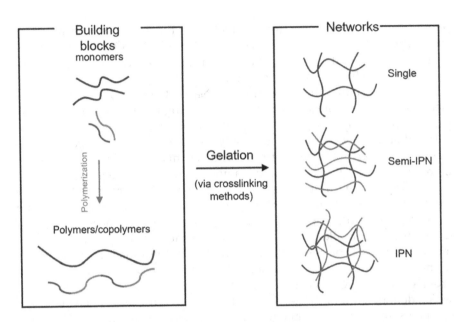

FIGURE 2.2
Synthesis schematic of building blocks and polymer networks of hydrogels. (Adapted with permission [11], Copyright 2020, American Chemical Society.)

hydrogen bonding, ionic bonding, π-π stacking, and intermolecular interactions based on polarities such as polar-polar or apolar-apolar. Other physical interactions that affect the strength of the gel are the two types of polymer chain entanglement. Each is a consequence of the type of crosslinking bond utilized. Alongside that, amorphous polymeric materials can entangle in two ways which can be topological or cohesional entanglement. In that sense, topology is based on the three-dimensional entanglement of different polymeric chains through a physical process. Cohesional entanglement is based on the interaction between the neighboring segments that present attractive forces between each other. Topologically entangled structures have greatly enhanced mechanical properties and swelling properties because the chains are roughly entangled. The only way to alter the structure is to break the bonds that make up the chains. For this reason, the mechanical strength that natural hydrogels lack is well supplemented by the implementation of an IPN, a form of topological entanglement.

There are several methods by which crosslinking can occur. They can be divided into two main types: physical and chemical. Physically crosslinked polymers can take place generally through hydrogen bonding, ionic interactions, and the anchoring effect of dandling groups in polymeric chains, among other methods. Yet, another method to obtain a physically crosslinked hydrogel can be based on the crystallization process. It consists of allowing a polymer to swell when immersed in a proper solvent which is followed by a freeze-thaw process. Also, some polymers such as L-polylactic acid and D-polylactic acid can develop a stereo complex when mixed which is a type of physical crosslinking. Chemical crosslinking can also occur in several ways. Some examples of that include chemical reactions with complementary groups such as –OH and –COOH, –NH$_2$ and COOH, –OH and –N=C=O, and –OH and –C(O)H, among many others [15].

2.3 Hydrogels for Supercapacitors

A capacitor consists of two parallel plates separated by a dielectric. A typical dielectric capacitor can deliver its stored charge in just a few milliseconds, but the magnitude of that charge is greatly limited, restricting its use to circuit components in electrical systems. The specific energy of dielectric capacitors is simply too low to be a viable grid-scale energy storage option, necessitating modifications to achieve a higher capacitance. Even so, the specific energies of supercapacitors remain outclassed by that of batteries. Supercapacitors can be classified into two main types based on the way they achieve their heightened capacitances over unmodified dielectric capacitors. There are electric double-layer capacitors (EDLCs) and redox-active pseudo-capacitors. Supercapacitors are suitable for any application in which a high number (>100,000) of a rapid charge-discharge cycle is required. Everyday uses of supercapacitors include consumer electronics and regenerative braking in electric vehicles. Supercapacitors, however, may suffer from passive self-discharge over long periods. Also, the output voltage of a supercapacitor decreases linearly with discharge time. This means that supercapacitors that are to be used for an application that requires a steady output voltage require further modification to the output, losing additional energy in the process. All supercapacitors benefit from a high surface area, small electrolyte ion size, and high ion mobility. Currently, the most common material used for supercapacitor electrodes is activated carbon materials due to their low cost, high surface area, chemical stability, and porosity [11]. Also, the electrolyte is the component responsible for allowing ionic transport from one electrode to the other while preventing the flux of electrons [16]. They can be aqueous or organic, a liquid compound containing free ions [17], a solid state, or redox-active [18]. The need for ion mobility and interfacial contact between the electrolyte and electroactive materials has made liquid electrolytes the standard choice. In this sense, the water adsorption of hydrogels may allow them to achieve similar ionic transport properties. The benefit of that is that a hydrogel electrolyte can eliminate the leakage of the device and avoid reacting with the electrode materials, increasing its lifetime, and avoiding some safety concerns. This makes hydrogel electrolytes a viable choice for high-performance supercapacitors.

2.3.1 Hydrogel-based Electric Double-Layer Capacitors

EDLCs exploit the electrostatic interactions on the interface between electrode and electrolyte to store charge. The specific capacitances of EDLCs are lower than that of redox-active pseudocapacitors but can be increased in primarily two ways: increasing the surface area and appropriately matching the size of the electrolyte ion to the pores of the electrode material. Because the performance of EDLCs is overwhelmingly affected by these two aspects, this section will focus on them. A high surface area benefits EDLCs by increasing the area of a double layer formed at the interface. Maximizing the number of pores is a reliable way to increase the surface area of a material but may limit how well the electrolyte ions can access the pores, especially if the sizes differ greatly. This makes the matching of ion and pore size an important aspect in maximizing the specific capacitance of the material [18]. The size of solvated, partially solvated, and desolvated ions are all viable options when choosing a pore size. Considerations must also be made for the interactions between the ion and solvent.

Hydrogels created from carbon-based materials, especially graphene, are a common electrode material for EDLCs with their high surface area and tunable porosity. In one study [19], a facile method was developed for creating agarose-based hydrogels as precursors to a porous carbon (PC) structure with tunable pore sizes. Agarose, a naturally sourced material, was especially attractive because of its eco-friendliness, low cost, and appreciable strength. When heated, it formed a sol state, which was a colloidal suspension of particles. After cooling, it formed a gel. This allowed easily tunable pore sizes based on the concentration of the colloid concerning the solvent or the temperatures at which the gel was carbonized. To synthesize the gel, agarose and an activator, potassium oxalate ($K_2C_2O_4$), were dissolved in water and heated. Freeze-drying the hydrogel converted it to an aerogel, which was subsequently carbonized to a PC structure. The primary purpose of the activator was to introduce functionality into the structure, but $K_2C_2O_4$ was also able to generate additional micropores by converting to CO and CO_2 when carbonized.

A series of optimizations were carried out concerning the agarose:$K_2C_2O_4$ ratio and carbonization temperature. The ratio varied between 1:1 and 1:6, and the temperature varied between 600 and 800°C. Another sample used no activator. To investigate the effects of the $K_2C_2O_4$ activator further, samples were also created using other activators such as KOH, K_2CO_3, and KCl. It was found that as the proportion of $K_2C_2O_4$ increased, the size of the pores approached a uniform distribution and the thickness of the pores' walls decreased. There was a limit, however. Too much of the activator in comparison with the agarose meant that gelation could not occur. When using no activator, the resulting structure had far fewer pores, especially micropores. When using KCl as the activator, all the pores were closed off from the others. From these trends, the authors concluded that $K_2C_2O_4$ was unique in that it coordinated the pore structure while also interconnecting the pores as it released its gases upon carbonization. Next, the carbonization temperature varied using samples of 1:6 agarose: $K_2C_2O_4$ ratios. Some of the analyses made from these observations consisted of scanning electron microscopy (SEM), transmission electron microscopy (TEM), cyclic voltammetry (CV), and galvanostatic charge-discharge (GCD). It was observed that the activation with $K_2C_2O_4$ promoted a drastic change in morphology when compared with the neat sample as relatively uniform pores were formed displaying a sponge-like structure. In that sense, KCl also promoted a variation in morphology, however, it seemed to have led to a rougher structure rather than a porous one. Hence, the thermal decomposition of $K_2C_2O_4$ into CO and CO_2 seemed to have played a critical role in optimizing the hydrogel's morphology to yield an open pore structure. Based on this phenomenon, at 600°C, the decomposition process of $K_2C_2O_4$ into K_2CO_3 and CO was incomplete, which resulted in fragmented pores. At 700°C, a fully activated interconnected structure with homogeneous pore sizes was seen. At 800°C, gaseous potassium forms resulted in the merging or collapse of pores, and therefore the decrease in surface area. Under this line, the agarose:$K_2C_2O_4$ ratio of 1:3 achieved a higher specific capacitance as well as cycling stability than the 1:6 sample. Surface areas of up to 1754.9 m^2/g and pore sizes of up to 2.643 cm^3/g were achieved when using an agarose:$K_2C_2O_4$ ratio of 1:3. Based on that, the optimization of morphology, surface area, and electrochemical performance could be observed for the 1:3 agarose:$K_2C_2O_4$, which displayed the best electrochemical performance among the other samples with 166 F/g of specific capacitance at a current density of 0.125 A/g. It was noticed that samples showed a typical electrochemical behavior for an EDLC profile due to the squared-shaped CV along with the triangular shape on the GCD. Such profiles suggested that the capacitance was predominantly occurring due to the (de)insertion of ions into the porous structure of the PC as would be expected for a carbon-based hydrogel.

2.3.2 Hydrogel-based Redox-Active Pseudocapacitors

Redox-active pseudocapacitors use fast faradaic reactions during charge and discharge to allow a greater number of electrons to participate in the redox process. A high surface area benefits redox pseudocapacitors by increasing the number of sites at which redox reactions can take place. One common example of materials incorporated in pseudocapacitors is conducting polymers. Yet, they require a supportive material to have good mechanical properties and avoid conductivity loss from the swelling and shrinking that takes place during each charge-discharge cycle. Polyaniline (PANI) and polypyrrole (PPy) are examples of such polymers. Their integration into a hydrogel structure, however, has challenges. Because these polymers are insoluble in most organic solvents and water, the synthesis route of a heated dissolution followed by cooled gelation is strongly unlikely. However, one way to counter that issue consists of an in-situ polymerization, where a pregelation dispersion is made, followed by polymerization. Through that, this limitation can be overcome. Carbon allotrope support structures including carbon nanofibers (CNF), carbon nanotubes (CNT), and graphene are used extensively as they adapt well to the volumetric changes during charge and discharge and provide additional electron-conductive pathways [20]. These carbon structures can also be doped with heteroatoms, typically N and O, to further increase redox activity. Also, transition metal oxides and phosphides are other popular materials because of their high surface areas and good reversibility but are entirely constrained to the domain of stationary nonflexible applications.

Based on the use of conducting polymers on hydrogels, Zhou and coworkers developed an oxidant-templating method for creating PANI hydrogels [21]. For that, $V_2O_5 \cdot nH_2O$ nanowires were used as a sacrificial template for PANI nanofibers throughout the structure formed through oxidative polymerization. The templating effect was extremely important to the formation of a hydrogel structure, as otherwise, the polymer would simply precipitate. As the $V_2O_5 \cdot nH_2O$ wires were reduced, they became soluble, which resulted in wires that were pure PANI. In this sense, FT-IR analysis confirmed the presence of PANI and EDX confirmed the absence of vanadium. The oxidative polymerization was extremely quick – complete within 10 seconds of mixing. Once the viability of the templating was realized, hydrogels were created using PPy and PEDOT. Three dispersions of $V_2O_5 \cdot nH_2O$ of volume fractions 0.43%, 0.72%, and 0.90% were obtained. It was also noticed that the mechanical strength increased as the volume fraction increased. The resulting PANI hydrogels were named PHG-1, PHG-2, and PHG-3, respectively. PHG-2 was utilized as a supercapacitor electrode without a binder simply being cut into the desired shape and pressed onto the current collector. It achieved a specific capacitance of 636 F/g at 2.0 A/g, and a cycling stability of 83.3% after 10,000 cycles. This was superior to drop-cast PANI films, which lost roughly 50% of their capacitance after 10,000 cycles. Two aspects of PANI films limited their cyclic stability: the electrochemical degradation and the volumetric changes during charge and discharge. The hydrogel network PHG-2 was able to cope with the swelling/shrinking and exhibited a lower ohmic drop compared with the PANI film (Figure 2.3). This research exemplifies oxidant-templating as a facile and reliable way to synthesize conducting polymer hydrogels.

2.3.3 Hydrogel-based Hybrid Supercapacitors

Asymmetric supercapacitors are defined as those whose electrodes differ in material. Hybrid supercapacitors are a subset of asymmetric devices that specifically combine the properties of a battery-like (faradaic electrode) and an ELDC-like (non-faradaic) electrode, making them a subset of the previously mentioned asymmetric devices [22]. Hybrid supercapacitors can

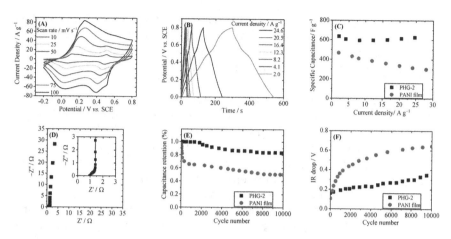

FIGURE 2.3
Electrochemical testing of PHG-2. (a) Cycling voltammetry testing at different scan rates. (b) Charge-discharge curves at different current densities. (c) Specific capacitance at different current densities of PHG-2 compared with PANI film. (d) Nyquist plot. € Cycling stability testing of PHG-2 compared with PANI film. (f) The ohmic drop of PHG-2 compared with PANI film. (Adapted with permission [21], Copyright 2018, American Chemical Society.)

operate within a larger voltage window, granting them higher capacitance and energy density compared with their symmetric counterparts. They also possess an EDLC-like lifetime and a lower rate of self-discharge.

Increasing the operating voltage window of a device has a great impact on maximizing its specific energy. This makes asymmetric supercapacitors (ASCs) an appealing option for this purpose as one can strategically overlap the voltage windows of two different electrode materials to achieve an overall wider potential range. Chen and coworkers [23] created an ASC consisting of a $Ti_3C_2T_x$@NiO-reduced graphene oxide (RGO) hydrogel anode and a defective RGO (DRGO) hydrogel cathode with a 1 M KOH electrolyte. It possessed a wide potential window of 1.8 V and achieved an energy density of 79.02 Wh/kg at 450 W/kg and 45.68 Wh/kg at 9000 W/kg. Stability testing showed capacity retention of 95.6% after 10,000 cycles at 10 A/g. Both hydrogels were produced hydrothermally, beginning with a sonicated dispersion of graphene oxide in water. To create the anode, the $Ti_3C_2T_x$@NiO particles were dispersed into the suspension. $Ti_3C_2T_x$@NiO particles were created by depositing NiO nanoflowers onto the $Ti_3C_2T_x$ sheets. This material was one of the first MXenes ever synthesized. It is a popular anode material because of its high conductivity and hydrophilicity [18]. Also, electrodes made from $Ti_3C_2T_x$ have been found to have good stability and high energy and power densities. However, pristine $Ti_3C_2T_x$ sheets and graphene oxide sheets tend to aggregate, resulting in lower surface areas and reduced ion transport [24]. Introducing the $Ti_3C_2T_x$ into a hydrogel structure helped to lessen the aggregation of the anode. Also, the defects present in the DRGO helped to lessen the aggregation of the graphene sheets in the cathode, as well as increase the size and number of pores. Through nitrogen absorption testing, the Brunauer–Emmett–Teller (BET) specific surface area of the RGO was found to be 198 m^2/g compared with the DRGO with a specific surface area of 591 m^2/g (Figure 2.4a,b). Aside from improving the dispersibility of RGO, the synthesized hydrogel dispensed the use of binders. The advancement of the materials was exemplified well in testing. The specific capacitances at 0.5 A/g were 112 F/g, 623 F/g, and 979 F/g for the pristine $Ti_3C_2T_X$, $Ti_3C_2T_X$@NiO, and $Ti_3C_2T_X$@NiO-RGO, respectively. Comparing the specific capacitance, also at 0.5 A/g, of the defect-free RGO and DRGO hydrogels, it was observed that there was an increase in the number of pores that could be correlated with an improvement in capacitance as it went from 178 to 261 F/g, respectively.

The DRGO hydrogel also showed an increase in cycling stability after 10,000 cycles to 95.6% from 94.4% of the RGO hydrogel. The CV testing of each electrode individually (Figure 2.4c) guided the tuning of the voltage window to 1.8 V. At 1.9 V, oxygen evolution began (Figure 2.4d), so the range

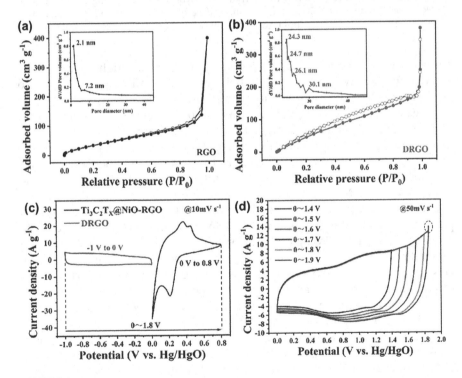

FIGURE 2.4
(a) N$_2$ adsorption of RGO hydrogel. (b) N$_2$ adsorption of DRGO hydrogel. (c) Individual voltage range testing for each ASC electrode, $Ti_3C_2T_X$@NiO-RGO, and DRGO. (d) CV curves of ASC at different potentials. (Adapted with permission [23], Copyright 2022, American Chemical Society.)

was set to 1.8 V to keep the device stable. The ASC was tested at different current densities, indicating highly reversible operation through its triangular shapes. Based on the electrochemical performance, it could be noted that the hydrogel structures reduced aggregation, removed the need for a binder, and by their wet, mesoporous structures, allowed greatly enhanced ion mobility and diffusion.

2.4 Hydrogels for Batteries

Batteries are one of the most common types of electrochemical devices today. When charged, the metal is intercalated within the anode. This intercalation physically degrades the battery, limiting the number of charge/discharge cycles it can endure. Discharging the battery splits the metal into an ion, which travels through the electrolyte, and an electron, which is delivered through the load. Unlike supercapacitors, they can supply a current at a steady voltage during discharge for extended periods, making them suitable for devices that have a similarly steady power draw. Electric vehicles are a relevant example, as the development of efficient and long-lifetime battery technologies is a major impediment to their widespread use.

The construction of sustainable and affordable rechargeable batteries requires abundant or otherwise cheap naturally sourced materials. High specific energy, high specific power, and long lifetimes are all desirable qualities for a battery to have [25]. The specific energy is determined by the voltage the cell can achieve and its capacity, while specific power depends on the rate capability of the cell, which is primarily determined by ion mobility, electron transport, and kinetic considerations. The lifetime of a cell is directly determined by the reversibility of its charge-discharge cycles. This quality is quantified by Coulombic efficiency, which describes the ratio of energy put into the system during charge versus the energy released during discharge. To achieve a long cell lifetime, Coulombic efficiency must be very close to 100%. No system is perfectly efficient and will inevitably degrade over many cycles. High internal resistance from corrosion or high temperatures, side reactions, and high charge-discharge currents hurt the Coulombic efficiency of the cell.

2.4.1 Metal-Ion Batteries

Metal-ion batteries are cells in which a single type of metal ion moves between the anode and cathode. Today, the majority of rechargeable batteries are of the lithium-ion type, which powers mobile phones and laptop computers. Li-ion batteries (LIBs) have a greater operating potential and boast a much higher energy density and a lower rate of self-discharge than other common cell types but suffer from a higher production cost and the risk of lithium dendrite formation on the anode, which can short circuit the system or result in thermal runaway, putting their safety into question. Zinc-ion batteries (ZIBs) exist as a prospectively safer, lower-cost, and more eco-friendly alternative to LIBs [26, 27]. They are compatible with aqueous electrolytes, while other multivalent metal ions such as magnesium and aluminum are not. Issues remain, however, with dendrite formation and side reactions. Gel electrolytes are applicable here for their higher ionic conductivity when compared with solid-state electrolytes. If LIBs are desired, however, a hydrogel electrolyte consisting of polyethylene oxides is useful as dendrite prevention because they tend to act as a host material for lithium ions [28].

A solid electrolyte interface (SEI) forms in most LIBs between the anode and electrolyte, acting as a passivation layer to prevent electrolyte decomposition. The stability of the SEI is greatly influenced by the type of binder used in the battery's construction [29]. Therefore, the optimization of the binder material is of importance to the cycling stability of a LIB. Sandu et al. [29] synthesized a poly(3,4-ethylene dioxythiophene):poly(styrene sulfonate) (PEDOT: PSS) hydrogel binder for LIB. It has several advantages over the organic binders typically found within batteries. Regular PEDOT: PSS has an adhesive strength of 55-65 N/cm^2, which is already greater than the strength of a typical polyvinylidene fluoride (PVDF) binder at 35–40 N/cm^2. The PEDOT: PSS hydrogel exhibited an even greater adhesive strength, but saw a decrease in conductivity from 0.69 ± 0.03 S/cm to 0.20 ± 0.02 S/cm. The authors attributed this loss of conductivity to the mechanical disruption of the PEDOT: PSS matrix during ball milling and the partial reduction of the material. When mixing additives into regular PEDOT: PSS solutions, the dispersion was found to be unstable. Silicon nanoparticles and carbon both failed to remain in the suspension, forming sediments (Figure 2.5a). When using the PEDOT: PSS hydrogel, however, viscosity was increased enough to keep the additives colloidally suspended, resulting in electrode slurries that were highly stable for months, remaining a single homogeneous phase. Graphite/Al particle and Sn/Si particle systems were both investigated. The incorporation of aluminum particles proved only possible with the hydrogel binder, and the silicon nanoparticle system showed a greatly improved cycling stability and rate performance when using the hydrogel formulation rather than the regular PEDOT: PSS (Figure 2.5d). The ability of a hydrogel matrix to increase adhesive strength and support the incorporation of nanoparticle additives and ensure their homogeneity was demonstrated.

2.4.2 Metal-Air Batteries

Metal-air batteries have attracted a lot of interest because of their theoretically high specific energies. Lithium, iron, and zinc are popular choices for their construction, but other options include aluminum, magnesium, and calcium. Most of these use aqueous electrolytes such as KOH, LiOH, or

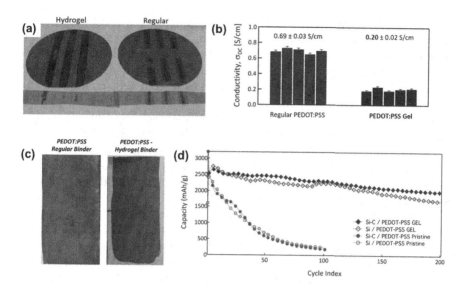

FIGURE 2.5
Mechanical and electrochemical comparison of regular PEDOT: PSS and hydrogel PEDOT: PSS. (a) Adhesion strength testing. (b) Conductivity testing. (c) Visual demonstration of sedimentation in a commercial binder and lack thereof in the hydrogel binder. (d) Cyclic stability testing comparing the binder types with a cycling rate of ~300 mA/g. (Adapted with permission [29], Copyright 2017, American Chemical Society.)

NaOH. Metal-air batteries differ from other cell types as they use a fully metal anode and an open-air cathode. During discharge, O_2, CO_2, and H_2O from the air are reduced, and the metal anode is oxidized. Like metal-ion batteries, the formation of metal dendrites on the anode is a complication that must be dealt with. Another aspect to consider is passivation: the natural formation of oxide films on the surface of metals. While this layer is useful in other applications to prevent corrosion, in batteries it only hinders the performance of the cell. When charging aluminum and zinc batteries, it is generally easier to produce H_2 gas than it is to produce the respective metal. The intentional heightening of the overpotential for hydrogen evolution is required to limit it as a side reaction.

Liu and coworkers [30] created several Al-air coin cell batteries using chitosan as a polymer membrane separator. Chitosan is a highly affordable and popular naturally sourced polymer that is derived from chitin by deacetylation. Chitin is a material sourced from the exoskeletons of crabs, shrimp, and other crustaceans [31]. Because it is dissolvable in low concentrations of weak organic acids, a facile synthesis method was possible, which was based on the digestion of chitin in a weak organic acid solution for one day. After a drying process, a chitosan hydrogel (CHG) membrane was obtained. Three additional gels were created, following an identical procedure with the exception that SiO_2, SnO_2, or ZnO were added to the aqueous solution to suppress the anodic corrosion rate. SEM imaging showed the existence of homogeneously dispersed particles throughout the structure. Elemental mapping of these particles proved they were SiO_2. Several coin cells were constructed from a Mn_3O_4/C air cathode, an Al metal anode, and a varied CHG membrane. The discharge performances of coin cells containing the three 10% by weight SiO_2, SnO_2, and ZnO CHGs were studied. SiO_2 maintained the highest voltage plateau and the longest discharge time. Two additional gels were obtained using SiO_2 with weight percentages of 5% and 15% for comparison. Again, SiO_2 performed best. It was found that SiO_2 and SnO_2 were originally introduced to inhibit corrosion as well as to enhance the discharge performance of the coin cells. The authors attributed this increase in performance to the formation of SiO_3^{2-} and SnO_3^{2-}.

The recyclability of the coin cells was demonstrated. First, a discharged cell was soaked in a diluted acetic acid solution to fully dissolve the hydrogel membrane. The cathode and other cell components like the waterproof carbon paper cover and stainless-steel shell can be reused for new cells. $Al(OH)_3$, the discharge product, was washed off from the aluminum anode with additional diluted acetic acid. Adding sodium bicarbonate caused the remaining aluminum ions to precipitate out of the solution. This recycling method process could be done with only food-grade ingredients. For example, vinegar and baking soda could be used as they are diluted acetic acid solution and sodium bicarbonate, respectively. Likewise, any required filtration steps can be done with a small funnel and a coffee filter. This research stands as a promising example of a sustainable, affordable, and facile fabrication of Al-air coin cells with a good specific energy of 288.5 mA/g at 1.0 mA/cm². They are completely recyclable and highly portable, pointing to small-scale consumer electronics such as hearing aids and wristwatches as suitable applications.

2.4.3 Metal-Sulfur Batteries

Lithium-sulfur batteries (LSBs) exist as a cheaper and potentially more powerful alternative to LIBs due to their high theoretical energy density of nearly 2600 Wh/kg and abundance on Earth. However, to match LIBs in specific capacitance, one would have to increase the sulfur loading of LSBs to at least 6 mg/cm², which is beyond the current stable limit. LSB technology is also still held back by sulfur's low electrical conductivity at room temperature, volumetric changes during charge and discharge, the formation of lithium dendrites, and the shuttle effect caused by the buildup of polysulfides in the electrolyte [32]. The shuttle effect describes the diffusion of these polysulfides between the anode and cathode, causing capacitive losses, poor cycling stability, and corrosion of the anode. Therefore, the mitigation of this effect is the most direct route to increasing the energy density and lifespan of LSBs. Research aimed at mitigating the shuttle effect is focused on the encapsulation of the sulfur on a host material, insertion of functional layers or additives, or modification to the separator or electrolyte [33].

In one study, a precursive conductive polymer hydrogel was synthesized from PANI, phytic acid, and Super P nanoparticles, a commercially available conductivity enhancer [34]. Gelation was carried out on melamine foam, which has a greatly porous structure and good mechanical stability. Pyrolysis of the sample at 1000°C for 3 h yielded a carbonized 3D monolithic structure. The schematic for the synthesis of the hydrogel composite is presented in Figure 2.6a and its visual aspect and dimensions are shown in Figures 6b and c. The SEM analysis (Figure 2.6d) revealed pore sizes ranging from 200–500 nm that were evenly distributed throughout the 3D network and TEM (Figure 2.6e) showed a hierarchically porous structure. Alongside that, the highly porous structure as well as the element distribution at the hydrogel's composite surface can be seen in Figure 2.6f and Figure 2.6g, respectively. The specific surface area of the material was determined to be 483 m^2/g through N_2 absorption testing. Using the Barrett–Joyner–Halenda (BJH) method, pore volume was found to be 0.213 m^3/g. Next, the incorporation of cobalt or copper metal-organic frameworks (MOFs) was carried out. It was found to influence the pore dimensions, thereby affecting sulfur-loading and sulfur-retention capabilities. The sulfur-loading capacity was increased from 15 g/cm^3 to 17.1 g/cm^3 by ZIF-67, a cobalt MOF, and increased to 18.8 g/cm^3 by HKUST-1, a copper MOF. HKUST-1, with a pore size of 0.9 nm, was

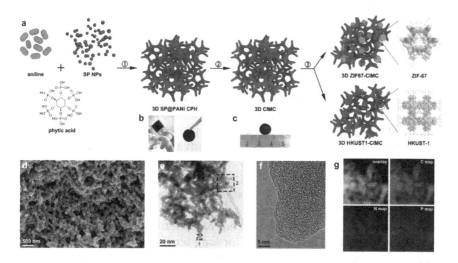

FIGURE 2.6
Scheme for the synthesis of a PANI and MOF-based hydrogel composite. (a) The synthesis scheme of monolithic ZIF-67 and HKUST-1 MOF electrodes. (b) Photograph of conductive polymer gel and cut 15 mm disc. (c) Post-carbonization 10 mm disc. (d) SEM image of undoped carbon electrodes. (e) TEM overview. (f) Area 1 from part e. (g) Area 2 from part e. (Adapted with permission [34], Copyright 2019, American Chemical Society.)

found to inhibit the shuttle effect more than ZIF-67, with a pore size of 0.34 nm. In this sense, the size of the polysulfide species was analyzed through UV-vis spectra. Hence, based on the measurements of surface area from BJH and polysulfide size determination through UV-Vis spectra, it was noted that S_8 polysulfide molecules were 0.68 nm in size, which enabled them to be accommodated by HKUST-1 but not ZIF-67.

The specific surface area was increased from the baseline 483 m^2/g to 634 m^2/g and 685 m^2/g by HKUST-1 and ZIF-67, respectively. Likewise, the BJH pore volume was increased from 0.213 cm^3/g to 0.438 cm^3/g by HKUST-1 and 0.488 cm^3/g by ZIF-67. The additional surface area and pore volume were attributed to the increased number of pores granted by MOF nanoparticles. The best-performing monolithic electrode (HKUST-1) had an aerial capacity of 16 mAh/cm^2 – roughly four times that of LIBs – and a volumetric capacity of 1231 mAh/cm^3. It retained 82% of its capacity after 300 cycles at 0.2 C up to a current density of 6.27 mA/cm^2. Overall, the monolith owed its high performance to its highly conductive N, P co-doped structure and hierarchal porosity. Also, the accessible macro-, meso-, and micropores functioned as excellent hosts for MOFs and soluble sulfur particles. The hosted MOFs themselves provided additional pores and further heightened the sulfur hosting and retention capability of the material. This research points to targeting a specific pore size as an avenue to increase sulfur loading and combat the shuttle effect, for which gel materials are an especially recognizable candidate.

2.5 Hydrogels for Fuel Cells

Fuel cells involve the reaction between a fuel such as hydrogen, natural gas, methanol, or ethanol, and an oxidizing agent, typically air or pure oxygen gas. This reaction is stoichiometrically identical to combustion, but the closed system of a fuel cell grants several advantages. Traditional combustion engines operate at an efficiency of about 33–35%, while fuel cells can reach over 60%, and in the case of carbon-based fuels, the carbon dioxide byproduct can be contained, isolated, and electrochemically reduced, nullifying its effects on the environment. Fuel cells can provide great amounts of energy, surpassing even batteries, but require much more time to deliver them. This makes fuel cells suitable for long-term energy generation, either to the grid, for remote locations, or for applications where an emission-less energy source is necessary – within indoor or otherwise sealed environments. Fuel cells are typically classified by the fuel they consume or by their electrolyte type. The four most common types of fuel cell electrolytes are the proton exchange membrane (PEM), anion exchange membrane (AEM), phosphoric acid, and solid oxide. Qualities important to the performance of a fuel cell include thermal stability and ion conductivity of the membrane, be it proton conductivity for PEMs or hydroxide conductivity for AEMs.

The most widely used PEM material is Nafion, an expensive and difficult-to-synthesize perfluorinated sulfonic acid (PFSA) ionomer. Nafion forms ion-conducting hydrophilic channel structures that provide excellent proton conductivity for PEMs. However, it is sensitive to changes in humidity, and when engineered into thin-film structures, it undergoes structural changes that hurt its proton conductivity and mass transport. One goal that has been a focus of research is to increase the conductivity by increasing the concentration of acid groups in the ionomer structure. Based on that objective, Katzenberg and coworkers developed an ester-bond crosslinked PAA-PVA hydrogel [35]. This sidesteps the inherent difficulties in engineering the complex morphology of a PFSA catalyst layer. For example, PFSA thin-film limitations are avoided due to the gel's homogeneous structure. Echoing the strategy employed with PFSA materials to achieve high conductivity, the material was doped with sulfuric acid. The schematic for the adsorption of the acid into the hydrogel's structure is provided in Figure 2.7a. The stability of the ester crosslinks under high acid concentrations was a concern, but FT-IR spectra taken after submersion in sulfuric acid showed a peak at 1145 cm^{-1}, an absorbance indicative of ester bonding as shown in Figure 2.7b. The signal persisted into further spectra taken at several time intervals up to 66 h of submersion, showing an acceptably slow rate of ester bond degradation at room temperature. The proton concentration of the electrolyte has two contributors: the acidic groups of the PAA-PVA polymer network and the acid dopant. The contribution of the polymer network, however, can be safely assumed to be negligible, as the pK_a of the acid dopant is significantly lower than the acidic polymer groups. This means the proton concentration is directly proportional only to the amount of acid the material has absorbed, allowing the assumption that the proton concentration of the electrolyte is simply twice the concentration of the diprotic acid dopant. The proton concentration at different acid dopant concentrations is shown in Figure 2.7c. When treated with 1 M H_2SO_4, the mass of the material increased by 3.4 times, meaning that the newly doped material's mass was over 75% sulfuric acid. The proton concentration was calculated to be more than 15 mmol g^{-1}, a value over 12 times higher than some Nafion electrolytes. This high concentration of sulfuric acid granted the material a proton conductivity as high as 350 mS cm^{-1}, independent of humidity as presented in the graphs from Figure 2.7d. The swelling ratio under humid conditions, however, was positively correlated with the concentration of sulfuric acid as observed in Figure 2.7e. These properties were maintained even when applied to thin-film structures (7–100 nm) and across a range of humidity levels (99% to <10% relative humidity).

The AEM of an anion exchange membrane fuel cell (AEMFC) requires high hydroxide ion conductivity and good stability in alkaline conditions. A hydrogel membrane, by its incredibly high-water uptake, is an attractive

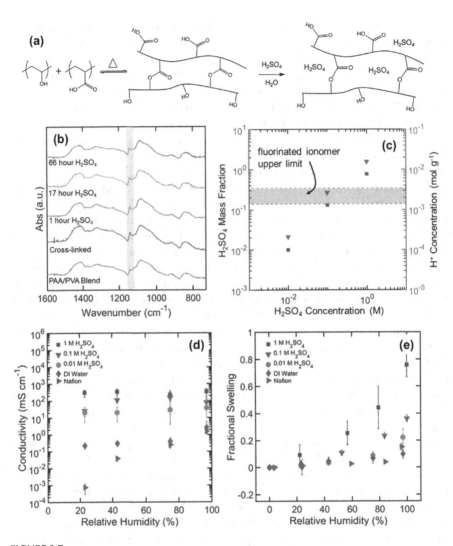

FIGURE 2.7
(a) The synthesis scheme of the PAA-PVA network. (b) FT-IR absorbances of PAA-PVA blend, acid-free crosslinked gel, and acid-submerged crosslinked gel after 1, 17, and 66 hours. (c) Acid concentration versus acid uptake/derived proton concentration. (d) Humidity versus conductivity of Nafion, DI water, and gels treated with 0.01 M, 0.1 M, and 1 M H_2SO_4. ϵ Humidity versus swelling ratio of Nafion, DI water, and gels treated with 0.01 M, 0.1 M, and 1 M H_2SO_4. (Adapted with permission [35], Copyright 2021, American Chemical Society.)

solution to both the needed hydroxide conductivity and alkaline stability. Because of their larger size, hydroxide ions have a transport coefficient that is roughly half that of protons, hurting hydroxide ion conductivity. Fortunately, the hydroxide ion conductivity is proportional to the water absorption in an AEM, making the optimization of the AEM's water uptake a great way to counteract the low transport coefficient and drive the success of AEMFCs. With a larger water uptake, however, accommodations must be made for the increased degree of swelling and risk of electrode flooding.

Yuan et al. [36] created a PVA hydrogel crosslinked with a multi-cation crosslinker (MC) chain that had a water uptake of 282–726 wt.% and a corresponding hydroxide conductivity of 29–70 mS/cm at 20°C and 50–150 mS/cm at 80°C. Additionally, at this temperature, the hydrogel's high-water content allowed it to be stable in 6 M NaOH conditions; after 240 h, the conductivity was still 65% of the baseline. To reduce dimensional swelling, a hydrophobic component, 4-(trifluoromethyl) benzaldehyde (TFBA), was added to the main PVA chain. Extensive AEM testing was done to determine an optimal ratio of each of the components, PVA, TFBA, imidazole-4-carbaldehyde (IM), and the MC, by measuring ion exchange capacity (IEC), water uptake (WU), swelling ratio (SR), elongation, strength, hydroxide conductivity, and gel fraction (GF). The tabulation of this data showed several trends. As the proportion of TFBA increased, the IEC, WU, and SR decreased. As the proportion of MC increased, IEC, WU, and SR increased. The swelling ratio was also shown to be positively correlated with temperature. PVA-TFBA$_{7.84}$-IM-MC$_{20}$ was chosen for the construction of a membrane electrode assembly (MEA), which showed an H$_2$/O$_2$ single-cell peak power density of 715 mW/cm² at 65°C. This result indicated that hydrogel is a viable alternative for the construction of efficient AEMs.

2.6 Hydrogels for Flexible Devices

Electronics are traditionally consigned to be as rigid as the silicon wafer substrates on which they are constructed [37]. This makes them unsuitable for devices that are likely to experience bending, stretching, or compression, such as those that are to be worn. Electrochemical devices are likewise constrained by the rigidity of their electrode and other components. For example, a redox pseudocapacitor utilizing a transition metal oxide cannot cope under stress like a conducting polymer can. Typically, during stress, the electrode of a device is damaged or otherwise lost.

Gel electrolytes, electrodes, and separators are used for flexible devices, as they can endure the strain and remain functional. Even when cut or broken, some can spontaneously self-heal and restore their capacitance. Dynamic covalent bonding and physical interactions are the keys to granting self-healing properties to a material. These kinds of bonds include Diels–Alder, disulfides, imines, borate-esters, olefin metathesis, and phthalocyanine bonds. Physical interactions include hydrogen bonding, metal coordination, and electrostatic interactions [13]. Hydrogels with multiple repair mechanisms are more resilient; if one repair mechanism is disrupted by its environment, the other can remain active. Designing devices that remain operable in a wider range of conditions, like temperature, will aid in the realization of its widespread use.

Tao et al. [38] created a self-healing hydrogel electrolyte by crosslinking sodium alginate with dynamic catechol-borate bonds. These bonds allowed a capacitor based on this electrolyte to spontaneously heal after being cut or broken, retaining 97.3% of its ionic conductivity after ten break-heal cycles. Electrochemical analysis in terms of CV (Figure 2.8a) and GDC (Figure 2.8b) was performed on the self-healing hydrogel. The capacitor delivered a specific capacitance of 97 F/g at 1 A/g and 75 F/g at 10 A/g. After a single cut/heal cycle, its specific capacitance was similar, at 93 F/g at 1 A/g and 55 F/g at 10 A/g. Cyclic stability testing done on a self-healed sample showed only a slight increase in the degradation of its capacitance per cycle. Catechol-borate bonds are also tolerant to high salt contents, avoiding damage to the gel's network (via the salt-out effect) and granting the capacitor the ability to operate in colder environments. Typically, aqueous hydrogel electrolytes, upon freezing, lose capacitance, ion conductivity, flexibility, and self-healing properties. At a temperature of −10°C, the capacitor retained 80% of its room-temperature capacitance. In this sense, Figure 2.8c demonstrates its self-healing mechanism and summarizes its electrochemical properties after subsequent cut/heal cycles. This self-healing mechanism was disrupted by the introduction of urea, and to a larger degree by fructose. It could also be observed that the hydrogel displayed a relative fluctuation in stress-strain properties when different healing times were applied as seen in Figure 2.8d. Also, the hindrance of self-healing properties was analyzed based on the addition of urea and fructose as their hydroxyl groups would react more promptly to the catechol hydroxyl groups, preventing the self-healing process. In this sense, fructose presented a stronger influence in hindering the self-healing process rather than urea as seen in Figure 2.8e.

In another study, Wang and coworkers [39] produced a $Zn-MgO_2$ rechargeable battery using a crosslinked polyacrylamide (PAAm) hydrogel electrolyte. It was capable of enduring large compressive forces, losing negligible amounts of capacity, and maintaining a stable voltage output independent of the compressive load. Interestingly, as the compressive force increased, the ionic conductivity increased. The authors attribute this to the increased contact between the electrode and electrolyte during compression. The battery containing the PAAm hydrogel electrolyte was compared to another with an aqueous electrolyte of the same chemistry. A wristband was constructed from the compressible batteries and a piezoresistive sensor was able to detect the intensity and frequency of pressures applied to the band in good resolution and with consistent output, showing promise as an energy storage device for wearable electronics (Figure 2.9). The device showed that electrochemical stability was negligibly influenced by mechanical stress-strain deformations. This effect suggested that there was likely an intimate contact between the hydrogel and electrodes that allowed it to undergo relatively high mechanical stress without considerable loss of performance.

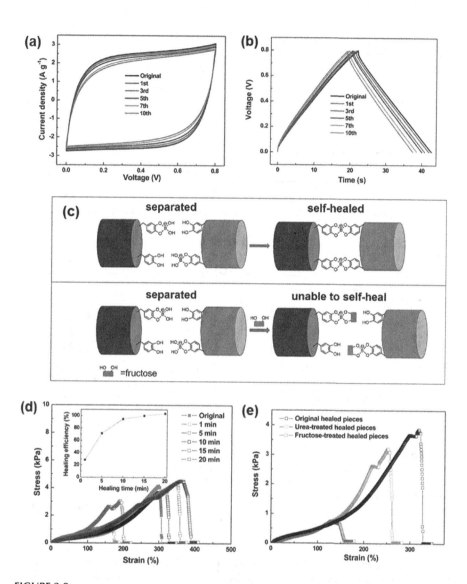

FIGURE 2.8
Hydrogel with self-healing properties based on catechol-borate bonds. (a) CV curves at 100 mV/s scan rate. (b) Charge-discharge curves at 1 A/g. (c) Self-healing mechanism of the catechol-borate bonding – interference of diol in the mechanism. (d) Stress-strain analysis of samples after different self-healing times. € Stress-strain analysis of urea and fructose treated samples. (Adapted with permission [38], Copyright 2017, American Chemical Society.)

Furthermore, the PAAm battery's specific capacity at 4 C was 136.2 mAh/g while the aqueous battery was 154.9 mAh/g. Although comparatively lower in capacity, cyclic stability testing showed that the PAAm electrolyte had greater capacity retention. After 500 charge-discharge cycles, the aqueous electrolyte maintained 90.7 mAh/g, a retention of 58.55%. The PAAm

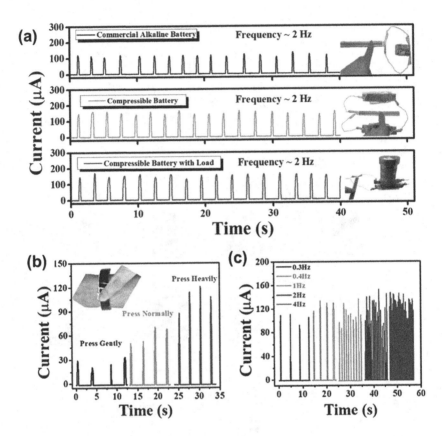

FIGURE 2.9
Pressure sensitivity characterizations. (a) Comparison of pressure frequency/resolution in commercially available pressure-sensitive battery and synthesized battery, with and without a compressive load. (b) Comparison of pressure intensities on a smart wristband by current output. (c) Comparison of pressure frequencies on a smart wristband by current output. (Adapted with permission [39], Copyright 2018, American Chemical Society.)

electrolyte, after 1000 cycles, maintained 94 mAh/g, a retention of 69.02%. This clearly shows that, while sacrificing specific capacity, the hydrogel electrolyte retains a significantly greater amount of its capacitance after many charge-discharge cycles. While the majority of hydrogel-based sensor technologies are, at best, tangentially related to electrochemical devices, they still warrant an aside, as the material's response to certain stimuli can be related to the modulation of its electrochemical characteristics or simply a detectable electrical signal, like through the piezoelectric effect. Hydrogels stand as an excellent platform for the integration of stimuli-responsive modifications. Unique changes in the hydrogel's properties can be brought about by both physical and chemical stimuli such as light, force, pressure, changes in temperature, pH, or magnetic fields, and the presence and/or concentration of molecular species [40].

2.7 Conclusion and Perspectives

This chapter provides some of the recent approaches for the development of hydrogels and their composites for energy storage applications in supercapacitors and batteries. Through that, it could be observed that the introduction of hydrogels into energy storage and flexible devices has been demonstrated to be an efficient strategy to further enhance their performance. Such improvement is mostly based on the inherent properties of hydrogels such as their highly porous morphology and their surface area. To obtain this type of structure, one of the most commonly used methods is freeze-drying or similar approaches that require using a matrix that can be either a polymer, MOF, carbon-based, or a composite that is soaked into a specific solvent, which can promote a satisfactory degree of swelling followed by its removal to form a highly porous matrix. For that, it is important to properly select the components for the synthesis, such as monomers, crosslinking agents, and proper solvents, and the rate of solvent evaporation to obtain a tailored structure suitable for the desired end.

Even though the synthesis of hydrogels is not necessarily a laborious process, it is often time-consuming and leads to a relatively low yield, which is a current challenge for this process. Yet the capability of obtaining a highly porous structure is extremely attractive for energy storage devices such as supercapacitors and batteries. In this sense, hydrogels can function, for example, as solid electrolyte components when obtained as ionically conducting materials based on the flux of ions through their interconnected network. Through that, an effective ion transport can occur, allowing relatively higher potentials to be applied when compared with aqueous and some organic media. Also, hydrogels have demonstrated appreciable mechanical robusticity and elasticity, which makes them suitable candidates to be used in flexible devices that can go through mechanical stress. Also, efficient hydrogel composites suitable as electrodes can be obtained by incorporating electroactive materials into their network structure that can give them appreciable electrochemical properties based on the percolation theory. Based on that, there have been some approaches to developing hydrogel-based composites that can display the inherently high surface area of hydrogels along with the redox activity of transition metal-based compounds that are embedded in their structure. In addition, conductive nanofiller can also be incorporated to facilitate electron transfer steps.

Another scope for hydrogels is their use as self-healing materials due to reversible chemical bonding formation and breakage as well as physical crosslinking. In the field of energy storage, such property can be particularly important in the use of Li-S batteries, for instance, which undergo a great mechanical strain due to the incorporation of bulky S atoms into the electrodes. Therefore, the use of hydrogels can be a feasible option to accommodate the volume expansion that is inherent to this technology. Yet, further work is needed as few hydrogel-based materials can properly function under such harsh conditions. Thus, throughout the discussion provided in this chapter, it is noticeable that hydrogels hold great potential as materials that can be used in the next generation of energy storage devices as they can counter some of the current issues faced by the scientific community.

References

1. R.B. Jackson, A. Ahlström, G. Hugelius, C. Wang, A. Porporato, A. Ramaswami, J. Roy, J. Yin, Human well-being and per capita energy use, *Ecosphere*. 13 (2022) 1–10.
2. EIA, International energy outlook 2021, Int. Energy Outlook. (2021) 1–202.
3. S. Bashir, M. Hina, J. Iqbal, A.H. Rajpar, M.A. Mujtaba, N.A. Alghamdi, S. Wageh, K. Ramesh, S. Ramesh, Fundamental concepts of hydrogels: Synthesis, properties, and their applications, *Polymers (Basel)*. 12 (2020) 1–60.
4. O. Wichterlie, D. Lim, Hydrophilic Gels for Biological Use, *Nature*. 185 (1960) 117–118.
5. X. Sun, S. Agate, K.S. Salem, L. Lucia, L. Pal, Hydrogel-based sensor networks: Compositions, properties, and applications – A review, *ACS Appl. Bio Mater*. 4 (2021) 140–162.
6. M. Greiner, X. Yin, L. Fernández-Díaz, E. Griesshaber, F. Weitzel, A. Ziegler, S. Veintemillas-Verdaguer, W.W. Schmahl, Combined influence of reagent concentrations and agar hydrogel strength on the formation of biomimetic hydrogel-calcite composites, *Cryst. Growth Des*. 18 (2018) 1401–1414.
7. C.S.A. Musgrave, F. Fang, Contact lens materials: A materials science perspective, *Materials (Basel)*. 12 (2019) 1–35.
8. B.D. Ratner, A.S. Hoffman, Synthetic hydrogels for biomedical applications, in: *Hydrogels for Medical and Related Applications*, American Chemical Society, Washington, DC, 1976: p. 1.
9. J.Y.C. Lim, Q. Lin, K. Xue, X.J. Loh, Recent advances in supramolecular hydrogels for biomedical applications, *Mater. Today Adv*. 3 (2019) 100021.
10. J. Zhang, L. Dai, Heteroatom-doped graphitic carbon catalysts for efficient electrocatalysis of oxygen reduction reaction, *ACS Catal*. 5 (2015) 7244–7253.
11. Y. Guo, J. Bae, Z. Fang, P. Li, F. Zhao, G. Yu, Hydrogels and hydrogel-derived materials for energy and water sustainability, *Chem. Rev*. 120 (2020) 7642–7707.
12. J. Duan, W. Xie, P. Yang, J. Li, G. Xue, Q. Chen, B. Yu, R. Liu, J. Zhou, Tough hydrogel diodes with tunable interfacial adhesion for safe and durable wearable batteries, *Nano Energy*. 48 (2018) 569–574.
13. H. Dai, G. Zhang, D. Rawach, C. Fu, C. Wang, X. Liu, M. Dubois, C. Lai, S. Sun, Polymer gel electrolytes for flexible supercapacitors: Recent progress, challenges, and perspectives, *Energy Storage Mater*. 34 (2021) 320–355.
14. J. Maitra, V.K. Shukla, Cross-linking in hydrogels – A review, *Am. J. Polym. Sci*. 4 (2014) 25–31.
15. M. Mahinroosta, Z. Jomeh Farsangi, A. Allahverdi, Z. Shakoori, Hydrogels as intelligent materials: A brief review of synthesis, properties and applications, *Mater. Today Chem*. 8 (2018) 42–55.

16. Y. Lv, Y. Lin, F. Chen, F. Li, Y. Shangguan, Q. Zheng, Chain entanglement and molecular dynamics of solution-cast PMMA/SMA blend films affected by hydrogen bonding between casting solvents and polymer chains, *RSC Adv.* 5 (2015) 44800–44811.
17. T. Zhou, X. Gao, B. Dong, N. Sun, L. Zheng, Poly(ionic liquid) hydrogels exhibiting superior mechanical and electrochemical properties as flexible electrolytes, *J. Mater. Chem. A.* 4 (2016) 1112–1118.
18. Y. Gogotsi, B. Anasori, The rise of MXenes, *ACS Nano.* 13 (2019) 8491–8494.
19. S. Hwang, J. Zhou, T. Tang, K. Goossens, C.W. Bielawski, J. Geng, Agarose-based hierarchical porous carbons prepared with gas-generating activators and used in high-power density supercapacitors, *Energy and Fuels.* 35 (2021) 19775–19783.
20. A. Gupta, S. Sardana, J. Dalal, S. Lather, A.S. Maan, R. Tripathi, R. Punia, K. Singh, A. Ohlan, Nanostructured polyaniline/graphene/Fe_2O_3 composites hydrogel as a high-performance flexible supercapacitor electrode material, *ACS Appl. Energy Mater.* 3 (2020) 6434–6446.
21. K. Zhou, Y. He, Q. Xu, Q. Zhang, A. Zhou, Z. Lu, L.K. Yang, Y. Jiang, D. Ge, X.Y. Liu, H. Bai, A hydrogel of ultrathin pure polyaniline nanofibers: Oxidant-templating preparation and supercapacitor application, *ACS Nano.* 12 (2018) 5888–5894.
22. Y. Shao, M.F. El-Kady, J. Sun, Y. Li, Q. Zhang, M. Zhu, H. Wang, B. Dunn, R.B. Kaner, Design and mechanisms of asymmetric supercapacitors, *Chem. Rev.* 118 (2018) 9233–9280.
23. W. Chen, C. Hao, Z. Qiu, X. Zhang, H. Xu, B. Yu, S. Chen, High-energy-density asymmetric supercapacitor based on free-standing $Ti_3C_2T_X$@NiO-reduced graphene oxide heterostructured anode and defective reduced graphene oxide hydrogel cathode, *ACS Appl. Mater. Interfaces.* 14 (2022) 19534–19546.
24. H. Su, C. Zhang, X. Li, L. Wu, Y. Chen, Aggregation prevention: Reduction of graphene oxide in mixed medium of alkylphenol polyoxyethylene (7) ether and 2-methoxyethanol, *RSC Adv.* 8 (2018) 39140–39148.
25. A. Van der Ven, Z. Deng, S. Banerjee, S.P. Ong, Rechargeable alkali-ion battery materials: Theory and computation, *Chem. Rev.* 120 (2020) 6977–7019.
26. S. Ponnada, M.S. Kiai, R. Krishnapriya, R. Singhal, R.K. Sharma, Lithium-free batteries: Needs and challenges, *Energy & Fuels.* 36 (2022) 6013–6026.
27. B. Wang, J. Li, C. Hou, Q. Zhang, Y. Li, H. Wang, Stable hydrogel electrolytes for flexible and submarine-use Zn-Ion batteries, *ACS Appl. Mater. Interfaces.* 12 (2020) 46005–46014.
28. Q. Lu, Y.B. He, Q. Yu, B. Li, Y.V. Kaneti, Y. Yao, F. Kang, Q.H. Yang, Dendrite-free, high-rate, long-life lithium metal batteries with a 3D cross-linked network polymer electrolyte, *Adv. Mater.* 29 (2017).
29. G. Sandu, B. Ernould, J. Rolland, N. Cheminet, J. Brassinne, P.R. Das, Y. Filinchuk, L. Cheng, L. Komsiyska, P. Dubois, S. Melinte, J.F. Gohy, R. Lazzaroni, A. Vlad, Mechanochemical synthesis of PEDOT:PSS hydrogels for aqueous formulation of Li-Ion battery electrodes, *ACS Appl. Mater. Interfaces.* 9 (2017) 34865–34874.
30. Y. Liu, Q. Sun, X. Yang, J. Liang, B. Wang, A. Koo, R. Li, J. Li, X. Sun, High-performance and recyclable Al-Air coin cells based on eco-friendly chitosan hydrogel membranes, *ACS Appl. Mater. Interfaces.* 10 (2018) 19730–19738.
31. J. Robertson, Chitin and chitosan: Production and application of versatile biomedical nanomaterials, *Am. Math. Mon.* 111 (2004) 915.

32. W. Ren, W. Ma, S. Zhang, B. Tang, Recent advances in shuttle effect inhibition for lithium sulfur batteries, *Energy Storage Mater.* 23 (2019) 707–732.
33. N. Hu, X. Lv, Y. Dai, L. Fan, D. Xiong, X. Li, SnO_2/Reduced graphene oxide interlayer mitigating the shuttle effect of Li-S batteries, *ACS Appl. Mater. Interfaces.* 10 (2018) 18665–18674.
34. B. Liu, R. Bo, M. Taheri, I. Di Bernardo, N. Motta, H. Chen, T. Tsuzuki, G. Yu, A. Tricoli, Metal-organic frameworks/conducting polymer hydrogel integrated three-dimensional free-standing monoliths as ultrahigh loading Li-S battery electrodes, *Nano Lett.* 19 (2019) 4391–4399.
35. A. Katzenberg, C. Muñoz Davila, B. Chen, T. Siboonruang, M.A. Modestino, Acid-doped hydrogel electrolytes for electrocatalyst interfaces, *ACS Appl. Polym. Mater.* 2 (2020) 2046–2054.
36. C. Yuan, P. Li, L. Zeng, H. Duan, J. Wang, Z. Wei, Poly(vinyl alcohol)-based hydrogel anion exchange membranes for alkaline fuel cell, *Macromolecules.* 54 (2021) 7900–7909.
37. M. Zou, Y. Ma, X. Yuan, Y. Hu, J. Liu, Z. Jin, Flexible devices: From materials, architectures to applications, *J. Semicond.* 39 (2018) 011010.
38. F. Tao, L. Qin, Z. Wang, Q. Pan, Self-healable and cold-resistant supercapacitor based on a multifunctional hydrogel electrolyte, *ACS Appl. Mater. Interfaces.* 9 (2017) 15541–15548.
39. Z. Wang, F. Mo, L. Ma, Q. Yang, G. Liang, Z. Liu, H. Li, N. Li, H. Zhang, C. Zhi, Highly compressible cross-linked polyacrylamide hydrogel-enabled compressible $Zn-MnO_2$ battery and a flexible battery-sensor system, *ACS Appl. Mater. Interfaces.* 10 (2018) 44527–44534.
40. J.Y.C. Lim, S.S. Goh, X.J. Loh, Bottom-up engineering of responsive hydrogel materials for molecular detection and biosensing, *ACS Mater. Lett.* 2 (2020) 918–950.

3
Approaches for Hydrogel Synthesis

Ramji Kalidoss and Radhakrishnan Kothalam

CONTENTS

3.1 Introduction 47
3.2 General Properties of a Hydrogel 49
    3.2.1 Swelling 49
    3.2.2 Mechanical Properties 50
    3.2.3 Degradation 50
    3.2.4 Porosity 51
    3.2.5 Self-Healing 51
    3.2.6 Ionic Conductivity 52
    3.2.7 Response to Stimulus 52
3.3 Crosslinking 53
    3.3.1 Physically Crosslinked Hydrogels 53
    3.3.2 Chemically Crosslinked Hydrogels 55
3.4 Graphene-based Hydrogels 57
    3.4.1 Templated Assistive Self-assembly Methods 57
    3.4.2 Mixed Solutions Methods 59
3.5 Polymer-based Hydrogels 59
3.6 Biomass-derived Carbon Hydrogels/Aerogels 60
3.7 Conclusion 62
References 62

3.1 Introduction

Multidimensional mesh networks of an interlinked polymer capable of swelling and resisting dissolving in water are called hydrogels. They are characterized by high flexibility and appear to be jelly-like structures with significant elasticity [1]. This type of material is commonly witnessed in our daily lives such as jiggle desserts, contact lenses, baby diapers, and air fresheners. Their ability to retain water arises from the presence of hydrophilic functional groups in polymeric chains and their resistance to dissolution comes from polymeric crosslinking [2]. Materials with these properties are natural and synthetic hydrogels. Bacterial biofilms, plant structures, gelatin, and agar are some examples of natural hydrogels apart from hydrogels derived from bodily fluids. However, the desired functionality depending on how the application is used cannot be adjusted and their composition may vary between batches as they are not controllable. With these hindrances in natural hydrogels, synthetic hydrogels with enhanced properties have been formulated. The possibilities of gelation chemistry to synthetically induce the desired properties of hydrogels are found in the applications of tissue engineering, energy storage and conversion, drug delivery, and chemical sensors [3]. Hydrogel characteristics such as its biocompatibility, biodegradability, viscoelasticity, and hydrophilicity are suitable for various biomedical applications. Also, the capability to respond to various stimuli such as electric fields, temperature, magnetic fields, ionic strength, and biological molecules makes them suitable for sensor applications [4]. Apart from these, hydrogels have recently found application in energy storage and conversion due to their semi-solid nature and elasticity [5]. Conventional energy storage devices are heavy and the liquid electrolytes possess a serious threat of leakage, which is expensive and toxic in nature, while hydrogels are semi-solid, elastic, and less expensive. Also, hydrogels and derived materials have favorable properties for storage applications such as electrolyte permeability, structural flexibility, and electronic conductivity [6, 7]. Their usage in electrochemical applications ensures the device is flexible and stretchable, enabling it to work in various mechanical deformations [8]. Hydrogels to be applicable for such applications should possess certain physical, chemical, and mechanical properties listed in Figure 3.1. Section 3.2 introduces the necessary properties of hydrogels for different applications. Section 3.3

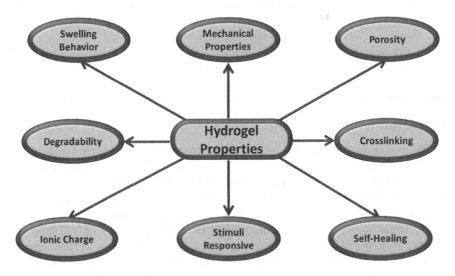

FIGURE 3.1
Significant properties of hydrogels for various applications.

discusses different crosslinking mechanisms to synthesize different types of hydrogels. Finally, in Sections 3.4–3.6 emphasis on graphene-derived hydrogels, hydrogels formed from polymers, and hydrogels derived from biomass is provided.

3.2 General Properties of a Hydrogel

3.2.1 Swelling

Hydrogels swell because networks of crosslinked polymers form when the material is immersed in a liquid solution instead of dissolving within them. Because the polymer network acts as a matrix to bind the liquid together, there is no lag time between the processes, and the ingested liquid acts as a selective filter to allow unrestricted diffusion of particular solute molecules [9]. Depending on the type of hydrogel, they can absorb water at a rate anywhere from 10%–20% up to hundreds of times their dry weight. This establishes a minimum value for the spectrum of feasible absorption rates.

The performance of a hydrogel in various applications can be greatly affected by the behavior of water absorption. Primary water binding occurs when the initial water molecules penetrate the matrix of a dry hydrogel and hydrate the most polar hydrophilic groups [10]. The network expands when the polar groups are moistened, revealing the hydrophobic groups. Water molecules become hydrophobically bound, or secondarily bonded, because of interactions between these hydrophobic groups and water molecules. Thus, primary and secondary bound water together is called total bound water [11]. Furthermore, due to the osmotic driving force of the network chains, water will continue to be absorbed by the network even after the hydrophobic and hydrophilic sites have interacted and bound with water molecules. Additional swelling occurs from the physical or covalent crosslink, leading to the retraction of desolvation of the elastic network. Therefore, the hydrogel will swell until it reaches a point of equilibrium [12].

Additionally, the water that exists outside of the network chains and/or in the center of bigger pores, macropores, or voids is referred to as free water, bulk water, or void water. When this occurs, ionic, polar, and hydrophobic groups have reached their maximum capacity of water absorption. After the water that is bound to ions, atoms, and molecules has reached its maximum capacity, the leftover water is taken into the body by this pathway. If the chains and crosslinks in the network are degradable, there is a chance that the hydrogel will eventually break down and disintegrate as the network expands.

This different category of water absorption by a hydrogel can be estimated using a variety of methods expressed as a percentage of the overall water content of the sample represented in Equation 3.1. Hydrogel water is typically characterized using differential scanning calorimetry (DSC), small molecular probes, and nuclear magnetic resonance (NMR) [2]. These characterizations provide an indirect measure of the degree of swelling that can again be used to assess a wide variety of hydrogel characteristics, such as crosslinking, mechanical strength, rate of degradation, and so on. One of the simplest, least expensive, and most reliable ways to tell a crosslinked gel apart from the uncrosslinked polymer it was originally made from is to evaluate its swelling and swollen state stability.

$$Water\ Uptake = \frac{Swollen\ Weight - Dry\ Weight}{Dry\ Weight} \times 100 \qquad (3.1)$$

3.2.2 Mechanical Properties

The hydrogel's mechanical properties can be tailored to meet the needs of the application. To increase the stiffness of a hydrogel, it can be heated, which increases the degree of crosslinking. Hydrogels have Young's modulus that is unmatched by any other material because of the water and gel matrix combination. The material, environmental factors, and research goals should all be taken into account while designing experiments to determine the causes and effects of changes in mechanical characteristics. The mechanical characteristics (Poisson modulus, young modulus, storage, and loss moduli) can be estimated with a rheometer or DMA tool using any of the numerous available methods [13].

3.2.3 Degradation

Hydrolysis of the crosslinks or the polymer backbone causes hydrogels to deteriorate. This can be accomplished by either incorporating degradable units into the polymer itself or by exposing the polymer network to an enzyme well-suited to the task. Therefore, hydrolysis or enzymatic activities may be responsible for degradation [14, 15]. Polymers undergo hydrolysis when they come in touch with an aqueous medium. A possibility of enzymatic degradation exists if proteolytically degradable domains are incorporated into the polymer's structure. Regardless of the application, the capacity to detect and measure the degradation of the hydrogels is crucial for assessing the performance of a system based on hydrogels. Methods that are commonly used to assess the degradation quality include measuring the hydrogel's swelling, analyzing its mechanical properties, and determining the amount of breakdown products produced. The results of these studies should be comparable and interconnectable with one another because they are all products of the same procedure.

3.2.4 Porosity

The pre-existing smaller gaps within the network or the phase separation during synthesis are two possible causes of porosity in hydrogels. A hydrogel matrix's average pore size, pore size distribution, and pore interconnectivity are all significant but challenging characteristics. In general, these properties are classified as "tortuosity", which describes the overall pliability of the hydrogel matrix [16]. The length of an effective diffusion channel across a barrier hydrogel film can be estimated by multiplying the film thickness by the pore volume fraction ratio and then dividing it by the tortuosity. The composition of the hydrogel polymer network and the density of the crosslinks are two of the most important factors that can influence these other characteristics.

Hydrogels have porous structures; hence, their properties are strongly reliant on the surrounding fluid. This is particularly true for dissolved ionic solutes, which cause Donnan effects, and dissolved uncharged solutes, which partition unequally between the gel and solution phases. These two kinds of solutes are dissolved (osmotic effects). For the objective of rational hydrogel design, knowledge of the pore size distribution is useful. This distribution is determined by hydrogel characterization and is often expressed as a proportion of C and T.

$$\%T = \frac{Weight\ of\ Monomer + Weight\ of\ crosslinkers}{Total\ Volume} \tag{3.2}$$

$$\%C = \frac{Weight\ of\ crosslinkers}{Weight\ of\ Monomer + Weight\ of\ crosslinkers} \tag{3.3}$$

3.2.5 Self-Healing

The reversible intermolecular interactions that take place within polymeric networks give hydrogels the added advantage of being able to self-heal. Despite suffering from extensive damage, hydrogels can revert to their original shape and properties. Hydrogels' self-healing ability can be qualitatively assessed in two ways [17]: either by viewing the scar left behind after the hydrogels have healed, or hydrogels can be suspended such that once they have healed, they can hang horizontally or vertically. Meanwhile, the self-healing ability can also be demonstrated using quantitative approaches such as time-dependent adhesion tests, rheology analysis, and tensile and compressive stress-strain measurements. The self-healing efficiency (ε_H) of a typical tough hydrogel can be defined by the ratio of elongation of the healed (λ_b) with respect to the pristine ($\lambda_{b,0}$) hydrogel represented in Equation 3.4.

$$\varepsilon_H = \frac{\lambda_b}{\lambda_{b,0}} \times 100\% \tag{3.4}$$

3.2.6 Ionic Conductivity

Hydrogels often incorporate functional components such as metal cross-linkers, conductive polymer backbones, solid electrolyte particles, and active electrode materials. Electron transmission over nanofillers, ion conduction in aqueous electrolytes, electron/ion conduction jointly across polymer backbones, and the effect of polymer-electrolyte interaction are just a few examples of the mobility that allows ions and electrons to travel through these materials (solvent and salt) [6]. Certain phenomena illustrate the ability of ions and electrons to move through these substances. Therefore, it is not simple to understand how ions and electrons are carried along the channel. A lack of ion and electron movement will prevent electrochemical energy storage and conversion devices from operating at their optimal levels.

The electrical conductivity of hydrogels can be greatly enhanced by using aqueous electrolytes with high ionic conductivity. Hydrogels can trap large amounts of aqueous electrolytes because of their hydrophilic functional groups in the polymer networks. Hydrogels use water as both the solvent and the charge carriers (also called cations and anions) [18]. The majority of electrolytes are aqueous. The conductivity (σ) of a hydrogel is proportional to the concentration of charge carriers (n_i), the valence of the mobile ion charges (Z_i), and the ionic mobility (μ_i).

$$\sigma = \sum_i n_i \mu_i Z_i e \qquad (3.5)$$

Due to water's high dielectric constant (= 80) and low viscosity (1 cP at 20°C), aqueous electrolytes often have a higher conductivity in hydrogels than organic electrolytes. Water is an effective solvent because it can solvate both cations and anions as a single water molecule (H–O–H) holds both acidity (–H) and Lewis basicity (–O–). Water's high dielectric constant also promotes salt dissociation, resulting in the formation of solvated cations and anions, which leads to an increase in charge carrier concentration (n_i). Meanwhile, the low viscosity of water contributes to an increase in ionic mobility (μ_i) which contributes to a rise in total conductivity (σ).

3.2.7 Response to Stimulus

Changes in light, pH, humidity, temperature, solvent, electrical field, pressure, and ionic strength are just some of the environmental stimuli that can cause hydrogels to shrink, swell, and undergo a sol-gel phase transition. Hydrogels with stimuli-responsive qualities have uses in the medical environment such as drug delivery and the electrical energy storage sector [19]. This is because of their high water content, extreme responsiveness to environmental cues, unique physicochemical characteristics, and the flexibility with which their structural and organizational aspects can be altered. Hydrogels can expand and contract in size due to changes in the water content of their networks, a process known as swelling and de-swelling. Several factors in the surrounding environment can trigger this shift. Polymers undergo structural changes when stimulated, making them more or less hydrophilic, respectively, which results in swelling and deflating. Hydrogels expand when they absorb water and contract when they release water from their bulk phase. This enables the use of hydrogels as stimulus-responsive materials that can respond to a wide range of environmental stimuli by expanding, contracting, deforming, or performing other desirable functions.

3.3 Crosslinking

The discussed properties of hydrogels are derived through types of low molecular weight chemical crosslinkers with polymeric chains. Depending on the application, the crosslinks might be made either physically or chemically (Table 3.1). Numerous processes, such as hydrophobic interactions within chains, complicate collages among a polycation and polyanion, and ionic interactions between multivalent cations and polyanion can result in physical crosslinking (Figure 3.2). Further, chemical crosslinking is the technique that produces gels with chemical bonds. Many different processes, including UV irradiation, heating, and chemical crosslinking via crosslinker, can cause the crosslinking to occur (Figure 3.2).

3.3.1 Physically Crosslinked Hydrogels

Ionic interaction, crystallization, stereocomplex formation, hydrophobization, and hydrogen bonding are all used to create physically crosslinked hydrogels.

Ionic crosslinking. Typically, this interaction occurs between two molecules or polyelectrolytes with opposite charges. The resulting system can be categorized as a polyelectrolyte or polyion complex or

TABLE 3.1

Comparison between the Properties of Physically and Chemically Crosslinked Polymers

Property	Physical crosslinking	Chemical Crosslinking
Bonding between polymer chains	Ionic bonding	Covalent bonding
Strength	Weak	Strong
Durability	Low	High
Self-healing	High	Low

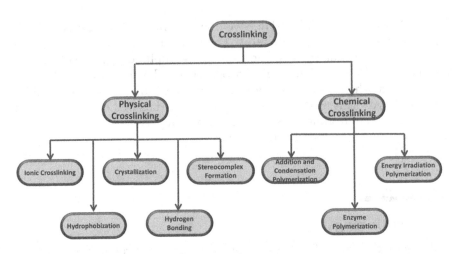

FIGURE 3.2
Commonly used mechanism for physical and chemical crosslinking of hydrogels.

even a complex coacervate. The conditions under which the ions form gels in a solution include the functional charge density of the ion, the ionic strength of the solution, the pH of the solution, and the type of the ion. Ionic interactions can crosslink hydrogels even at mild environmental conditions like room temperature and physiological pH [20]. Even if the polymer in question does not contain any ionic groups, this crosslinking is still a distinct possibility. And because metallic ions are incorporated throughout the production process, the resulting hydrogel has more durability. Crosslinking of acid groups, such as those produced from carboxylic, phosphonic, or sulfonic acid-type monomers, is facilitated by multivalent metal ions (aluminum, zirconium, titanium, chromium, zinc).

Crystallization. Hydrogel is created when polymer chain crystallites physically crosslink with one another at specific nodes in the network. For instance, the hydrogel can be formed by repeatedly freezing and thawing an aqueous PVA solution [21]. Hydrogel characteristics are affected by factors such as freezing period, freezing temperature, molecular weight, aqueous solution concentration, and the number of freeze-thaw cycles.

Stereocomplex formation. In this scenario, two polymers with different topologies come into contact and stimulate one another, resulting in a new stereocomplex with a different chemical composition [22]. Optimization will determine whether the stereocomplexity is a homo- or hetero-type. Complexes are formed between identical molecules with distinct stereochemistry in the first case. Contrarily, hetero-stereocomplexation involves interactions between molecules with different stereochemistry. No toxic organic solvents or chemical crosslinkers are needed for stereocomplexation. The main drawback of the stereocomplexation method is that it can only be employed with a specific set of polymer compositions. Therefore, even a small change in the stereochemistry or the composition can drastically decrease or nullify the stereochemical interaction.

Hydrophobization. Hydrophobic interactions can result in the development of hydrogel crosslinks in water-soluble polymers with hydrophobic end groups, side chains, or monomers. A hydrophobic physical crosslinking can be established using high-frequency sound waves and thermal induction at the lower critical solution temperature (LCST) or the upper critical solution temperature (UCST) [23]. Under the right circumstances, either method can induce a sol-to-gel transition, which is necessary for the formation of hydrogels.

Hydrogen bonding. The hydrogen bond is an extremely important example of a non-covalent interaction. Electron-donor groups, such as pyridine and imidazole, could react with amide, urea, carboxylic acid, pyrrole, and carbazole groups, as well as hydroxyl groups, to create hydrogels. One example is the utilization of a hydrogen bond to stabilize a secondary structure in a hydrogel made from agarose or peptide [24]. However, a hydrogel cannot be formed simply because hydrogen bonds exist between the sub-groups. Multiple multivalent hydrogen bonds can lead to the creation of a strong crosslinking network [23].

3.3.2 Chemically Crosslinked Hydrogels

Hydrogels with chemical crosslinking can be made using a variety of gelation methods. Addition polymerization, condensation polymerization, enzyme polymerization, gamma polymerization, and electron beam polymerization are all examples of this type of polymerization technique.

Addition and condensation polymerization. Polymerization requires the sequential addition of monomer functional groups to polyfunctional crosslinking agents. A hydrogel can be made by combining water-soluble monomers with crosslinking agents like tetramethylethylenediamine (TEMED). The synthesis of a hydrogel is possible due to the crosslinking of polymer chains in the presence of water. Water fills the voids in the hydrogel's network, creating its distinctive surface topography. Polyurethanes, polyesters, and nylon polymers are frequently used in the synthesis of hydrogels [25].

Chain growth polymerization. This procedure includes free radical polymerization, controlled free radical polymerization, anionic polymerization, and cationic polymerization. The precursor follows initiation, dissemination, and culmination for chain growth polymerization. During the initiation phase, a free radical active site is formed that can add monomers in a chain-like fashion. The creation of poly(N-isopropyl acrylamide) hydrogel is a by-product of free radical polymerization. Hydrogels made from PVA can be made using the free radical copolymerization method. Ethylene glycol di-methacrylate (EGDMA) serves as the crosslinking agent, while benzoyl peroxide kicks off the reaction, allowing polyvinyl alcohol (PVA) to chemically crosslink with monomer (methacrylic acid) in an aqueous media [26]. Ethylene glycol di-methacrylate (EGDMA) was used in the procedure because of its effectiveness as a crosslinking agent.

Energy irradiation polymerization. In this method, powerful electromagnetic radiation is used as the crosslinker. Ionizing-radiation techniques are highly effective when used in conjunction with a sterilizing process to create hydrogels. Water-soluble monomer or polymer molecules can have their chain ends crosslinked by these high-energy radiations without the need for any extra crosslinker. To ionize even the most elementary molecules, such as those found in air or water, ionizing radiation must possess a very high energy level. Electron beams and gamma rays are two kinds of such radiation. When irradiating a polymer solution, several reactive sites are produced alongside the polymer chains. These radicals are then combined, leading to the formation of several crosslinks [27]. The three-stage process of free radical polymerization is also followed by irradiation polymerizations: initiation, propagation, and termination. The generation of hydroxyl radicals kickstarts the free radical polymerization of the vinyl monomers, which spreads like rapid chain addition. Hydrogel formation is finalized when the network reaches the crucial gelation point. This method of polymerization eliminates the use of toxic chemicals and a controlled environment. The experiments can be performed at room temperature and a physiological pH without any need for optimization of environmental conditions. Procedures for irradiation are also simple, and the crosslinking point can be modified simply by altering the dose of radiation used [27].

Enzyme polymerization. Enzyme-mediated crosslinked hydrogels are of particular interest as they undergo relatively benign chemical reactions, and the majority of the relevant enzymes are abundant in nature. Enzymes such as tyrosinase, transglutaminase, phosphopantetheinyl transferase, plasma amine oxidase, lysyl oxidase, thermolysin, phosphatases, β-lactamase, and phosphatase/kinase are used to initiate crosslinking in an aqueous solution at neutral pH and moderate temperatures. By manipulating enzyme activity, the polymerization reaction can be controlled directly [28]. The gelation rate can also be adjusted to suit different applications.

3.4 Graphene-based Hydrogels

3.4.1 Templated Assistive Self-assembly Methods

The self-assembly process refers to the phenomenon in which a non-covalent link triggers a spontaneous transition of graphene oxide's (GO) two-dimensional structure into a stable 3D graphene structure. It is widely acknowledged as an effective approach for producing graphene-based hydrogels. Several unique 3D graphene hydrogels' self-assembly processes have been developed.

Soft templates. Soft templates are fluid-like states comprising flexible organic polymers, surfactants, and block copolymers. The precursor's interaction with these templates is established by weak non-covalent bonds. Soft templates can be used to create graphene-based hydrogels with variable porosities and properties. As a result, they make it possible to create more durable graphene monoliths. Using a GO-hexane droplet emulsion technique, Li et al. developed microporous graphene monoliths [29]. The hexane drops were used to create an extremely porous, 3D matrix. Hexane droplets were used as soft templates to make hydrogel; when mixed with GO solution, these droplets, ranging in size from 10 mm to 200 mm, swiftly formed a homogeneous suspension. Following that, they undergo hydrothermal reduction, which causes the GO sheets to self-assemble by wrapping them around sphere-shaped droplets. The hexane is then removed by dialyzing the hydrogels in water at 80°C. The hydrogels created by this template-assisted technique featured spherical pores with wrinkled rGO as walls.

Hard templates. Hard templates are often made of solid-state substances like polymers, silica, and carbon that have specific structures and shapes. Du et al. used hard templates as N-doping agents [30]. To produce a stable N-doped graphene hydrogel, amino-functionalized silica was utilized as a template that has the potential to generate significant changes in the porosity, surface area, and quality of the carbon. Also, the hard template avoids the restacking of graphene sheets because of their strong p-p interactions. A modified Stober approach was used to create functionalized silica particles with (3-aminopropyl) triethoxysilane (APTES), tetraethyl orthosilicate, and silica solution as precursors. Graphene oxide was successfully synthesized using Hummers' technique. Various weight ratios of functionalized silica and GO were hydrothermally reduced, and the obtained product is freeze-dried followed by pyrolysis as illustrated in Figure 3.3. The silica particles embedded with the hydrogel were then etched to obtain uniformly distributed pores.

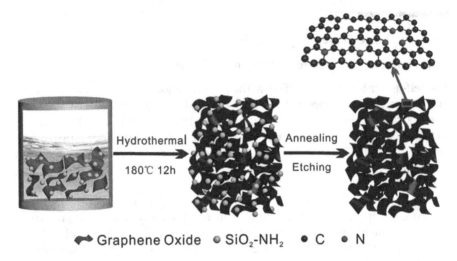

FIGURE 3.3
Schematic illustration of the preparation procedure of N-doped graphene hydrogel using amino-functionalized silica as a hard template. (Adapted with permission [30], Copyright [2018], Elsevier.)

In general, templated synthesis approaches such as hydrothermal process, chemical reduction, and metal ion-driven procedure are utilized for unique 3D graphene-based hydrogel synthesis. Using a hydrothermal synthesis, it is demonstrated that producing graphene-based hydrogels is simple and effective. The hydrogel has the potential to produce novel topologies and features such as porous network structures, extremely low densities, outstanding thermal properties, and thermal stability. In addition, they possess relatively high moisture content and a storage modulus that was several orders of magnitude more than that of typical hydrogels, as well as better electrical conductivity. The hydrothermal synthesis technique demands conditions of high pressure and high temperature, and the mixtures must be completely hermetically sealed. It is beneficial to boost the gelling rate and retain gel integrity. High temperatures and pressures, on the other hand, not only call for specialized and pricey equipment and a significant amount of power, but they also are not appropriate for production on a large scale.

Therefore, to develop a strategy that is both simple and practical, the chemical reduction method was developed. This method involves the use of moderately strong chemical reducing agents, such as hydrazine hydrate, oxalic acid, ascorbic acid, and others. These agents were added to the GO solution to increase the reduction degree of graphene to improve its performance [31]. In the mild chemical reduction process, unlike hydrothermal synthesis, neither chemical crosslinkers nor high pressures are required. Furthermore, the shapes of 3D graphene hydrogels can be controlled by varying the type of reactors that are used.

3.4.2 Mixed Solutions Methods

The fact that multilayer graphite can be created unintentionally during preparation and application is a big issue. GO has a far higher concentration of oxygen-containing groups on its surface than graphene. Because of these oxygen-containing groups, graphene oxide disperses efficiently in polar solvents such as water, ethanediol, and dimethylformamide [32]. This aids strong hydrogen bonds between GO and aqueous solution, thus forming stable colloidal dispersion systems. The mixed GO solution approach is an excellent tool for creating graphene-based hydrogel in this scenario.

The graphene/gelatin hydrogel and sulfonated graphene (SG)/poly(vinyl alcohol) (PVA) hydrogel were synthesized using the mixed solution method. Graphene/gelatin hydrogel nanocomposite was prepared by the dispersion of graphene in sodium dodecyl sulfate (SDS) solution and the dissolution of gelatin in distilled water. The solutions were then mixed and transferred to a petri dish to form a hydrogel film at room temperature [33]. Mixed solution methods, unlike other methods, do not demand high temperatures and pressure for hydrogel synthesis. However, they deliver higher-quality hydrogels with better storage modulus response and sensitivity.

3.5 Polymer-based Hydrogels

Chemical or electrochemical methods can be used to oxidize the monomers of conducting polymers such as polyaniline and polypyrrole. Hydrogels are frequently created using a chemical oxidation process. Aniline was oxidized into polyaniline salt using iron (III) chloride in water, and polypyrrole was made using ammonium peroxydisulfate [34, 35]. Both procedures were carried out in an aqueous media. This is only one example of how selecting an oxidant could be beneficial. It is assumed, at least in the most basic case, that the oxidant will infiltrate the hydrogel and react with the monomer to form a conducting polymer. As a result, the hydrogel that has been infused with a conducting polymer must be created. The term "interfacial polymerization" has been commonly used to describe this type of technique. Similarly, polyaniline has been reported to be produced by pre-treating the hydrogel with an oxidant solution and then exposing it to a monomer solution. According to research, water or aqueous acid solutions can dissolve the oxidant and the monomer [36].

Conductive polymers are often encased in a matrix of crosslinked water-soluble polymers, which are subsequently expanded with water or aqueous electrolyte solutions to form a conductive hydrogel. A hydrogel is created at the start of the synthesis by crosslinking a water-soluble polymer. A polymer solution, such as polypyrrole, polyaniline, or PEDOT, is then diffused into a hydrogel matrix. The oxidant solution is then poured over the hydrogel containing the monomer.

3.6 Biomass-derived Carbon Hydrogels/Aerogels

According to recent studies, biomass may be a useful ingredient in the synthesis of carbon aerogels because of its biological degradation capabilities. As the precursor is abundant and environmentally friendly, it facilitates limiting the adverse effects, while delivering efficient physical, chemical, and electrical properties. Biomass precursors such as cellulose, agarose, bagasse, chitosan, and wood pulps were used to prepare carbon hydrogels for electrochemical applications [37–39].

Bagasse. The cellulose and bagasse precursors were bleached with sodium chlorite/glacial acetic acid to remove lignin and hemicellulose. Subsequently, to create hierarchical porous carbon aerogels, bagasse aerogels were freeze-dried, then carbonized and activated by NaOH and KOH solution (Figure 3.4). Subsequently, the mixture was undisturbed for 12 hours for gelation and was regenerated. The created hierarchical carbon aerogels showed impressive electrochemical abilities due to their high specific surface areas and suitable pore size distributions. These characteristics made the material permeable to electrolyte ions facilitating faster electron transport [40].

Lignin. Lignin is an organic polymer extracted from cell walls and is thus abundantly available. Also, sulfite pulping of wood leads to a by-product, lignosulfonate. Graphene hydrogels functionalized with lignosulfonate showed a 3D interconnected microporous structure [41]. More electroactive sites were created, and the aggregation of reduced graphene oxide (rGO) was suppressed as a result of the uniform distribution of lignosulfonate over the material.

Chitosan. Chitosan precursor was used to synthesize nitrogen-doped graphene-based hierarchical porous carbon aerogels by precisely controlling the formation-carbonization-activation process. The chitosan aerogels were produced by freeze-drying using a lyophilizer, followed by carbonization in a tube furnace and activation using a KOH solution. The N-doped graphene-based carbon aerogels were composed of amorphous carbon with micro, meso, and macropores that were produced during the aerogel preparation-carbonization process. The micropores were created during the chemical activation process [42]. Carbon aerogels are a good choice for use as electrode

Approaches for Hydrogel Synthesis

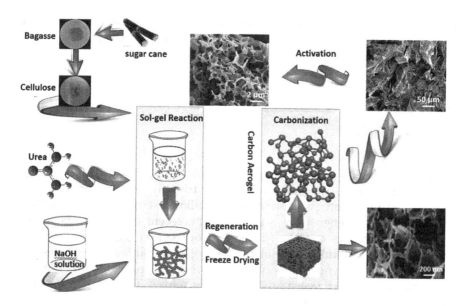

FIGURE 3.4
Schematic illustration of the synthesis of hierarchical porous carbon aerogels with Bagasse and cellulose as precursors. (Adapted with permission [40], Copyright [2014], American Chemical Society.)

material in supercapacitors because heat induces amorphous carbon to transform into the graphene skeleton, which makes the structure of these materials more stable. This resulted in a high specific capacitance and good electrochemical performance.

Cellulose. Bacterial cellulose-lignin resorcinol formaldehyde (BC-LRF) hydrogels were synthesized by crosslinking lignin resorcinol formaldehyde (LRF) and then impregnating with bacterial cellulose [43]. It was further supercritical dried with CO_2. Carbon aerogels prepared with BC-LRF can undergo significant reversible deformations due to the toughening effect of bacterial cellulose networks. The aerogel also possessed a unique structure resembling a blackberry and a significant number of mesopores. These characteristics encourage ion transit and adsorption, which leads to higher capacitance.

A carbon aerogel with a large surface area was prepared by pyrolyzing the cellulose aerogels and then activating them with carbon dioxide. Initially, cellulose microcrystals were dissolved in aqueous NaOH solutions to create cellulose aerogels. After that, supercritical carbon dioxide was used to gelate, regenerate, and dry the mixture [44]. After further CO_2 activation, the carbon aerogels' specific surface area increased with significant pore volume distribution. The resulting activated carbon aerogels demonstrated exceptional electrochemical capabilities when utilized as supercapacitors due to the very porous and interconnected nanostructure, which allowed for the efficient transport of electrolyte ions and electrons.

3.7 Conclusion

The need for hydrogels for a variety of applications in the past decade demands different preparation procedures to tune the properties. Also, the synthesis procedure is environmentally friendly, does not require complicated optimization procedures, and facilitates large-scale production. Herein, we comprehensively articulated the recent developments in the synthesis procedures of carbon-based hydrogels from graphene, polymers, and biomass. Crosslinking of polymer chains using different methods was emphasized. Graphene and polymer precursors to prepare carbon-based hydrogels require expensive preparation procedures, trained manpower to operate the equipment, and toxic chemicals. Thus, the research toward the synthesis of hydrogels is focused on sustainable biomass precursors along with simple and cost-effective procedures.

References

1. K. Kabiri, H. Omidian, M. J. Zohuriaan-Mehr, S. Doroudiani, Superabsorbent hydrogel composites, and nanocomposites: A review. *Polymer Composites*, 32 (2011) 277–289.
2. E. M. Ahmed, Hydrogel: Preparation, characterization, and applications: A review. *Journal of Advanced Research*, 6 (2015) 105–121.
3. S. Correa, A. K. Grosskopf, H. Lopez Hernandez, D. Chan, A. C. Yu, L. M. Stapleton, E. A. Appel, Translational applications of hydrogels. *Chemical Reviews*, 121 (2021) 11385–11457.
4. F. Ullah, M. B. H. Othman, F. Javed, Z. Ahmad, H. Md. Akil, Classification, processing and application of hydrogels: A review. *Materials Science and Engineering: C*, 57 (2015) 414–433.
5. J. Anjali, V. K. Jose, J. M. Lee, Carbon-based hydrogels: Synthesis and their recent energy applications. *Journal of Materials Chemistry A*, 7 (2019) 15491–15518.
6. Y. Guo, J. Bae, Z. Fang, P. Li, F. Zhao, G. Yu, Hydrogels and hydrogel-derived materials for energy and water sustainability. *Chemical Reviews*, 120 (2020) 7642–7707.
7. Y. Shi, L. Peng, G. Yu, Nanostructured conducting polymer hydrogels for energy storage applications. *Nanoscale*, 7 (2015) 12796–12806.
8. L. Li, P. Wu, F. Yu, J. Ma, Double network hydrogels for energy/environmental applications: Challenges and opportunities. *Journal of Materials Chemistry A*, 10 (2022) 9215–9247.

9. L. Racine, I. Texier, R. Auzély-Velty, Chitosan-based hydrogels: Recent design concepts to tailor properties and functions: Chitosan-based hydrogels: Tailoring properties and functions. *Polymer International*, 66 (2017) 981–998.
10. A. S. Hoffman, Hydrogels for biomedical applications. *Annals of the New York Academy of Sciences*, 944 (2006) 62–73.
11. H. Ismail, M. Irani, Z. Ahmad, Starch-based hydrogels: Present status and applications. *International Journal of Polymeric Materials*, 62 (2013) 411–420.
12. S. Eichler, O. Ramon, Y. Cohen, S. Mizrahi, Swelling and contraction driven mass transfer processes during osmotic dehydration of uncharged hydrogels. *International Journal of Food Science and Technology*, 37 (2002) 245–253.
13. M. L. Oyen, Mechanical characterisation of hydrogel materials. *International Materials Reviews*, 59 (2014) 44–59.
14. Z. Pan, L. Brassart, Constitutive modelling of hydrolytic degradation in hydrogels. *Journal of the Mechanics and Physics of Solids*, 167 (2022) 105016.
15. F. J. Vernerey, E. C. Greenwald, S. J. Bryant, Triphasic mixture model of cell-mediated enzymatic degradation of hydrogels. *Computer Methods in Biomechanics and Biomedical Engineering*, 15 (2012) 1197–1210.
16. H. Dechiraju, M. Jia, L. Luo, M. Rolandi, Ion-conducting hydrogels and their applications in bioelectronics. *Advanced Sustainable Systems*, 6 (2022) 2100173.
17. W. Wang, R. Narain, H. Zeng, Rational design of self-healing tough hydrogels: A mini review. *Frontiers in Chemistry*, 6 (2018) 497.
18. I. Cunha, R. Barras, P. Grey, D. Gaspar, E. Fortunato, R. Martins, L. Pereira, Reusable cellulose-based hydrogel sticker film applied as gate dielectric in paper electrolyte-gated transistors. *Advanced Functional Materials*, 27 (2017) 1606755.
19. M. C. Koetting, J. T. Peters, S. D. Steichen, N. A. Peppas, Stimulus-responsive hydrogels: Theory, modern advances, and applications. *Materials Science and Engineering: R: Reports*, 93 (2015) 1–49.
20. J. Maitra, V. K. Shukla, Cross-linking in hydrogels – A review. *American Journal of Polymer Science*, 4 (2014) 25–31.
21. K. H. Hong, G. Sun, Photoactive antimicrobial PVA hydrogel prepared by freeze-thawing process for wound dressing. *Journal of Applied Polymer Science*, 116 (2010) 2418–2424.
22. H. Cao, X. Chang, H. Mao, J. Zhou, Z. L. Wu, G. Shan, Y. Bao, P. Pan, Stereocomplexed physical hydrogels with high strength and tunable crystallizability. *Soft Matter*, 13 (2017) 8502–8510.
23. W. Hu, Z. Wang, Y. Xiao, S. Zhang, J. Wang, Advances in crosslinking strategies of biomedical hydrogels. *Biomaterials Science*, 7 (2019) 843–855.
24. X. Cui, J. Li, Y. Hartanto, M. Durham, J. Tang, H. Zhang, G. Hooper, K. Lim, T. Woodfield, Advances in extrusion 3d bioprinting: A focus on multicomponent hydrogel-based bioinks. *Advanced Healthcare Materials*, 9 (2020) 1901648.
25. W. E. Hennink, C. F. van Nostrum, Novel crosslinking methods to design hydrogels. *Advanced Drug Delivery Reviews*, 54 (2002) 13–36.
26. H. M. El-Husseiny, E. A. Mady, W. A. El-Dakroury, M. B. Zewail, M. Noshy, A. M. Abdelfatah, A. S. Doghish, Smart/stimuli-responsive hydrogels: State-of-the-art platforms for bone tissue engineering. *Applied Materials Today*, 29 (2022) 101560.
27. I. M. El-Sherbiny, M. H. Yacoub, Hydrogel scaffolds for tissue engineering: Progress and challenges. *Global Cardiology Science and Practice*, 2013 (2013) 38.

28. L. S. Moreira Teixeira, J. Feijen, C. A. van Blitterswijk, P. J. Dijkstra, M. Karperien, Enzyme-catalyzed crosslinkable hydrogels: Emerging strategies for tissue engineering. *Biomaterials*, 33 (2012) 1281–1290.
29. Y. Li, J. Chen, L. Huang, C. Li, J. D. Hong, G. Shi, Highly compressible macroporous graphene monoliths via an improved hydrothermal process. *Advanced Materials*, 26 (2014) 4789–4793.
30. Y. Du, L. Liu, Y. Xiang, Q. Zhang, Enhanced electrochemical capacitance and oil-absorbability of N-doped graphene aerogel by using amino-functionalized silica as template and doping agent. *Journal of Power Sources*, 379 (2018) 240–248.
31. G. Liao, J. Hu, Z. Chen, R. Zhang, G. Wang, T. Kuang, Preparation, properties, and applications of graphene-based hydrogels. *Frontiers in Chemistry*, 6 (2018) 450.
32. J. I. Paredes, S. Villar-Rodil, A. Martínez-Alonso, J. M. D. Tascón, Graphene oxide dispersions in organic solvents. *Langmuir*, 24 (2008) 10560–10564.
33. T. Tungkavet, N. Seetapan, D. Pattavarakorn, A. Sirivat, Graphene/gelatin hydrogel composites with high storage modulus sensitivity for using as electroactive actuator: Effects of surface area and electric field strength. *Polymer*, 70 (2015) 242–251.
34. H. Ahmad, M. M. Rahman, M. A. Ali, H. Minami, K. Tauer, M. A. Gafur, M. M. Rahman, A simple route to synthesize conductive stimuli-responsive polypyrrole nanocomposite hydrogel particles with strong magnetic properties and their performance for removal of hexavalent chromium ions from aqueous solution. *Journal of Magnetism and Magnetic Materials*, 412 (2016) 15–22.
35. J. Stejskal, M. Trchová, P. Bober, P. Humpolíček, V. Kašpárková, I. Sapurina, M. A. Shishov, M. Varga, Conducting polymers: Polyaniline. In John Wiley & Sons, Inc. (Ed.), *Encyclopedia of Polymer Science and Technology* (2015), John Wiley & Sons, Inc, pp. 1–44.
36. J. Stejskal, Conducting polymer hydrogels. *Chemical Papers*, 71 (2017) 269–291.
37. J. Jiang, Q. Zhang, X. Zhan, F. Chen, Renewable, biomass-derived, honeycomb-like aerogel as a robust oil absorbent with two-way reusability. *ACS Sustainable Chemistry & Engineering*, 5 (2017) 10307–10316.
38. M. S. Soffian, F. Z. Abdul Halim, F. A. Aziz, M. Rahman, M. A. Mohamed Amin, D. N. Awang Chee, Carbon-based material derived from biomass waste for wastewater treatment. *Environmental Advances*, 9 (2022) 100259.
39. S. Zhou, L. Zhou, Y. Zhang, J. Sun, J. Wen, Y. Yuan, Upgrading earth-abundant biomass into three-dimensional carbon materials for energy and environmental applications. *Journal of Materials Chemistry A*, 7 (2019) 4217–4229.
40. P. Hao, Z. Zhao, J. Tian, H. Li, Y. Sang, G. Yu, H. Cai, H. Liu, C. P. Wong, A. Umar, Hierarchical porous carbon aerogel derived from bagasse for high performance supercapacitor electrode. *Nanoscale*, 6 (2014) 12120–12129.
41. F. Li, X. Wang, R. Sun, A metal-free and flexible supercapacitor based on redox-active lignosulfonate functionalized graphene hydrogels. *Journal of Materials Chemistry A*, 5 (2017) 20643–20650.
42. P. Hao, Z. Zhao, Y. Leng, J. Tian, Y. Sang, R. I. Boughton, C. P. Wong, H. Liu, B. Yang, Graphene-based nitrogen self-doped hierarchical porous carbon aerogels derived from chitosan for high performance supercapacitors. *Nano Energy*, 15 (2015) 9–23.

43. X. Xu, J. Zhou, D. H. Nagaraju, L. Jiang, V. R. Marinov, G. Lubineau, H. N. Alshareef, M. Oh, Flexible, highly graphitized carbon aerogels based on bacterial cellulose/lignin: Catalyst-free synthesis and its application in energy storage devices. *Advanced Functional Materials*, 25 (2015) 3193–3202.
44. G. Zu, J. Shen, L. Zou, F. Wang, X. Wang, Y. Zhang, X. Yao, Nanocellulose-derived highly porous carbon aerogels for supercapacitors. *Carbon*, 99 (2016) 203–211.

4

Architecture of Hydrogels

Tamara D. Erceg and Nevena R. Vukić

CONTENTS

4.1 Introduction 67
4.2 Classification of Hydrogels 69
    4.2.1 Classification of Hydrogels Based on the Nature of Crosslinking 69
    4.2.2 Classification of Hydrogels According to Microstructural Properties 72
    4.2.3 Classification of Hydrogels According to Composition and Structure 73
    4.2.4 Classification of Hydrogels Based on Pore Size 73
    4.2.5 Classification of Hydrogels According to Sensitivity to External Stimuli 73
4.3 Parameters of Hydrogel Architecture 75
4.4 Conclusions 78
Acknowledgment 78
References 79

4.1 Introduction

Polymer networks represent one of the most important classes of polymer materials, from theoretical and application points of view. These systems represent three-dimensional structures that are created by establishing chemical (covalent) or physical bonds between segments of linear polymer chains. Linking of macromolecular chains, usually via covalent junction points, provides these materials with unique properties such as reduced creep, increased modulus of elasticity, improved chemical resistance, and resistance to abrasion [1]. Polymer networks form the basis of materials such as hydrogels, elastomers, thermosetting resins, and adhesives. Specially designed polymer networks can absorb large amounts of solvent, which are regulated by their chemical composition and microstructure. Chemical composition implies the nature of the polymer chains that ensures their affinity toward the liquid phase. Microstructure implies the degree of crosslinking, the distribution of pores, and the elasticity of segments between junction points. Hydrogels are defined as hydrophilic three-dimensional polymer networks that swell in water but do not dissolve in it. They are obtained by the reaction of one or more monomers, resulting in homo- and copolymer hydrogels. The ability of hydrogels to absorb water comes from their thermodynamic compatibility with water and the presence of hydrophilic groups (carboxylic, hydroxyl, amino, etc.) attached to the main chain. On the other hand, the existence of crosslinks between polymer chains prevents their dissolving in water in such a way that they create an elastic polymer network that expands as the result of polymer chain-solvent interaction. According to Hoffman, the amount of water present in hydrogels ranges from 10% of the mass of the xerogel to a mass that is 1000 times greater than the mass of the xerogel [2]. Xerogel represents a dry three-dimensional polymer network created by the evaporation of the solvent from the gel structure, while hydrogel is a two-component system consisting of the xerogel and the water that fills the spaces between polymer chains. The most important property of a hydrogel is its ability to swell and retain water, which depends on the presence of hydrophilic groups, crosslinking density, and glass transition temperature (T_g). A greater number of hydrophilic groups contributes to a greater swelling ability, while it decreases with an increase in crosslinking density because increasing the number of crosslinks increases the hydrophobicity and decreases the elasticity of the polymer networks. In contact with a compatible solvent, the xerogel swells; there are interactions of polymer chains and solvent molecules that are strong enough to elongate the chains between network nodes, but not to break them. The behavior of neutral hydrogels in the corresponding medium is the result of the action of two opposing forces – the osmotic force pressure that leads to the stretching of the chains and the expansion of the network and the elastic force that is caused by this expansion resists. When the osmotic pressure is higher, hydrogels swell, and when it is lower than the elastic force, they shrink. When they are equalized, a characteristic equilibrium state is established, which is described by the equilibrium degree of swelling.

The architecture of the polymer network strongly influences the swelling behavior of hydrogels. It has two levels of organization: molecular and supramolecular. The molecular level implies monomer structure – functional groups that are employed in the reaction of polymerization and crosslinking as well as functional groups that affect the swelling behavior (amino, carboxyl, hydroxyl, etc.). Supramolecular structures imply a 3D organization of monomers and chain segments, which results in measurable properties such as pore size, mesh size, and crosslinking density. The structure of hydrogels is not homogeneous but consists of places with higher and lower degrees of crosslinking, which differ in their ability to swell. Swelling is strongly influenced by the porosity of the hydrogel, that is, the size and distribution of the pores in it, determining the speed of swelling and the maximum amount of absorbed water. As the degree of crosslinking of the polymer network increases, the glass transition temperature (T_g) also increases, which is reflected in numerous properties of hydrogels, including the degree of swelling.

In order to define and understand the hydrogel architecture, it is necessary to make their classification according to where these differences come from. Classification of hydrogels from hydrogel structure and architecture was carried out according to the origin, composition, and structure of the polymer network, the pore size, and the type of crosslinking.

4.2 Classification of Hydrogels

The nature of crosslinking, microstructural properties, composition, structure, and sensitivity of hydrogels toward external stimuli are closely related to hydrogel architecture. Therefore, the classification of hydrogels was carried out according to these factors.

4.2.1 Classification of Hydrogels Based on the Nature of Crosslinking

Crosslinking of hydrogels is a crucial factor that affects their structure. According to the nature of crosslinking, hydrogels are divided into physical and chemical. Physical hydrogels are formed via reversible interactions such as hydrogen, electrostatic, hydrophobic interactions, crystallization, molecular-specific binding, host-guest interactions, and π–π stacking.

1. Hydrogen bonds are formed between positively charged hydrogen atoms and negatively charged atoms (oxygen, nitrogen, fluorine, etc.). These kinds of bonds can be classified as strong, primary bonds in charge of the formation of 3D hydrogel structures (in polyvinyl alcohol (PVA) – tannic acid (TA) shape memory hydrogels, between PVA and TA), or weak – between PVA chains in the same example. The formation of hydrogels via hydrogen bonds is schematically presented in Figure 4.1a [3].
2. Electrostatic interactions are formed between cations and anions via different arrangements:
    - between positively charged amino groups and negatively charged carboxylic groups in amino acids as constitutive units of proteins;
    - between polyanion and polycation such as γ-polyglutamic acid and chitosan; Figure 4.1b represents schematically the formation of hydrogels via electrostatic interactions between polyanions and polycations) [4];

A) Hydrogel formed by hydrogen bonding

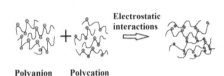

Hydrogen donor ◇   Hydrogen acceptor ◆

B) Hydrogel formed by electrostatic interaction

Polyanion   Polycation

C) Hydrogel formed by hydrophobic interactions

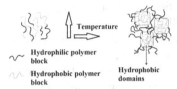

~ Hydrophilic polymer block
⋏⋏ Hydrophobic polymer block
Hydrophobic domains

D) Hydrogel formed by host-guest interactions

Host   Guest

FIGURE 4.1
Physical crosslinking.

- between charged polymers and multivalent ions (sodium alginate and calcium ion) via the formation of cationic bridges which results in an "egg-box" structure of the hydrogel [5].

3. Hydrophobic association hydrogels are formed in aqueous solutions via micellar copolymerization, combining the hydrophilic monomers with the micelles based on hydrophobic monomers. Figure 4.1c represents the formation of hydrogels via hydrophobic interactions [6]; the unique network structure of these hydrogels comes from the micelles that act as physical crosslinks connecting external hydrophilic and inner hydrophobic polymer chains [7, 8].

4. Crystallization: crystalline polymer domains form during the freezing/thawing process and act as crosslinks [9].

5. Molecular-specific binding implies the self-assembling of small molecules into a 3D matrix of structures such as nanospheres, nanorods, and nanofibers, which can hold water molecules in their cavities [10]. Deoxyribonucleic acid (DNA) supramolecular hydrogels are in the form of self-assembled hydrogels where DNA can serve as a reversible crosslinker [11].

6. Host-guest interactions: enable the creation of unique architectures of self-assembled hydrogel systems; cyclic oligosaccharides with a hydrophilic outer surface and hydrophobic inner cavities such as cyclodextrins (CDs) act as host molecules for constructing these supramolecular systems with different guest molecules such as poly(ethylene glycol) and poly(ε-caprolactone), which pass through their inner cavity forming supramolecular polymers – polyrotaxanes with a series of CDs threaded on the polymer chains, blocked

on the ends with large groups. Aggregates of multiple CDs between neighboring chains make physical crosslinks in the polyrotaxane hydrogels. Hydrogels formed by host-guest interactions are schematically presented in Figure 4.1d [12, 13].

7. π-π stacking is a type of binding between two molecules that are rich and poor in electrons. It usually occurs among aromatic compounds. Such interaction occurs in DNA/graphene oxide hydrogels between the unwound DNA chain and the graphene oxide sheet [14, 15].

Chemical or permanent hydrogels are polymer networks where polymer chains are interconnected by covalent bonds. These hydrogels have greater importance due to their stable structure and good mechanical properties. They are obtained in one of the following ways:

1. By crosslinking linear chains with low-molecular components – crosslinking agents (reaction of esterification, amidation polyaddition), for example, preparation of hydrogels based on carboxymethyl cellulose crosslinked using different kinds of polycarboxylic acids [16].
2. By the reaction of side groups of multifunctional monomers (reaction of esterification, amidation polyaddition, such as preparation of polyurethane (PU) hydrogels based on multifunctional isocyanate and polyethylene glycol diol [17].
3. By applying radiation (UV, ionizing), which initiates the creation of free radicals capable of causing simultaneous polymerization and crosslinking of polymer chains.
4. Simultaneous polymerization and crosslinking by the radical polymerization mechanism in the presence of initiators using conventional heating or microwaves [18].
5. By the two-step method, where the first step includes the polymer/biopolymer grafting via esterification, etherification, or amidation, and crosslinking in the second step via free-radical polymerization [19].

Covalent networks can be formed by stepwise reactions (polycondensation, polyaddition) or radical polymerization. For the covalent network to be formed by the reaction of polycondensation requires the presence of a tri- or multifunctional monomer that will represent a branching point of the network. A crosslinker with two double bonds is required for radical polymerization. The hydrogel can also be synthesized by polymerization in an aqueous solution, where the generation of free radicals can be induced by radiation. In addition to nuclear radiation in the synthesis of hydrogels, lower energy radiation can also be applied, such as microwaves.

4.2.2 Classification of Hydrogels According to Microstructural Properties

According to the physical structure of the polymer network, hydrogels can be amorphous, semi-crystalline, with hydrogen bonds, and supramolecular and hydrocolloid aggregates. Amorphous hydrogels are hydrogels in which water is the main constituent of the dispersed phase [20]. They are in a semi-liquid state and represent physically crosslinked polymer networks, with randomly dispersed junction points. Semi-crystalline hydrogels contain crystalline domains in an amorphous environment. Crystalline domains can consist of side chains of hydrophobic monomers in hydrogels obtained by a combination of hydrophilic and hydrophobic monomers. Also, the formation of crystalline regions can occur by the orientation of polymer chains in ordered domains by means of hydrogen bonds between the chain segments composed of hydrophilic monomers. Hydrogen-bonded hydrogels are those in which at a certain temperature (lower critical temperature) of an aqueous solution, the establishment/breaking of hydrogen bonds between water molecules and polymer side groups occurs, which is followed by hydration or dehydration of the polymer network [21]. Supramolecular hydrogels are three-dimensional crosslinked structures that contain large proportions of water (above 97%). Unlike polymer hydrogels, supramolecular hydrogels consist of networks of nanofibers created by the self-organization of small molecules (hydrogelators) (Figure 4.2). These molecules can be self-organized into complex structures using different non-covalent mechanisms, including hydrogen bonds, ionic interactions, and metal-ligand complexes, forming a three-dimensional network within which water is encapsulated. Hydrocolloid aggregates are formed by the establishment of secondary bonds between long chains of macromolecules (proteins, polysaccharides) in aqueous solutions. These can be ionic interactions, Van der Waals forces, and hydrogen bonds.

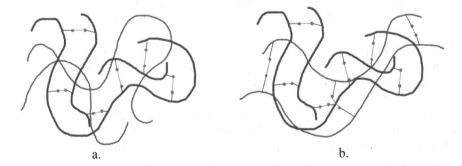

FIGURE 4.2
Multipolymer interpenetrating hydrogels: (a) interpenetrating, (b) semi-interpenetrating polymer hydrogels.

4.2.3 Classification of Hydrogels According to Composition and Structure

Based on the monomers included in their composition, hydrogels are divided into:

1. Homopolymeric: composed of only one type of monomer [22].
2. Copolymer: composed of at least two types of monomers [23].
3. Multipolymer interpenetrating hydrogels: composed of two natural and/or synthetic, mutually independent polymer networks (Figure 4.2a), or semi-interpenetrating hydrogels – composed of polymer network and linear polymer synthesized in juxtaposition (Figure 4.2b).

A subtype of these systems is semi-penetrating polymer networks, where one component is a polymer network, and the other a linear polymer that permeates it where the components can be of the same or different origin (natural or synthetic).

4.2.4 Classification of Hydrogels Based on Pore Size

The microarchitecture of hydrogels can be modified by adding porogen, such as salt leaching and gas foaming, and also by controlling the crosslinking conditions. For manipulating the hydrogel geometry and desired topography, other methods can be applied like micropatterning, micro-molding, and 3D printing [24].

According to the size of the pores, hydrogels are divided into:

1. Macroporous (pore size: 10^{-7}–10^{-6} m)
2. Microporous (pore size: 10^{-8}–10^{-7} m)
3. Non-porous (pore size: 10^{-9}–10^{-8} m) [25]

4.2.5 Classification of Hydrogels According to Sensitivity to External Stimuli

According to the sensitivity to external stimuli, hydrogels are divided into conventional and smart. Conventional hydrogels are not sensitive to external changes, while smart hydrogels show significant changes in the degree of swelling in response to small changes in pH solution, temperature, ionic strength, light, and electric and magnetic fields, which in their specific structure makes it possible. The ability to react to external stimuli implies a specific structure of hydrogels, that is, the incorporation of active compounds or using a design of nanocomposite hydrogels. Hydrogels sensitive to pH and electrical fields possess ionic groups (polyelectrolyte character). Light-sensitive hydrogels are obtained by incorporating "light switches" into the polymer network, in the form of side branches. It is known that the inclusion complex is β-CD and light-sensitive azo compounds. The azo group can be included in the cavity of β-CD, building an inclusion complex under the visible light that dissociates under UV irradiation due to a mismatch of β-CD and cis-Azo [26, 27]. Hydrogels sensitive to visible light can be obtained by incorporating chromophoric groups that absorb light from the visible part of the spectrum and emit it in vision heat, leading to an increase in the temperature of the hydrogel. An increase in temperature leads to a change in the degree of swelling of temperature-sensitive hydrogels. UV-responsive hydrogel light is obtained by introducing leuco groups into the structure that ionize under the influence of UV rays, which leads to their discontinuous swelling, as a result of an increase in osmotic pressure inside the hydrogel. After the cessation of exposure to UV rays, they shrink rapidly. The sensitivity of hydrogels to a magnetic field can be achieved by in-situ synthesis in the presence of magnetic particles, such as magnetite.

Since most polymer networks are heterogeneous and do not have a controlled structure, a study was recently published on ideal covalent polymer networks. Ideal covalent networks have well-defined polymer chain lengths, node functionality, and network architecture. Ideal polymer networks are prepared using low molecular weight end-functionalized macromers that are mixed near the overlap concentration and make crosslinks via the reaction of polycondensation or polyaddition. The ideal architecture of these networks is manifested in gels with controllable properties such as elasticity, mesh size, toughness, and stretchability. Ideal polymer networks are prepared using low molecular weight end-functionalized macromers that are mixed near the overlap concentration and make crosslinks via the reaction of polycondensation or polyaddition. These networks are not viscoelastic, as a consequence of the permanent nature of the crosslinks, but for the viscoelastic properties, reversible crosslinks can be introduced. Ideal reversible polymer networks have well-controlled polymer network architectures but feature reversible crosslinks. Also, there are no entanglements or other topological and point defects such as crosslinks with different functionality. Because of the association and dissociation of the reversible crosslinks in them, these networks show mechanical properties such as viscoelasticity. According to the theory published in the study of Parada and Zhao [28], the viscoelastic responses with the network architecture and properties of the reversible crosslinks can be quantitatively correlated. In this correlation, the ideal reversible polymer networks are formed of small end-functionalized 4-arm macromers near the overlap concentration. One-half of them have functionalized with group A and the other with group B. All arms were below critical entanglement molecular weight and of equal length. Functional groups A and B are complementary and cannot form reversible crosslinks with themselves but can with each other. A pair of complementary functional groups can be at equilibrium at the unbound or the bound state (A + B or C). This corresponds to a crosslink in the network.

4.3 Parameters of Hydrogel Architecture

The architecture of hydrogels is characterized by mesh size, crosslinking density, and porosity. Hydrogel porosity affects the swelling rate and the mechanical and functional properties of hydrogels. Increasing the porosity leads to a faster swelling response and increased swelling capacity of the hydrogel. The shear storage modulus of hydrogel in swollen state decreases with an increase in porosity. Pore size and architecture strongly affect hydrogel application, especially in tissue engineering. The overall porosity of hydrogels can be controlled by solvent casting/particle leaching, gas foaming, freeze drying, and electrospinning [29]. The solvent casting/particle leaching method implies dispersing of porogen (sodium chloride, sugars, paraffine) with controlled particle size into a polymer solution. After the leaching of solute particles using a selective solvent, the highly porous and interconnected matrix can be obtained [30]. The overall porosity and pore structure (which can be cubic or spherical) of hydrogel is affected by porogen type and geometry, particle size, and the concentration of prepolymer [31]. A freeze-drying method is also known as lyophilization, implying a rapid cooling of prepolymer/crosslinker solution and removal of the solvent by sublimation in the vacuum. Samples that are frozen between −20 and −80°C result in open-pore structures and a linear network of inner channels [32, 33]. Interconnected structures with certain pores size can be obtained using a technique of gas foaming that includes the addition of a foaming/blowing agent in a prepolymer mixture (sodium bicarbonate, ammonium bicarbonate), which is able to chemically decompose, releasing gasses such as CO_2 and NH_3 that generate a porous structure [34]. In a study by Erceg et al., highly porous hydrogels were obtained by simultaneous polymerization and water evaporation in the microwave field [35, 36]. The amount of water present in the monomer solution and the ratio of monomers strongly affect the hydrogel porosity. The scanning electron microscopy (SEM) pictures of this type of hydrogel are presented in Figure 4.3. The porous, cellular structure of the hydrogel induced by solvent evaporation is achieved during the preparation of polyurethane hydrogels using tetrahydrofuran (THF) as a solvent. Depending on whether the synthesis is carried out in open or closed systems, hydrogels with different microstructures are obtained [17]. Simultaneous polyaddition and THF evaporation in an open vessel give hydrogel with a sponge-like structure (Figure 4.4a) while preparation in closed, at a lower temperature, systems give more compact hydrogels (Figure 4.4b). Gas foaming with dense CO_2 enables the formation of a highly porous structure of hydrogel with stable pores using emulsion templating [37]. The formation of the ultrafine fibrous hydrogel with interconnected interfiber pores can be achieved using the electrospinning process [38].

Hydrogel porosity can be determined using a liquid displacement method, using the hexane, toward hydrogels, and does not show affinity and easily

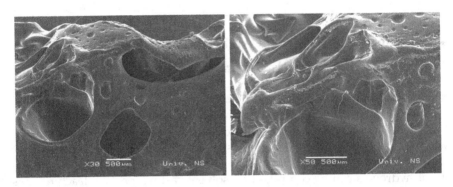

FIGURE 4.3
SEM image of acrylamide/acrylic acid hydrogel prepared in the microwave field. (Adapted with permission [36], Copyright [2020], Springer Nature.)

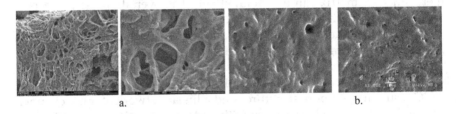

FIGURE 4.4
SEM image of poly[(phenyl isocyanate)-co-formaldehyde]/polyethylene glycol hydrogels prepared in (a) open and (b) closed systems. (Adapted with permission [17], Copyright [2021], Elsevier.)

penetrates hydrogel pores. To measure the porosity of hydrogel, xerogel is immersed in a graduated cylinder with a predetermined volume of hexane, designated as V_1. After a certain time interval (up to 24 h), xerogel is taken out from the cylinder, and the residual volume of hexane in the graduated vessel is recorded as V_2. The porosity of hydrogel (xerogel) is calculated according to Equation 4.1 as a ratio between a volume of hexane retained in the pores of xerogel (V_p) and a total volume of xerogel (V_T) [39]:

$$Porosity = \frac{V_p}{V_T} \cdot 100\% = \frac{V_2 - V_2}{V_T} \cdot 100\% \qquad (4.1)$$

The mesh size (ξ) is a linear distance between two crosslinks and can be calculated using two methods/equations. The first method implies using Equation 4.2 [40]:

$$\xi = \alpha \left(\overline{r_0^{-2}} \right)^{\frac{1}{2}} \qquad (4.2)$$

where α is the extension of the polymer chains calculated from the swollen polymer volume fraction (v_p) using the following Equation 4.3 and $\overline{r_0}^{-2}$ is the root-mean-square of the end-to-end distance of the polymer chains between two successive crosslinks. A swollen polymer volume fraction is evaluated experimentally, as the solvent parameters affect the hydrogel swelling behavior.

$$\alpha = (v_p)^{-\frac{1}{3}} \tag{4.3}$$

$\overline{r_0}^{-2}$ is calculated using Equation 4.4:

$$(\overline{r_0}^{2})^{\frac{1}{2}} = l \cdot \left(\frac{2C_n \overline{M_c}}{M_r} \right)^{\frac{1}{2}} \tag{4.4}$$

Where l is the length of the bond in the polymer backbone, C_n is the Flory characteristic ratio (given in the literature data), $\overline{M_c}$ is the average molecular weight between two crosslinks, and M_r is the molecular weight of repeat units of the hydrogel. This method implies the existence of literature data for the listed parameters, therefore exposing certain limitations. The second method of mesh size determination implies the application of Equation 4.5 based on the measurements of rheological properties of hydrogel in the equilibrated swollen state in a linear viscoelastic region [41]:

$$\xi = \left(\frac{k_p \cdot T}{G_e} \right)^{\frac{1}{3}} \tag{4.5}$$

Where k_p is the Boltzmann constant, T is an absolute temperature, and G_e is the plateau value of shear elastic modulus. The mesh size values of hydrogels calculated in this way are in the range between 50 and 1000 Å [19].

Crosslinking density can be calculated using the Mooney–Rivlin equation, which correlates the deformation behavior of a polymer network with its crosslinking properties. The general form of this theory is given by Equation 4.6:

$$\frac{\sigma_e \cdot v_p^{1/3}}{\left(\lambda - \frac{1}{\lambda^2} \right)} = 2C_1 + 2C_2 \frac{1}{\lambda} \tag{4.6}$$

Where σ_e is the applied tensile stress, v_p is the polymer volume fraction, and λ is the extension ratio. Plotting of the left side of Equation 4.6 against $\frac{1}{\lambda}$ gives a straight line with an intercept of $2C_1$ and slope of $2C_2$. Based on the determined C_1 value, crosslinking density (N, mol/m^3) can be calculated using Equation 4.7 [42]:

$$N = 2C_1 / kT \tag{4.7}$$

A qualitative comparison of hydrogel crosslinking density can be carried out by determination and comparison of their swelling and thermal and rheological properties. Therefore, the equilibrium swelling ratio decreases with increasing crosslinking density; the glass transition temperature of xerogel increases with increasing crosslinking density; and the shear elastic modulus increases with increasing crosslinking density.

4.4 Conclusions

Hydrogel architecture is affected by its chemical composition, synthesis conditions, microstructural characteristics, and type of crosslinking. Controlling the synthesis conditions as well as the hydrogel individual and pores clusters is important for the gaining of engineered properties. Since most polymer networks are heterogeneous and do not have a controlled structure, some studies confirm the existence of ideal polymer networks – covalent/irreversible and reversible. Ideal covalent networks have well-defined polymer chain lengths, node functionality, and network architecture, giving gels controllable properties such as elasticity, mesh size, toughness, and stretchability. Ideal polymer networks are prepared using low molecular weight end-functionalized macromers that are mixed near the overlap concentration and make crosslinks via the reaction of polycondensation or polyaddition. These networks are not viscoelastic because of the permanent nature of the crosslinks. Ideal reversible polymer networks have well-controlled polymer network architectures and reversible crosslinks. Because of the association and dissociation of the reversible crosslinks in them, these networks show viscoelasticity. The parameters of hydrogel architecture are porosity, mesh size, and crosslinking density, which can be quantitatively expressed using the equations based on parameters calculated for hydrogels in a swollen state. Therefore, the architecture of hydrogels is most important from the point of regulation of their main property – swelling behavior.

Acknowledgment

The study is supported by the Minister of Science, Innovation and Technological Development of the Republic of Serbia (contracts no. 451-03-68/2022-14/ 200134).

References

1. A. K. Whittaker, The structure of polymer networks, G. A. Webb (ed.), *Modern magnetic resonance*, Dordrecht, Netherlands, Springer, 2006.
2. A. S. Hoffman, Intelligent polymers, K. Park (ed.), *Controlled drug delivery, challenges and strategies*, American Chemical Society, Washington, DC, 1997.
3. Y. Chen, L. Peng, T. Liu, Y. Wang, S. Shi, H. Wang, Poly(vinyl alcohol)–tannic acid hydrogels with excellent mechanical properties and shape memory behaviors, *ACS Appl. Mater. Inter.* 8 (2016) 27199–27206.
4. T. L. Sun, T. Kurokawa, S. Kuroda, A. B. Ihsan, T. Akasaki, K. Sato, M. A. Haque, T. Nakajima, J. P. Gong, Physical hydrogels composed of polyampholytes demonstrate high toughness and viscoelasticity, *Nat. Mater.* 12 (2013) 932–937.
5. M. F. Akhtar, M. Hanif, N. M. Ranjha, Methods of synthesis of hydrogels. A review, *Saudi Pharm. J.* 24 (2016) 554–559.
6. S. Abdurrahmanoglu, V. Can, O. Okay, Design of high-toughness polyacrylamide hydrogels by hydrophobic modification, *Polymer.* 50 (2009) 5449–5455.
7. Y. Liu, Z. Li, N. Niu, J. Zou, F. Liu, A simple coordination strategy for preparing a complex hydrophobic association hydrogel, *J. Appl. Polym. Sci.* 135 (2018) 46400.
8. G. Jiang, C. Liu, X. Liu, G. Zhang, M. Yang, F. Liu, Construction and properties of hydrophobic association hydrogels with high mechanical strength and reforming capability, *Macromol. Mater. Eng.* 294 (2009) 815–820.
9. P. Guo, J. Liang, Y. Li, X. Lu, H. Fu, H. Jing, S. Guan, D. Han, L. Niu, High-strength and pH-responsive self-healing polyvinyl alcohol/poly 6-acrylamidohexanoic acid hydrogel based on dual physically cross-linked network, *Colloids Surf. A.* 571 (2019) 64–71.
10. X. Zhang, X. Chu, L. Wang, H. Wang, G. Liang, J. Zhang, J. Long, Z. Yang, Rational design of a tetrameric protein to enhance interactions between self-assembled fibers gives molecular hydrogels, *Angew. Chem. Int. Ed.* 51 (2012) 4388–4392.
11. J. Liu, Oligonucleotide-functionalized hydrogels as stimuli responsive materials and biosensors, *Soft Matter.* 7 (2011) 6757–6767.
12. Q. Hu, G. Tang, P. K. Chu, Cyclodextrin-based host–guest supramolecular nanoparticles for delivery: From design to applications, *Acc. Chem. Res.* 47 (2014) 2017–2025.
13. K. L. Liu, Z. Zhang, J. Li, Supramolecular hydrogels based on cyclodextrin-polymer polypseudorotaxanes: Materials design and hydrogel properties, *Soft Matter.* 7 (2011) 1129–11297.
14. Y. Xu, Q. Wu, Y. Sun, H. Bai, G. Shi, Three-dimensional self-assembly of graphene oxide and DNA into multifunctional hydrogels, *ACS Nano.* 4 (2010) 7358–7362.
15. J. Li, X. Jia, L. Yin, Hydrogel: Diversity of structures and applications in food science, *Food Rev. Int.* 37 (2021) 313–372.
16. T. Erceg, A. Stupar, M. Cvetinov, V. Vasić, I. Ristić, Investigation the correlation between chemical structure and swelling, thermal and flocculation properties of carboxymethylcellulose hydrogels, *J. Appl. Polym. Sci.* 138, 10 (2021) 50240.

17. J. Tanasić, T. Erceg, Lj. Tanasić, S. Baloš, O. Klisurić, I. Ristić, The influence of reaction conditions on structural properties and swelling kinetics of polyurethane hydrogels, intended for agricultural purposes, *React. Funct. Polym.* 169 (2021) 105085.
18. T. Erceg, S. Cakić, M. Cvetinov, T. Dapčević-Hadnađev, J. Budinski-Simendić, The properties of conventionally and microwave synthesized poly(acrylamide-co-acrylic acid) hydrogels, *Polym. Bull.* 77, 4 (2020) 2089–2110.
19. T. Erceg, G. Brakus, A. Stupar, M. Cvetinov, M. Hadnađev, I. Ristić, Synthesis and characterization of chitosan-acrylic acid based hydrogels and investigation the properties of bilayered design with incorporated alginate beads, *J. Polym. Environ.* 30 (2022) 3737–3760.
20. S. Bale, V. Banks, S. Haglestein, K. G. Harding, A comparison of two amorphous hydrogels in the debridement of pressure sores, *J. Wound Care.* 7, 2 (1998) 65–68.
21. E. Díez-Peña, P. Frutos, G. Frutos, I. Quijada-Garrido, J. M. Barrales-Rienda, The influence of the copolymer composition on the diltiazem hydrochloride release from a series of pH Sensitive poly[(N-isopropylacrylamide)-co-methacrylic acid)] hydrogels, *AAPS Pharm. Sci. Tech.* 5, 2, (1998) 69–76.
22. L. Yang, J. S. Chu, J. A. Fix, Colon-specific drug delivery: New approaches and in vitro/in vivo evaluation, *Int. J. Pharm.* 235, 1–2 (2002) 1–15.
23. Z. Maolin, L. Jun, Y. Min, H. Hongfei, The swelling behavior of radiation prepared semi-interpenetrating polymer networks composed of polyNIPAAm and hydrophilic polymers, *Radiat. Phys. Chem.* 58 (2000) 397–400.
24. F. D. Martinez-Garcia, T. Fischer, A. Hayn, C. T. Mierke, J. K. Burgess, M. C. Harmsen, A beginner's guide to the characterization of hydrogel microarchitecture for cellular applications, *Gels.* 8 (2022) 535.
25. N. Chirani, L. H. Yahia, L. Gritsch, F. L. Motta, S. Chirani, S. Fare, History and applications of hydrogels, *J. Biomed. Sci.* 4, 2 (2015) 1–23.
26. Y. Takashima, S. Hatanaka, M. Otsubo, M. Nakahata, T. Kakuta, A. Hashidzume, H. Yamaguchi, A. Harada, Expansion–contraction of photoresponsive artificial muscle regulated by host–guest interactions, *Nat. Commun.* 3 (2012) 1270.
27. Y. Y. Xiao, X. L. Going, Y. Kang, Z. C. Jiang, S. Zhang, B. J. Li, Light-, pH- and thermal responsive hydrogels with the triple-shape memory effect, *Chem. Commun.* 52 (2016) 10609–10612.
28. G. A. Parada, X. Zhao, Ideal reversible polymer networks, *Soft Matter.* 14 (2018) 5186–5196.
29. N. Annabi, J. Nichol, X. Zhong, C. Ji, S. Koshy, A. Khademhosseini, F. Dehghani, Controlling the porosity and microarchitecture of hydrogels for tissue engineering, *Tissue Eng. Part B Rev.* 16, 4 (2010) 371–83.
30. S. W. Suh, J. Y. Shin, J. Kim, J. Kim, C. H. Beak, D. I. Kim, H. Kim, S. S. Jeon, I. W. Choo, Effect of different particles on cell proliferation in polymer scaffolds using a solvent-casting and particulate leaching technique, *ASAIO J.* 48, 5 (2002) 460–4.
31. J. Zhang, L. Wu, D. Jing, J. Ding, A comparative study of porous scaffolds with cubic and spherical macropores, *Polymer.* 46 (2005) 4979–4985.
32. X. Wu, L. Black, G. Santacana-Laffitte, C.W. Jr. Patrick, Preparation and assessment of glutaraldehyde-crosslinked collagen-chitosan hydrogels for adipose tissue engineering, *J. Biomed. Mater. Res. Part A.* 81, 1 (2007) 59–65.

33. S. Stokols, M. H. Tuszynski, The fabrication and characterization of linearly oriented nerve guidance scaffolds for spinal cord injury, *Biomaterials.* 25 (2004) 5839–5846.
34. K. M. Huh, N. Baek, K. Park, Enhanced swelling rate of poly(ethylene glycol)-grafted super porous hydrogels, *J. Bioact. Compat. Polym.* 20, 3 (2005) 231–243.
35. T. Erceg, V. Teofilović, N. Vukić, I. Ristić, Labelling as an incentive for the development of superabsorbent polymer materials obtained by energy-efficient polymerization method, Rzeszow, Rzeszow, Politechnika Rzeszowska im. Ignacego Lukasiewicza (2021) 69–85.
36. T. Erceg, T. Dapčević, M. Hadnađev, Hadnađev, I. Ristić, Swelling kinetics and rheological behaviour of microwave synthesized poly(acrylamide-co-acrylic acid) hydrogels, *Colloid. Polym. Sci.* 299 (2021) 11–23.
37. Z. Bing, J. Y. Lee, S. W. Choi, J. H. Kim, Preparation of porous $CaCO_3$=PAM composites by CO_2 in water emulsion templating method, *Eur. Polym. J.* 43 (2007) 4814–4820.
38. L. Li, Y.-L. Hsieh, Ultra-fine polyelectrolyte hydrogel fibers from poly(acrylic acid)=poly(vinyl alcohol), *Nanotechnology.* 16 (2005) 2852.
39. H. V. Chavda, R. D. Patel, I. P. Modhia, C. N. Patel, Preparation and characterization of superporous hydrogel based on different polymers, *Int. J. Pharm. Investig.* 2 (2012) 134–139.
40. J. Peters, M. Wechsler, N. Peppas, Advanced biomedical hydrogels: Molecular architecture and its impact on medical applications, *Regen. Biomater.* 8, 6 (2021) 1–21.
41. L. Pescosolido, L. Feruglio, R. Farra, S. Fiorentino, I. Colombo, T. Coviello, P. W. Matricardi, W. E. Hennink, T. Vermonden, M. Grassi, Mesh size distribution determination of interpenetrating polymer network hydrogels, *Soft Matter.* 8 (2012) 7708–7715.
42. Z. Xia, M. Patchan, J. Maranchi, J. Elisseeff, M. Trexler, Determination of cross-linking density of hydrogels prepared from microcrystalline cellulose, *J. Appl. Polym.* 127 (2012) 4537–4541.

5

Characteristics of Hydrogels

Jagadeeshwar Kodavaty, Suresh Kumar Yatirajula,
Abhishek Kumar Gupta, and Rejeswara Reddy Erva

CONTENTS

5.1 Introduction 83
    5.1.1 Hydrogel Systems and Their Importance 84
    5.1.2 Natural and Synthetic Polymers Used in Hydrogel Drug Delivery Systems..86
    5.1.3 Transportation Properties in Hydrogels 86
    5.1.4 Rheology of Gels 87
5.2 Rheology as a Tool to Characterize Hydrogels 89
5.3 SAOS 91
5.4 Gel Strength...92
5.5 Gelation Time 93
5.6 Characterization of Hydrogels Using Dynamic Light Scattering 95
5.7 Surface Morphology Using Environmental Scanning Electron Microscopy...96
5.8 Conclusion 97
References...98

5.1 Introduction

Hydrogels are potential materials for drug delivery systems. Hydrogels have significant importance in various applications like medicine and chemical, biological, petrochemical, and other allied fields of engineering. As the name suggests, a hydrogel contains 80 to 90% water and the rest is other constituents like natural or synthetic polymers, fibers, particles, and animal extracts [1]. These hydrogels are the result of physical or chemical interactions among the constituents present in them [2]. Some organogels can show gel characteristics upon the polymerization reaction from monomers. In this type, the reaction time, temperature, and pH become important as the onset concentration of the formed polymer by polymerization reaction decides the strength and gelation time of the organogel [3]. However, other organogels result from secondary forces like physical interactions or hydrogen bonding [4].

DOI: 10.1201/9781003351566-6

Drug delivery techniques include formulations, methods, and technology that deliver a pharmaceutical compound to a particular place in the human body, where its therapeutic treatment is needed. The release of insulin in the human body when glucose levels are critical is the best example of a drug delivery system. The pharmaceutical molecule when swallowed should release the necessary therapeutics into the body and it controls the glucose levels based on the delivery that follows a zero-order system. In these circumstances, the release of the drug becomes continuous, and the immediate control of glucose levels is achieved.

Hydrogels made of polymers have good biocompatible properties and are ideal for drug targeting. Numerous formulations through hydrogel systems have been suggested, which include economical, stable, and biocompatible. The three-dimensional networks of polymers in hydrogels exhibit distinct properties and structures because of covalent, ionic, hydrogen bonding, and crosslinking. Hydrogels' crucial role was discovered in the late 1950s when creating a polymer contact lens material (2-hydroxyethyl methacrylate). Because they are porous, hydrogels can absorb a lot of water or biological fluids and have a biocompatible nature. The distribution and release kinetics of the medicinal component to the desired location is greatly influenced by the size of these polymer networks. These gels include a range of mesh sizes, including macro (0.1–1 μm), micro (10–100 nm), and nano (1–10 nm). In general, morphology, swelling, and elasticity define hydrogels [5–7].

The kinetics of drug release will be determined by swelling or the amount of water absorbed depending on the mesh size. As these hydrogels are typically subjected to deformations, elasticity is a property of the mechanical strength of the polymer network and stability of the material. Hydrogels are categorized as pH or ionic-sensitive, thermosensitive, or occasionally protein, peptide, or biomolecule (which is created by an organism) sensitive depending on the specific usage and formulation. These hydrogels have been used in several pharmaceutical applications, such as eye medication delivery, cosmetics, and wound healing, and are widely used in the fields of medication delivery systems and tissue engineering [8, 9].

5.1.1 Hydrogel Systems and Their Importance

Conventionally, a hydrogel is a jelly kind of substance that embeds a maximum quantity of water. Because of this reason, these hydrogels are more compatible with the human body as the human body is made of 80 to 90% water. Although they contain a huge amount of water, they act like solid materials that are reluctant to flow. The formation of hydrogels and organogels is very important as their internal structure of formation leads to parameters like gelation time and gel strength. A schematic of the gel is shown in Figure 5.1.

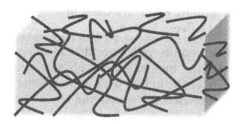

FIGURE 5.1
Schematic of a hydrogel containing polymer entangled molecules and water.

The traditional way of showing the gelation process is to keep the solution in a test tube and allow the gelation to continue. After the solution becomes a gel, the test tube is inverted, and if the contents do not fall, then the gel has been formed, as shown in Figure 5.2 [10].

In characterization, neither the gel strength nor the gelation time is given importance. This type of characterization will not give any clues about the gelation process. We cannot evaluate the extent of hydrogen bonding, the crosslinking reaction, or the polymerization reaction that could be responsible for the formation of a gel. We can also not say at what exact time the solution transforms to a gel and the duration of time the gelation process takes to complete, as these things are important during the design of a hydrogel used in drug delivery systems. It is, therefore, significant to find complete characteristics of hydrogel materials, beginning from the solution to a complete hydrogel state, to estimate gelation properties like hydrogel strength, gelation time, and structural features [11, 12].

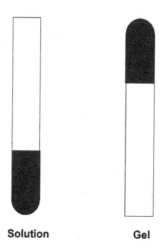

FIGURE 5.2
Solution to gel formation indicated by inverting the test tube.

5.1.2 Natural and Synthetic Polymers Used in Hydrogel Drug Delivery Systems

Drug delivery systems require polymers that meet several requirements, including chemical stability, impurity levels, physical structure, and processing simplicity. Biopolymers including polysaccharides and natural gums are widely used in medication delivery systems and biomaterials. Agar, alginate, chitosan, dextrin, gellan gum, guar gum, karaya gum, locust bean gum, pectin, starch, hyaluronic acid, and xanthan gum are a few examples of the various natural polymers employed. Natural or biopolymers have several advantages over synthetic polymers, but their main benefits are that they are non-toxic, less expensive, and readily available from renewable sources [13].

These polymers contain microbial contamination and vary in composition from batch to batch since they are produced from microbial sources. Natural polymers have limited application options due to their inferior mechanical qualities to those of synthetic polymers. Synthetic polymers have recently begun to take the place of natural polymers. These offer the advantages of being simple to prepare, having good functional control, being microbially uncontaminated, and having excellent mechanical qualities. Poly(vinyl pyrrolidone), poly(methyl methacrylate), poly(vinyl alcohol), poly(acrylic acid), poly(acrylamide), poly(ethylene glycol), and poly(methacrylic acid) are a few examples of synthetic polymers that are utilized in drug delivery systems. The fact that these synthetic polymers break down into biocompatible molecules and can be removed via metabolic routes is a significant advantage [14, 15].

5.1.3 Transportation Properties in Hydrogels

The crosslinkages of hydrogels, which are swelling networks of polymers, allow them to retain water while still maintaining structural integrity. The effectiveness of the gel matrix in a hydrogel is dependent on the permeation of dissolved gases, nutrients, and so on, and the pace at which pharmaceuticals are released while retaining their original structure. These hydrogels are created by injecting solutions – with or without medicines or biologics – into various applications. The shape, rheology, swelling, and transport characteristics of hydrogel materials are determined by their functions and network structure. For the system's total study, biological characterization is just as crucial as material characterization.

The transport of solvent or solute through and within the gel are important aspects to be studied in the design of a hydrogel. Crosslinking density and the size of the hydrogels will be used to determine how a model drug diffuses through the swollen gel composed of solvent channels. Solvent transport within the hydrogel is found by swelling experiments. However, solute transfer through the swollen gel is important and usually analyzed with standard diffusional laws [16–18].

Enzymatically biodegradable polymers are required for novel biomedical technology such as regenerative medicine, tissue engineering, controlled drug delivery, and biotechnology. Despite the long history of the usage of natural polymers, recent years have seen an increase in the use of synthetic polymers in biological applications. Different hydrologically and enzymatically degradable polymers employed as biomaterials were described by Nair and Laurencin. Scaffolds built of polymeric materials that temporarily replace injured tissue and disintegrate at a rate corresponding to cell growth are polymeric materials for bone regeneration. Hyaluronan hydrogel has also been mentioned as a prospective contender due to its natural abundance, hydrolysis stability, biodegradable qualities, and numerous biological features. It was also underlined that these materials are completely biodegradable and immunologically inactive [19].

5.1.4 Rheology of Gels

Polymer gels can absorb biological fluids or solvents during swelling. Due to physical or chemical crosslinks, the entangled polymer gel that holds vast amounts of solvent will not dissolve. The hydrophilic groups are attracted to the solvent, and the gel networking gives off a dissolution, creating the structure of the hydrogel with characteristics like those of bodily cells. By making the gel matrix swell and putting stress on the polymer networks, the solvent enters the gel matrix. Given that the human body contains 70% water, any hydrogel employed as a biomaterial must be able to absorb significant volumes of bodily fluids to become increasingly biocompatible [20].

The number of polymer gels made of either natural or synthetic polymers is determined by the swelling ratio, which is the ratio of the weight or volume of the network that has expanded to the equilibrium of the weight of the original gel. The volume increase reported for a gel used in a biomaterial system typically ranges from 10 to 20 times the initial volume. These swelling gel networks can hold drug molecules and release them as necessary. The transport properties of the gel are therefore critical. Through the application of certain criteria, the engineering point of view is often employed to access the swelling kinetics of the solute as well as the medicine release kinetics. Since these elements often depend on the mesh size and effective crosslinking density of the network, methods have been developed to measure these characteristics using swelling data [21, 22].

During the gel formation through a solution state, the polymer molecules bind to each other using hydrogen or covalent bonding, resulting in physically and chemically crosslinked gels, respectively. During gelation, the molecules might interact locally forming a microgel that grows to form a macro-gel, or they might have a certain way of forming the gel that leads to the final structure of the gel. These structural features are apparent in the surface morphology of the gels that were formed. Hydrogels made of unique sizes and structures have significant applications in various fields of biomedical applications. It is important to analyze the structure and morphology of the materials.

Gelation is the process of converting a solution into an elastic solid by the utilization of covalent, ionic, or atomic bonding during the sol-gel transition. The creation of a continually interconnected network is the result of an increase in the number of physical or chemical links between molecules or particles during the gelation process. Rheology is the most effective way to assess reaction kinetics since gelation is a kinetic process with a measurable modulus [22].

Rheological behavior provides structural formation hints throughout the dynamic process of gel formation from solutions, and phenomenological studies may shed light on the evolution of complex structures. In other words, while covalent bonding causes chemical gelation, the gel's molecular weight keeps rising and its flexibility develops until its viscosity reaches a specific level. The resulting gel has a viscoelastic character but is still in the form of a liquid. The gel's characteristics will undergo drastic changes as the molecular weight continues to rise; these changes are commonly referred to as the gelation time or gelation temperature. The gel point, according to Winter, is the point at which the weight average molecular weight reaches infinity, or the greatest molecular cluster crosses the sample. Very gradually and with limited behavior, the transition through the gel point occurs. It is stressed that the real gel point cannot be located, and the substance is both a viscoelastic liquid and a viscoelastic solid. The mobility of the chain segments affects the material's gel strength [23].

The variables viscoelasticity, viscosity, and gel strength are crucial when analyzing hydrogels. The real portion of G and an imaginary component of G are measured when the material under consideration is exposed to sinusoidal oscillatory shear at a specific angular frequency and strain percentage; these quantities are referred to as the storage modulus G' and loss modulus G", respectively. In general, a hydrogel's mechanical characteristics depend on the characteristics of each of its constituent parts. The contribution of each property in a mixture of polymers will have an impact on the material's overall qualities. In most cases, the polymers are combined to create hydrogels, which have better qualities and functions. Therefore, it is important to characterize these materials using rheology. Gel rheology can be used to evaluate the strength of gels generated, even though the rheology during gelation can give hints about structural aspects [24].

During the analysis of hydrogels, parameters such as viscoelasticity, viscosity, and gel strength are important. When the material under consideration is subjected to sinusoidal oscillatory shear, at a particular angular frequency and strain percentage, the real part of G and an imaginary part of G are measured, which are known as storage modulus G' and loss modulus G", respectively. In general, the mechanical properties of hydrogels depend on the properties of their components. In a blend of polymers, the contribution of the properties will affect the overall properties of the material under investigation. In most cases, the polymers are blended and made in the form of hydrogels that have improved properties and functionalities. Hence, the characterization of these materials using rheology is significant. Although the structural features can be hinted at by rheology during gelation, the strength of the gels formed can be estimated using gel rheology [25].

5.2 Rheology as a Tool to Characterize Hydrogels

Rheology is the study of how fluids move and change when forces are applied. Contrary to water, polymer solutions have non-Newtonian properties. Depending on the concentration of entanglement present in the solutions, these solutions also have some degree of elasticity. As a result, these polymer solutions are viscoelastic liquids in nature, meaning that both their viscosity and elasticity affect how they behave. Similar to this, the viscoelastic properties of these crosslinked aqueous polymer solutions, which exhibit features of both viscosity and elasticity, can be assessed in the form of modulus utilizing rheology as a technique. When a solution is viscous, strain rate determines stress, and when a material is elastic, strain determines stress. Both strain and strain rate affect stress for viscoelastic liquids and solids.

It is common practice in rheological analysis to analyze polymer blends and solutions, soft materials, emulsions, polyelectrolytes, biological molecules, and more using dynamic oscillatory shear. Even though small amplitude oscillatory shear rheology (SAOS) is a popular test method, it is ineffective when processing activities include significant deformations. When the deformations involved are significant, it is vital to have a complete understanding of the materials. With the right test protocols for complicated materials, large amplitude oscillatory shear rheology (LAOS) therefore becomes prominent [26].

It is common practice in rheological analysis to analyze polymer blends and solutions, hydrogels, soft materials, emulsions, polyelectrolytes, and biological molecules using dynamic oscillatory shear. Despite being a standard test technique, SAOS is ineffective when processing operations include substantial deformations [27–29]. In these situations, having a complete understanding of the materials' characteristics is important. With the right test techniques for complicated materials, LAOS therefore became prominent in the investigation of Ramya et al. [30].

One method of characterization of these hydrogels is to find the response of the hydrogel for a given change in some input. A hydrogel is neither a liquid nor a solid and the properties lie in between solid and liquid, and is viscoelastic in nature. To check the viscoelastic nature, the stress-strain relation is obtained and analyzed [26]. To obtain the relation, the fluid is placed between two plates and sheared, as shown in Figure 5.3.

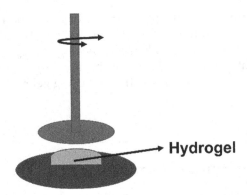

FIGURE 5.3
A hydrogel was placed between two plates for the measurement of viscoelastic properties.

The top plate is usually moving, and the bottom plate will be stationary in an instrument, called a rheometer, used to characterize hydrogels. The plate surfaces are flat or curved, differing in dimensions. The top moving plate will be moved such that it will not undergo any friction. The bottom plate is used to vary the temperature by means of a Peltier connected to it. The movement of the top plate strains the hydrogel placed in between the plates and the stresses are estimated by carefully measuring the torque that appears from the top plate. The strains and stresses are captured by various sensors connected to the rheometer and the software present will give us the stress-strain relations directly [18, 24, 28, 30–32].

To examine how a material responds mechanically to viscoelastic qualities, oscillatory shear rheology is used. This will give an understanding of the physical processes and molecule structures that result in special qualities. The fundamental idea behind this technique is to subject the sample to sinusoidal shear deformation and then evaluate the ensuing stress response. A common experiment involves placing a sample between parallel plates, inducing a time-dependent strain (t) = o sin(t), and measuring the time-dependent stress (t) by calculating the torque the sample produces. For a viscoelastic material, the time-dependent stress is given by

$$\sigma(t) = \gamma_o \left[G'(\omega)\sin(\omega t) + G''(\omega)\cos(\omega t) \right]$$

where G′ is the storage modulus, which is characteristic of the elastic nature of the material, and G″ is the loss modulus, which is characteristic of the viscous nature of the material [33, 34]. The ratio of the loss modulus to the storage modulus, or G″/G′, is known as the loss tangent tan, and it serves as a measure of damping. It is crucial to investigate the material by analyzing

Characteristics of Hydrogels

the frequency-dependent characteristics G' and G" in either an undisturbed or disturbed state. Techniques for examining materials at strains where the materials' characteristics fluctuate are called linear and non-linear oscillatory rheology. Linear oscillatory rheology is a methodology that uses modest strains to probe materials without changing their physical characteristics, whereas non-linear oscillatory rheology uses larger strains that change the materials' physical characteristics. Such data provides additional insights into the structure and underlying characteristics that lead the material to become viscoelastic. A non-linear stress response shows a fundamental shift in material behavior from the equilibrium. Another name for this test procedure is large amplitude oscillatory shear rheology.

5.3 SAOS

The stress-strain relation will provide us with elements like storage and loss modulus for the characterization of gels, solutions, paints, and suspensions. The variables observed will be strain dependent when the material is sandwiched between two plates and strain is imparted by observing the resultant stresses in oscillatory shear rheology. A test procedure is referred to as small amplitude oscillatory shear rheology if the strains applied in the experiment are sufficiently tiny so as not to impact the properties of the material. When the strain is continuously varied while the viscoelastic properties, such as the storage and loss modulus, are observed, the modulus values will eventually diverge. The strain percentage at which a material maintains the initial values of its attributes is known as the linear viscoelastic range [35, 36].

During the estimation of the viscoelastic regime and finding the extent of strain percentage that the material could be probed, the solutions immediately after the addition of the crosslinker were placed between the plates for analysis. The moving plate is positioned such that the gel formed between the moving and stationary plates takes the shape of a gap between the plates. After lowering and adjusting the rotating plate and after the formation of the gel, a strain sweep is introduced by slowly increasing the strain on the material at a constant frequency [33, 37]. The modulus, usually the storage modulus G', is followed during the strain sweep. Since the hydrogel does not break its structure at low strains but only deforms, at this stage, the storage modulus will become constant. At higher strains, the deformations become too long, leading to the breakage of the structure of the hydrogel and the storage modulus will fall due to the breakage of the structural feature of the hydrogel, as evident from Figure 5.4.

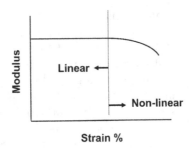

FIGURE 5.4
Finding the linear viscoelastic regime to obtain the linear strain percentage for a hydrogel, which can be probed further.

5.4 Gel Strength

An important characteristic of a hydrogel is the extent of gelation. This requires firstly the conformation of the formation of gel and then characterization using some physical parameter. As discussed earlier, the storage modulus is the characteristic parameter that tells us about the extent of the solidness of a material. Meaning, the higher the value of storage modulus G′, the stronger the gel. To confirm the formation of gel, the sample is placed between the two plates in a rheometer and the moving plate is lowered and the solution is allowed to form the gel. After the gel has been formed (leaving the contents of the hydrogel for a period that is approximated using the crude test upside-down method of gel characterization), a frequency sweep is conducted by fixing the strain at a value that lies in the linear viscoelastic regime [38, 39]. If the contents are jellified, then the storage modulus becomes independent of frequency and will give us the same value, as shown in Figure 5.5 [1, 28, 29, 34]. This is an indication of the gel. The following steps need to be followed to characterize the hydrogel for gel strength.

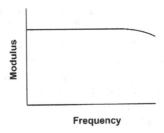

FIGURE 5.5
Conformation of the formation of a hydrogel.

1. Place the solution in between the fixed and moving plates of a rheometer.
2. Position the moving plate by lowering it carefully.
3. Allow the contents to form a gel and take the shape of the gap between the plates.
4. Impose a frequency sweep on the hydrogel by the varying frequency at a strain found using a linear region.
5. Follow the storage modulus and observe the value.
6. It should be independent of the frequency change for the conformation of the gel.
7. To compare two different hydrogels for their strength, compare the storage modulus values.

5.5 Gelation Time

Winter explained that sol-gel transition incorporates the idea of molecular weight in the study of rheological expression to chemical gelation [15, 16]. The solution starts with a molecular weight that increases in accordance with the crosslinking procedure. When the elasticity kicks in and the viscosity reaches a certain degree throughout the process, the polymer is still liquid and possesses viscoelastic qualities. The molecular weight continues to rise with more crosslinking reactions until a point known as the "gel point", where a sizable molecular cluster forms and the molecular weight diverges to infinity, is reached. The equilibrium modulus increases exponentially at the gel point until the crosslinking reaction ends and the phase change results from the solid's viscoelastic nature. Measuring the equilibrium modulus or steady shear viscosity are two typical techniques for determining the gel point [40].

The measurement of steady shear viscosity has some drawbacks, including the following:

1. The relaxation time increases dramatically during the gelation phase, preventing the achievement of steady shear.
2. Network structure deteriorates and is prone to failure.
3. The gel appears to be shear thinning before it gels.
4. After the gelation point, the stress and strain rates can exceed the instrument's detectable limits.

Because of this, Winter noted that the mechanical experiments with small amplitude oscillatory shear are ideal for figuring out the gelation point and modulus progression [24]. When examining the dynamic mechanical characteristics of a curing system, which are a function of reaction, Laza et al. attempted to analyze the gel point based on a number of factors. Based on the greatest difference between the gel's elastic and viscous properties, they detected a maximum peak in tan, with crossing points of the elastic and viscous moduli signifying the storage and release of energy. The cross line drawn at the G' curve as G' shoots up corresponds to the point on the tangent line to the G' curve. The dynamic viscosity reaches a point when it is several orders of magnitude [41].

In their research, Winter found that the storage modulus and loss modulus are equal at the gel point for any significant frequency or temperature. When the relaxation exponent is 0.5, the gel point will represent the intersection of the storage and loss moduli. In terms of the damping factor [25], Raghavan et al. examined the legation point. According to their crosslinking system observation, the G' and G" curves reveal the same slope as crosslinking progresses, which corresponds to the development of crucial gel and meets the Winter criterion [26]. Hayaty et al. provided the following criteria for determining the gel point based on rheological studies considering the aforementioned literature [40].

Criterion 1. The baseline and the tangent drawn from the turning point of the G' curve are crossed to determine the gel time.

Criterion 2. The gel time is the amount of time needed for the viscosity to attain a very high value or toward infinity.

Criterion 3. The period at which the G' and G" curves cross is referred to as the gel time or tan δ = 1.

Criterion 4. The point at which tan δ becomes frequency independent is considered to be the gel time.

Based on the above discussion, it is important for us to find the gelation time. In most cases, the hydrogel sol-gel transition time can be found by carefully following the changes in modulus according to the change in time during a fixed strain percentage and frequency, where the hydrogel is probed for gelation time, as shown in Figure 5.6.

The following steps need to be followed to find the gelation time of a hydrogel:

1. Place the solution in between the fixed and moving plates of a rheometer.
2. Position the moving plate by lowering it carefully.
3. Adjust the strain percentage and frequency such that the hydrogel falls into a linear regime.
4. Immediately after positioning the moving plate, start following the modulus at varying times.

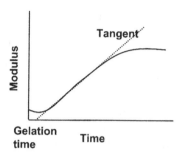

FIGURE 5.6
Finding the gelation time by following the storage modulus changes with time.

5. The storage modulus starts increasing with time as the gelation process continues.
6. The increase in storage modulus seizes at some time, indicating complete gel formation, as shown in Figure 5.6.
7. The gelation time can be found by simply drawing the tangent to the storage modulus curve and indicating the onset of gelation time.
8. To compare two different hydrogels for their strength, compare the storage modulus values.

By following the above characterization methods for a hydrogel, gel properties can be completely evaluated and compared with other kinds of important hydrogels.

5.6 Characterization of Hydrogels Using Dynamic Light Scattering

A polarizer directs laser light through it before it hits the scattering medium in a typical experiment involving light scattering. By picking out only polarized light, the analyzer examines the light that was scattered. This is then put into the detector. The scattering angle is determined by the detector's position. As a result, the detector area's intercepted beam and the incident beam's intersection are used to determine the scattered volume. The contemporary DLS instrument also allows for the measurement of the dispersed light's spectral distribution. To be precise and accurate, these instruments also use photomultipliers. Intramolecular interference, which depends on the mass distribution of the molecules and is the foundation for determining the radius of gyration, occurs when molecules are large enough in comparison with the wavelength of light. In DLS measurements, the radii of gyration of polymer solutions are assessed using intramolecular interference and scattered light frequency dependent on integrated light intensities [42].

Due to the Brownian motion of the molecules, when laser light is shone through a diluted polymer solution, the light intensity will change. I(t) provides the intensity at any moment t. The intensity of light varies and is given by I(t+τ) at any time larger than t, let's say (t+ τ). Instead of using discrete intensities, an autocorrelation function will be created based on changes in these intensities as sensed by the detector. The autocorrelation function describes the time-dependent relationship between two measurements. A 100% autocorrelation exists at zero time. As the measurement is nearly zero at this point and the particles may not have moved all that much, I(t+ τ) coincides with I(t). Because of the disparity in intensities between the initial state and the current state, the measured intensities will eventually stop correlating. When there is no similarity in intensities from the starting function, the correlation function eventually hits 0. The relation between diffusion coefficient D and the measurement vector q is given by

$$g(\tau) = g_0 e^{-Dq^2\tau}$$

and the measurement vector q is given by

$$q = \frac{4\pi n}{\lambda} \sin\frac{\theta}{2}$$

where n is the refractive index of the solution, λ is the scatting wavelength, and θ is the angle of scattering measurement to the ending state.

When the particle is smaller and has more Brownian motion, this occurs more quickly. Exponential decay can be used to describe how the autocorrelation function degrades. Measurements of dynamic light scattering have been utilized to examine the aqueous solution gelation behavior. Using the Horiba nanoparticle SZ 100, dynamic light scattering (DLS) from samples (samples used in rheology) was examined. At various points during gelation, the first-order correlation function was measured for the solutions. A disposable cuvette that was open on all four sides was used to transmit laser light with a 532 nm wavelength, and the intensity of the scattered light was measured at 90.

5.7 Surface Morphology Using Environmental Scanning Electron Microscopy

The morphological characteristics of hydrogels offer intricate details about their structural characteristics. The structural properties of these polymer molecules' three-dimensional structures can occasionally be examined utilizing their morphology. Environmental scanning electron microscopy (ESEM) was used to examine the morphology of the hydrogel. It was discovered that the surface characteristics were missing. The morphology of these gels also explains why the hydrogel does not have a porous structure when the crosslinker-to-polymer ratio is used to create the gel. In the three-dimensional network, covalent crosslinking was strict, leaving no possibility for pores to develop.

Protons are not employed in electron microscopes because electrons have shorter wavelengths and can observe nanostructure details at atomic resolution scales. Scanning electron microscopy is used to study the surface. Secondary electrons, backscattered electrons, transmitted electrons, X-rays, and light are produced when an electron beam interacts with atoms on the surface. Using a detector with a superior resolution of up to 1 nm, SEM creates a built image based on secondary electrons. When the inner shell electrons are lost and replaced by beam electrons, energy is released, and X-rays are produced as a result.

The elements in the sample are identified and measured using this X-ray emission. The backscattered electrons are produced because of the elastic interactions between the beam electrons and the atoms. As heavy atoms with high atomic numbers scatter electrons more intensely than lighter atoms, it is possible to determine the composition based on backscattered electrons. The creation of environmental scanning electron microscopes has given researchers the flexibility to study surface morphology on non-conductive samples (those without a gold coating) [43]. Gels were studied in an FEI Quanta 400 FEG (ESEM mode, gaseous secondary electron detector (GSED) at 0.96 Torr vacuum) at various drying times. The solution was put into Petri dishes once the crosslinker was added. At various stages of the drying process, a little piece of the gel was cut and placed on the stub for microscopic examination. The surface features were viewed at various magnifications, as the sample's whole surface was scanned.

5.8 Conclusion

The characterization of hydrogels can be accomplished based on scientific methods rather than traditional inversion methods. The rheology of gels can give more insights during characterization and is an appropriate tool to show gelation time and gel strength using the SAOS rheology method. This method of characterization has the advantage of accurately comparing different kinds of hydrogels and their structural features, which could enable us to distinguish a weak or strong hydrogel.

References

1. J. Kodavaty, "Poly (vinyl alcohol) and hyaluronic acid hydrogels as potential biomaterial systems-A comprehensive review," *J Drug Deliv Sci Technol*, 71(2022)[103298]
2. A. Vintiloiu and J.-C. Leroux, "Organogels and their use in drug delivery: A review," *J Control Release*, 125(2008)[179–192]
3. D. A. Loy, B. M. Baugher, C. R. Baugher, D. A. Schneider, and K. Rahimian, "Substituent effects on the sol–gel chemistry of organotrialkoxysilanes," *Chem Mater*, 12(2000)[3624–3632]
4. R. Tamate, K. Hashimoto, T. Horii, M. Hirasawa, X. Li, M. Shibayama, and M. Watanabe, "Self-healing micellar ion gels based on multiple hydrogen bonding," *Adv Mater*, 30(2018)[1802792]
5. S. Amin, S. Rajabnezhad, and K. Kohli, "Hydrogels as potential drug delivery systems," *Scientific Research and Essays*, 4(2009)[1175–1183]
6. S. R. Stauffer and N. A. Peppast, "Poly(vinyl alcohol) hydrogels prepared by freezing-thawing cyclic processing," *Polymer*, 33(1992)[3932–3936]
7. A. T. Metters, K. S. Anseth, and C. N. Bowman, "Fundamental studies of a novel, biodegradable PEG-b-PLA hydrogel," *Polymer*, 41(2000)[3993–4004]
8. L. Brannon-Peppas and N. A. Peppas, "Equilibrium swelling behavior of pH-sensitive hydrogels," *Chem Eng Sci*, 46(1991)[715–722]
9. T. Miyata, T. Uragami, and K. Nakamae, "Biomolecule-sensitive hydrogels," *Adv Drug Deliv Rev*, 54(2002)[79–98]
10. D. F. Argenta, T. C. dos Santos, A. M. Campos, and T. Caon, "Hydrogel nanocomposite systems: Physico-chemical characterization and application for drug-delivery systems," in *Nanocarriers for Drug Delivery*, Elsevier, (2019)[81–131]
11. F. G. C. Tessarolli, S. T. S. Souza, A. S. Gomes, and C. R. E. Mansur, "Influence of polymer structure on the gelation kinetics and gel strength of acrylamide-based copolymers, bentonite and polyethylenimine systems for conformance control of oil reservoirs," *J Appl Polym Sci*, 136(2019)[47556]
12. A. S. Hoffman, "Hydrogels for biomedical applications," *Adv Drug Deliv Rev*, 54(2002)[3–12]
13. G. W. M. Vandermeulen and H.-A. Klok, "Peptide/protein hybrid materials: Enhanced control of structure and improved performance through conjugation of biological and synthetic polymers," *Macromol Biosci*, 4(2004)[383–398]
14. P. A. Gunatillake and R. Adhikari, "Biodegradable synthetic polymers for tissue engineering," *e Cells Mater J*, 5(2003)[1–16]
15. M. Vert, "Polymeric biomaterials: Strategies of the past vs. strategies of the future," *Prog Polym Sci*, 32(2007)[755–761]
16. J.-Y. Lai, "Relationship between structure and cytocompatibility of divinyl sulfone cross-linked hyaluronic acid," *Carbohydr Polym*, 101(2014)[203–212]
17. Y. Ding, S. Zheng, X. Meng, and D. Yang, "Low salinity hot water injection with addition of nanoparticles for enhancing heavy oil recovery," *J Energy Resour Technol*, 141(2019)[7]
18. X. Yang, E. Bakaic, T. Hoare, and E. D. Cranston, "Injectable polysaccharide hydrogels reinforced with cellulose nanocrystals: Morphology, rheology, degradation, and cytotoxicity," *Biomacromolecules*, 14(2013)[4447–4455]

19. L. S. Nair and C. T. Laurencin, "Biodegradable polymers as biomaterials," *Progress in Polymer Science*, 32(2007)[762–798]
20. S. H. Kim, K. Hyun, T. S. Moon, T. Mitsumata, J. S. Hong, K. H. Ahn, and S. J. Lee, "Morphology–rheology relationship in hyaluronate/poly(vinyl alcohol)/borax polymer blends," *Polymer*, 46(2005)[7156–7163]
21. K. Hyun, S. H. Kim, K. H. Ahn, and S. J. Lee, "Large amplitude oscillatory shear as a way to classify the complex fluids," *J Non-Newtonian Fluid Mech*, 107(2002)[51–65]
22. X. Zhao, Q. Zhang, D. Chen, and P. Lu, "Enhanced mechanical properties of graphene-based poly (vinyl alcohol) composites," *Macromolecules*, 43(2010)[2357–2363]
23. Y.-F. Tang, Y.-M. Du, X.-W. Hu, X.-W. Shi, and J. F. Kennedy, "Rheological characterisation of a novel thermosensitive chitosan/poly(vinyl alcohol) blend hydrogel," *Carbohydr Polym*, 67(2007)[491–499]
24. H. H. Winter, "Evolution of rheology during chemical gelation," in *Permanent and Transient Networks*, Springer, (1987)[104–110]
25. H. H. Winter, "Can the gel point of a cross-linking polymer be detected by the G'–G" crossover?," *Polym Eng Sci*, 27(1987)[1698–1702]
26. S. R. Raghavan, L. A. Chen, C. McDowell, S. A. Khan, R. Hwang, and S. White, "Rheological study of crosslinking and gelation in chlorobutyl elastomer systems," *Polymer*, 37(1996)[5869–5875]
27. Q. Yuan, X. Lu, K. H. Khayat, D. Feys, and C. Shi, "Small amplitude oscillatory shear technique to evaluate structural build-up of cement paste," *Mater Struct*, 50(2017)[1–12]
28. J. Kodavaty and A. P. Deshpande, "Regimes of microstructural evolution as observed from rheology and surface morphology of crosslinked poly (vinyl alcohol) and hyaluronic acid blends during gelation," *J Appl Polym Sci*, 131(2014)[22].
29. J. Kodavaty and A. P. Deshpande, "Self-assembly and drying assisted microstructural domain formation in poly (vinyl alcohol) and hyaluronic acid gels," *Polym Bulletin*, 74(2017)[3605–3617]
30. K. A. Ramya, J. Kodavaty, P. Dorishetty, M. Setti, and A. P. Deshpande, "Characterizing the yielding processes in pluronic-hyaluronic acid thermoreversible gelling systems using oscillatory rheology," *J Rheol*, 63(2019)[215–228]
31. M. Harini and A. P. Deshpande, "Rheology of poly (sodium acrylate) hydrogels during cross-linking with and without cellulose microfibrils," *J Rheol*, 53(2009)[31–47]
32. A.-L. Kjøniksen and B. Nyström, "Effects of polymer concentration and cross-linking density on rheology of chemically cross-linked poly(vinyl alcohol) near the gelation threshold," *Macromolecules*, 29(1996)[5215–5222]
33. J. Kodavaty and A. P. Deshpande, "Mechanical and swelling properties of poly (vinyl alcohol) and hyaluronic acid gels used in biomaterial systems: A comparative study," *Def Sci J*, 64(2014)[3]
34. J. Kodavaty and A. P. Deshpande, "Evaluation of solute diffusion and polymer relaxation in cross-linked hyaluronic acid hydrogels: Experimental measurement and relaxation modeling," *Polym Bulletin*, 78(2021)[2605–2626]
35. N. Mahinpey, A. Ambalae, and K. Asghari, "In situ combustion in enhanced oil recovery (EOR): A review," *Chem Eng Commun*, 194(2007)[995–1021]
36. M. K. Nguyen and D. S. Lee, "Injectable biodegradable hydrogels," *Macromol Biosci*, 10(2010)[563–579]

37. S. K. Yatirajula, A. Shrivastava, V. K. Saxena, and J. Kodavaty, "Flow behavior analysis of Chlorella Vulgaris microalgal biomass," *Heliyon*, 5(2019)[01845]
38. D. Saraydin and Y. Çaldiran, "In vitro dynamic swelling behaviors of polyhydroxamic acid hydrogels in the simulated physiological body fluids," *Polym Bulletin*, 46(2001)[91–98]
39. D. Saraydin, E. Karadağ, and O. Güven, "The releases of agrochemicals from radiation induced acrylamide/crotonic acid hydrogels," *Polym Bulletin*, 41(1998) [577–584]
40. M. Hayaty, M. H. Beheshty, and M. Esfandeh, "A new approach for determination of gel time of a glass/epoxy prepreg," *J Appl Polym Sci*, 120(2011)[1483–1489]
41. J. M. Laza, C. A. Julian, E. Larrauri, M. Rodriguez, and L. M. Leon, "Thermal scanning rheometer analysis of curing kinetic of an epoxy resin: 2. An amine as curing agent," *Polymer*, 40(1999)[35–45]
42. L. Fang and W. Brown, "Decay time distributions from dynamic light scattering for aqueous poly(vinyl alcohol) gels and semidilute solutions," *Macromolecules*, 23(1990)[3284–3290]
43. O. Jeon, S. J. Song, K. J. Lee, M. H. Park, S. H. Lee, S. K. Hahn, and B. S. Kim "Mechanical properties and degradation behaviors of hyaluronic acid hydrogels cross-linked at various cross-linking densities," *Carbohydr Polym*, 70(2007) [251–257]

Part II

Hydrogels for Energy Applications

6
Hydrogels and Their Emerging Applications

Atta Rasool, Muhammad A. u. R. Qureshi,
Atif Islam, and Shumaila Fayyaz

CONTENTS

6.1 Introduction 104
    6.1.1 Hydrogels as 3D Material 104
    6.1.2 Properties of Hydrogels 104
    6.1.3 Crosslinking 105
        6.1.3.1 Physical Crosslinking 105
        6.1.3.2 Chemical Crosslinking 106
6.2 Materials for 3D Hydrogel Fabrication 106
    6.2.1 Collagen-based 3D Hydrogels 106
    6.2.2 Gelatin-based 3D Hydrogels 107
    6.2.3 Chitosan-based 3D Hydrogels 107
    6.2.4 Carrageenan-based 3D Hydrogels 108
    6.2.5 Pectin-based 3D Hydrogels 109
    6.2.6 Alginate-based 3D Hydrogels 110
    6.2.7 Hyaluronic Acid-based 3D Hydrogels 110
6.3 Synthetic Polymers Incorporated in 3D Hydrogels 111
6.4 Reinforcement Materials 112
    6.4.1 Graphene and Its Derivatives 113
    6.4.2 Natural Clays 113
    6.4.3 Dendrimers 113
6.5 Applications of 3D Hydrogel Frameworks 114
    6.5.1 Drug Delivery 114
    6.5.2 Tissue Engineering 114
    6.5.3 Wound Healing 115
    6.5.4 Cancer Treatment 115
    6.5.5 Eye Treatments 115
    6.5.6 Intelligent 3D Biosensors 116
        6.5.6.1 Optical Sensors 116
        6.5.6.2 Mechanical Sensors 116
        6.5.6.3 Electrical Sensors 116
    6.5.7 Agriculture 117
    6.5.8 Supercapacitors 118

6.5.9 Cosmetics 119
6.5.10 Wastewater Treatment 119
6.6 Recent Developments 119
6.7 Conclusions and Outlook 121
Acknowledgment 121
References 122

6.1 Introduction

6.1.1 Hydrogels as 3D Material

Hydrogels are 3D frameworks composed of hydrophilic polymers that are crosslinked by inter/intramolecular forces. They can be designed in variable forms, sizes, and shapes that can absorb water readily. Hydrogel-based materials have significance in drug delivery, tissue engineering, cosmetics, dentistry, and ocular and cancer treatments [1]. The swelling behavior is a unique response of hydrogels in water, ionic liquids, and biological fluids. Swelling action depends on pH, temperature, the nature of the polymer, and the extent of crosslinking, which enables them to be used in agricultural, biological, industrial, and biomedical fields. Covalent bonding, secondary forces, and chain enlargement are responsible for their swelling actions. Hydrogel's dehydrated form is known as the glassy state, while the hydrated form is termed the rubbery state. Swelling involves water diffusion, and the loosening up of polymeric chains followed by expansion, thus the glassy state is converted into a rubbery state [2].

6.1.2 Properties of Hydrogels

The properties of hydrogels are dependent on the polymer and their pendant groups. Hydrogels are hydrophilic, flexible, easily shapeable, porous, high swelling, soft, and elastic in nature. These act as intelligent materials that respond to temperature, light, ionic strength, pH, osmotic pressure, and magnetic and electric fields. Biocompatibility and immune compatibility are significant properties of hydrogels that generate specific responses in particular conditions without inflammatory responses. Degradation, mechanical stability, and non-toxicity of hydrogel materials are some of the prime features of hydrogel frameworks. Natural polymers like chitosan, alginate, gelatin, hyaluronic acid, and cellulose not only present good in vivo and in vitro biocompatibilities but also demonstrate biodegradation, mechanical stability, and non-toxicity. The physicochemical properties of hydrogels are significantly improved by the crosslinking and incorporation of fillers or reinforcement materials. In contrast to chemical crosslinking, physical crosslinking control over the microenvironment is limited because it is dependent on the inherent properties of the polymer.

6.1.3 Crosslinking

Hydrogels are prepared by crosslinking polymeric chains in an aqueous medium by physical crosslinking, chemical crosslinking, or ionic interactions, as illustrated in Figure 6.1.

6.1.3.1 Physical Crosslinking

In physical crosslinking, polymeric chains are linked to each other by secondary interactions [2]. Physical crosslinking is attained in 3D hydrogel frameworks by thermal condensation, ionic interactions, and self-assembly. Mainly polysaccharides undergo thermal condensation governed by temperature changes. Phase transitions that take place by a decrease in temperature are termed upper critical solution temperature (UCST), while a phase transition directed by a rise in temperature is known as low critical solution temperature (LCST). UCST and LCST are enthalpy- and entropy-derived effects, respectively. However, enthalpy-derived polymers are very rare. Hydrogels that are formed by electrostatic or chelating interactions of charged polymeric backbone make ionotropic hydrogels. Calcium alginate is a well-known example of ionotropic hydrogels. Self-assembly is another method of physical crosslinking that is comprised of hydrophobic, hydrogen,

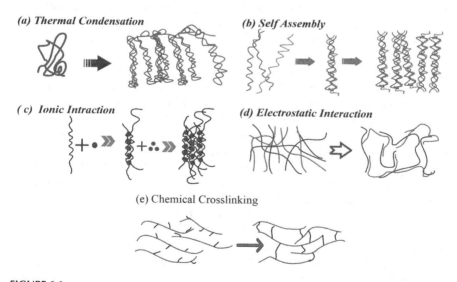

FIGURE 6.1
Different types of crosslinking in hydrogels. (Adapted with permission [3], Copyright [2020], Elsevier.)

and electrostatic interactions, which exists in protein-based hydrogels. These proteinaceous hydrogels exhibit shear thinning under stress and self-healing upon removal of stress, which is employed in ligand pair interaction and the development of protein structures [3].

6.1.3.2 Chemical Crosslinking

In chemical crosslinking, more stable, tunable, and covalently bound hydrogels are prepared. It is carried out by radical polymerization, aldehydes, complementary groups, enzymes, addition, and condensation reactions. However, radical polymerization is the most common mechanism.

6.2 Materials for 3D Hydrogel Fabrication

Natural polymers have extraordinary varieties in their structures that could be used in the fabrication of 3D hydrogels. Some well-known natural polymers will be the focus of this section.

6.2.1 Collagen-based 3D Hydrogels

Collagen is an important constituent of connective tissues, which has been extensively studied. It possesses a triple helical structure estimated at up to 25% of total dry weight in mammals [4]. Living cells are the source of collagen; however, it is also extracted from tendons, rat tails, bovine skin, and porcine skin. There are 29 different types of collagens but collagen type I, II, III, V, and XI are used for the preparation of collagen-based nanofibers and hydrogels. Among all types, collagen type I is widely used in 3D fibrous hydrogel hybrids owing to its inherent ability to undergo self-assembly. It exists in a liquid phase at low temperatures, at 37°C, and in neutral pH mediums, and it generates fibrous hydrogels. Slower gelation is the main drawback related to collagen type I, which needs 30 minutes for gelation compromising stability, uniformity, and homogeneity. Therefore, collagen type I is blended with support materials such as fillers and natural and synthetic polymers. For example, the group of Cho's prepared collagen type I and calcium alginate hydrogels for chondrocyte encapsulation that presented osteochondral regeneration and 90% cell compatibilities [5]. The addition of calcium alginate is expected to improve mechanical stability. Likewise, high-density collagen can also be employed to overcome lesser stabilities in collagen-derived hydrogel frameworks. Collagen type I hydrogels are promising 3D substrates not only for cell culturing but also used as scaffolds in tissue and tumor engineering.

6.2.2 Gelatin-based 3D Hydrogels

Gelatin is a water-soluble and partly hydrolyzed form of collagen. It is a biodegradable and biocompatible biopolymer. Its properties depend on the source from which it is extracted. Primarily it is obtained from mammals and fish. However, gelatin-based 3D hydrogels are limited in tissue engineering due to faster enzymatic biodegradation, lower mechanical strengths, slower gelation, and higher solubility. Gelatin-based 3D hydrogels have been devised with the addition of fillers and synthetic or natural polymers. To overcome slow gelation, functional groups have been increased in gelatin's backbone to regulate the design of the hydrogel matrix. For instance, methacrylate moiety was introduced in 2000 by the combination of gelatin and methacrylic anhydride. The product is known by various names such as gelatin methacryloyl, gelatin methacrylate, and gelatin methacrylamide. Gelatin methacrylate exhibited poor mechanical performance which affected long-term load-bearing capabilities in tissue engineering, repairment of cartilage, and bone healing. Poly(caprolactone) (PCL) networks of microfibers were evaluated to address the problem of poor mechanical performances. The results indicated stiffness, responsiveness, and higher mechanical loading in gelatin-based 3D frameworks. The group led by Zeeshan et al. also fabricated gelatin/poly(vinyl alcohol) (PVA) hydrogels by using a 3-aminopropyltriethoxysilane (APTES) crosslinker. It was reported that homogeneity and tensile strengths were significantly increased by increments in APTES quantity.

6.2.3 Chitosan-based 3D Hydrogels

Chitosan is a naturally occurring cationic copolymer obtained from chitin by the de-acetylation of the acetamido group. It is formed by β-(1-4) linked 2-amino-2-deoxyglucopyranose and 2 acetamido-2-deoxy-D-glucopyrnose groups. It is a well-known biodegradable, biocompatible, and non-toxic biopolymer for medico-biological applications [6]. It possesses two hydroxyl groups and one amino group per unit, which could be modified for desired properties. For instance, Rasool and his team fabricated chitosan/PVP biodegradable pH-sensitive hydrogels in the presence of poly(acrylic acid) (PAA). The extent of swelling was inversely related to the amount of PVP. These hydrogels were further explored for in vitro release of Ag sulfadiazine in phosphate buffer saline (PBS) solution for wound healing that depicted 91.2% drug release in one hour [7].

Another research group fabricated halloysite reinforced 3D hydrogel frameworks established by electrospun fibers of chitosan/PVA for cephradine release, which were characterized by thermogravimetric analysis (TGA), Fourier transform infrared spectroscopy (FTIR), and scanning electron microscope (SEM). The stability of reported hydrogels was directly related to the amount of halloysite; 90% of the drug was released in a PBS

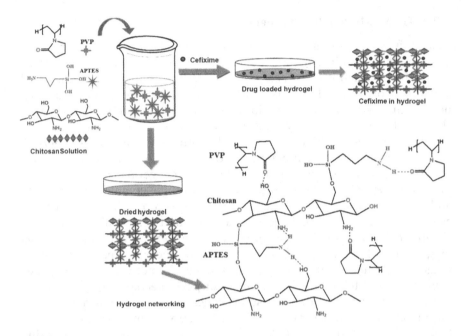

FIGURE 6.2
Proposed synthetic scheme for chitosan/PVP by APTES. (Adapted with permission [9], Copyright [2020], Elsevier.)

solution in 160 minutes [8]. Ata et al. fabricated chitosan/PVP hydrogel crosslinked by APTES for controlled release of cefixime in gastric pH medium depicted in Figure 6.2. The maximum swelling was reported at pH 2, which decreased upon a further increase in pH [9].

One research group designed chitosan-based 3D silane crosslinked hydrogels for benzocaine release. Chitosan/PVA/halloysite hydrogels were prepared by electrospinning technique using (3-glycidyloxypropyl)trimethoxysilane (GOPS). The fabricated hydrogel system not only represented significant antimicrobial properties but also demonstrated thermal stability that directly related to the amount of crosslinker. Benzocaine release was studied in PBS at 7.4 pH [10].

6.2.4 Carrageenan-based 3D Hydrogels

Carrageenan belongs to sulfated linear polysaccharides obtained from seaweeds. It has a high molecular weight and flexible anionic polysaccharide that produces a coiled structure. It has the characteristics of a geller, stabilizer, and thickener. Carrageenan is mainly employed in drug delivery, tissue engineering, and wound coverings. Based on sulfate content, it is categorized into kappa (k), iota (ι), and lambda (λ) carrageenan containing one, two, and three sulfate groups per disaccharide, respectively. k-carrageenan and

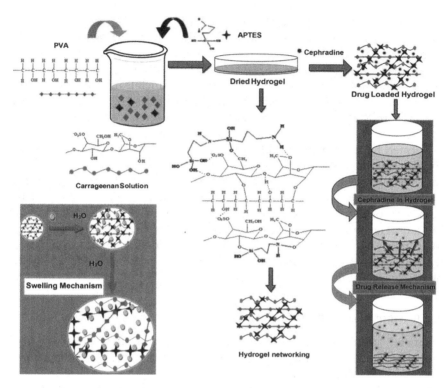

FIGURE 6.3
Diagrammatic presentation of saline crosslinked carrageenan/PVA hydrogel for cephradine release. (Adapted from [11]. Open Access Article licensed under a Creative Commons Attribution-Non-Commercial 3.0 Unported License.)

ι-carrageenan are employed for hydrogel preparations. Carrageenan possesses sulfate and hydroxyl groups that can be modified by oxidation, methylation, sulfation, and acetylation to improve its physicochemical properties. A novel, biodegradable, and pH-sensitive 3D carrageenan-based hydrogel was reported by Rasool et al. using PVA and APTES, which is represented in Figure 6.3. Higher swelling volumes were noticed with a lower quantity of APTES crosslinker. Cephradine was loaded on fabricated hydrogels for controlled release studies in simulated intestinal fluids, which showed 85.5% drug release in 7.5 hours [11].

6.2.5 Pectin-based 3D Hydrogels

Pectin is a complex heterogeneous polysaccharide that is present in the cell wall and middle lamella of terrestrial plants. Its chief constituent is galacturonic acid. Pectin is extracted from citrus fruit in the form of light brown powder that is used to thicken jams and jellies. The quantity of pectin found in citrus peel, rose hips, carrots, and apples is 30, 15, 1.4, and 1.5%, respectively. It is mainly used as a stabilizing, gelling, and viscosity agent. In the medical sector, it is involved in the preparation of medical adhesives and wound healing applications. However, Sriamornsak stated that pectin could be an appropriate choice for controlled and reproducible drug delivery applications in colon and gastro-retentive release systems [12]. A research team led by Morello prepared chitosan/pectin thermo-sensitive hydrogels for culturing turmeric spheroids while gelation was carried out by a weak base. Fabricated hydrogel presented thermo-responsiveness at 37°C and six-week stability. Hydrogels were claimed as appropriate 3D materials for oncological treatments [13]. Recently, pectin and acrylamide hydrogels crosslinked with N, N-bisacrylamide were reported by Siddiqua et al. for colon-specific deliveries of biopharmaceutical drugs using ammonium persulfate as an initiator. The authors inferred that the fabricated 3D framework proved nontoxic in in vivo and in vitro cytotoxicity assays. Further, gel fraction, porosity, and swelling were increased by an increase in pectin content; 90% of the drug release profile was detected [14].

6.2.6 Alginate-based 3D Hydrogels

Alginate's name comes from their gelling behavior. It is extracted from brown seaweeds. It is a hydrophilic, water-soluble, degradable, biocompatible, and low-cost polymer.

Alginate absorbs water readily, hence it is used in waterproofing, fireproofing, dehydrating products, and as a thickener. It contains a carboxylic group that can be negatively charged to incorporate metal ions like Ca^{2+}, Ba^{2+}, Sr^{2+}, and Ni^{2+} to construct polyanionic chains. Sodium, potassium, and calcium alginate are commercially available.

Alginate-based hydrogels were reported to stimulate regeneration in brain tissues and bone recovery by improvement in mechanical properties and porosity levels [15]. Alginate can be fabricated with natural as well as synthetic polymers like chitosan/carrageenan and PAA. For instance, Rasool et al. formulated carrageenan/sodium alginate and PEG 3D hybrid hydrogels using APTES for sustained release of lidocaine. The effect of variable molecular mass PEG and antimicrobial properties were also investigated. Drug release was carried out in PBS solution, which is indicative of controlled release behavior illustrated in Figure 6.4, while SEM micrographs are depicted in Figure 6.5 [16].

6.2.7 Hyaluronic Acid-based 3D Hydrogels

Hyaluronic acid (HA) is a linear and non-sulfated glycosaminoglycan with disaccharide repeating units of β-1,3-N-acetyl-D-glucosamine and β-1,4-D-glucuronic acid. It is found in all connective tissues, extracellular matrix, and cartilage. For 3D hybrid hydrogel systems, it is imperative to improve

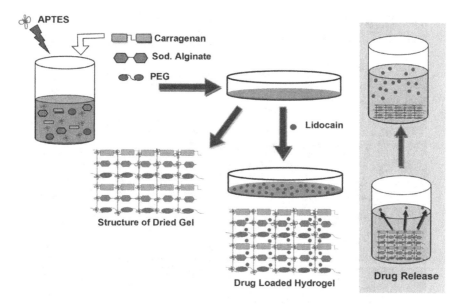

FIGURE 6.4
Synthetic illustration of carrageenan/sodium alginate/ poly(ethylene glycol) (PEG) for lidocaine release. (Adapted with permission [16], Copyright [2020], Elsevier.)

mechanical and rheological properties of HA, which can be achieved by chemical modifications or mixing of some appropriate polymer. Gelatin and HA-based 3D hydrogels were used to deliver human cardiac-derived progenitor cells for cardiac patch printing in mice which resulted in considerable improvement in vascularization, differentiation, and uniform distribution of human cardiac-derived progenitor cells. Similarly, Doyle and coworkers reported thermo-responsive HA and methylcellulose for the delivery of mesenchyme cells obtained from sheep adipose tissues.

6.3 Synthetic Polymers Incorporated in 3D Hydrogels

Synthetic polymers are chosen in the biomedical sector due to their easy processing, degradability, modifiability, and good mechanical strengths. Furthermore, these are hydrophilic, non-immunogenic, and biocompatible. The structural modifications of biopolymers are carried out by crosslinking them with synthetic polymers to acquire collective advantages and upgraded physicochemical characteristics. For example, PEG is employed in 3D hydrogels for drug delivery, tissue engineering, wound healing, and bone regenerative applications. Its structure contains two hydroxyl moieties

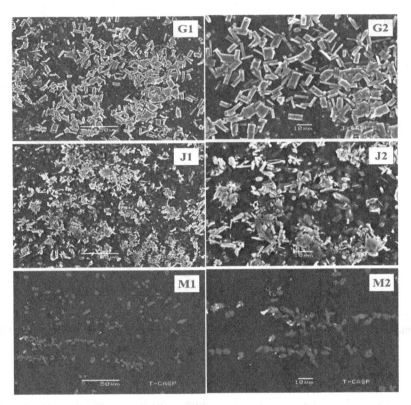

FIGURE 6.5
SEM images of GAP^{-10} (G1 × 50 μm), (G2 × 10 μm), GAP^{-60} (J1 × 50 μm), (J2 × 10 μm), and drug-loaded sample (M1 × 50 μm), (M2 × 10 μm). (Adapted with permission [16], Copyright [2020], Elsevier.)

that could be transformed to carboxyl, thiol, azide, amine, and acrylate functionalities. Consequently, gelatin acrylate/calcium alginate and PEG-tetra acrylate-based hollow tubes can be synthesized and show higher compression modulus and mechanical stability. Some commonly employed synthetic polymers for hydrogel fabrication are PVA, PEG, PVP, PAA, PCL, polymethyl methacrylate (PMMA), polyacrylamide (PAM), and polyurethane (PU).

6.4 Reinforcement Materials

Mechanical properties of hydrogels can be improved by the integration of reinforcement material that may impart responsiveness to biological, physical, or chemical stimuli. Several reinforcement materials have been utilized with a variety of polymers to achieve desired features in hydrogel matrices. Fillers are categorized as organic and inorganic depending on their origin. Metal particles are also incorporated as fillers. Fillers like silica, cellulose, and clay are known as inert fillers and improve the mechanical strength of the hydrogel matrix but do not induce stimuli-responsive behaviors. On the other hand, fillers that direct stimuli-responsive capability and increase mechanical performance are termed active nanofillers.

6.4.1 Graphene and Its Derivatives

It is desirable to devise electrically conductive hydrogels that are capable of bearing high loads coupled with good mechanical performance for biotechnological applications. Yi et al. stated that graphene not only improves stability and mechanical features but also enriches drug loading/release profiles [17]. Furthermore, stability, pore sizes, and degradation rates can be controlled. Graphene is an ideal reinforcement in 3D hybrid hydrogels due to outstanding mechanical and electrical properties coupled with facile conjugation reactions inherited from multiple functionalities such as hydroxyl, epoxy, and carboxyl present on its surface.

6.4.2 Natural Clays

Recently, researchers reported that the incorporation of clay such as halloysite, hydroxyapatite, and sepiolite filler extraordinarily improves mechanical strength in hydrogels [18]. For example, sepiolite is a soft, white, fibrous, hydrous clay of magnesium silicate that enhances drug loading, surface area, and dissolution of sparingly soluble drugs [19]. Likewise, halloysite strengthened 3D chitosan/PVA hydrogel architectures were fabricated for sustained release of cephradine by electrospinning technique. The stability of the stated hydrogels increased with the increase in the quantity of halloysite [8].

6.4.3 Dendrimers

Dendrimers are well-ordered, branched, and polymeric compounds that are symmetric at the core but exhibit 3D shapes. Their solubilities can be changed by the functionalization of exterior functional groups. Dendrimers such as poly(propylene imine) (PI) and poly(amidoamine) (PAAM) are versatile fillers in hydrogel formulations. Dendrimers are polyfunctional, multivalent, and biocompatible, and direct hydrogels as potential candidates for a targeted drug delivery carrier. For example, Holden et al. prepared PAAM/PEG acrylate 3D hydrogel for the efficient delivery of brimonidine and timolol maleate consecutively, which exhibited higher safety, solubility, and mucoadhesive properties [20].

6.5 Applications of 3D Hydrogel Frameworks

This section is dedicated to the application of 3D hydrogel frameworks in different sectors. Initially, biomedical applications such as drug delivery, tissue engineering, wound healing, cancer, and eye treatments are discussed followed by a brief overview of non-medical applications.

6.5.1 Drug Delivery

Conventional drug delivery systems have shown complete and rapid release of drugs that have severe physicochemical effects on normal cells. Therefore, researchers have shown an interest in designing polymeric 3D hydrogel blends for drug delivery applications.

Hydrogels are regarded as perfect polymeric 3D materials for drug delivery applications owing to their modest gelation, biocompatibility, biodegradability, renewability, safety, and elastic nature. For example, L-carnosine was loaded in a k-carrageenan/HA hydrogel fabricated by an ECH crosslinker. The drug release followed zero-order kinetics. An inverse relation of crosslinker and swelling volume was detected [21]. Quercetin release was also assessed in gelatin/ι-carrageenan 3D porous hydrogels. Drug release was improved by an increase in the quantity of carrageenan [22]. The obtained 3D hydrogel was explored for diclofenac sodium. Hydrogels depicted a higher rate of swelling in acidic pH as compared with neutral or basic media [23]. Jabeen et al. formulated chitosan/alginate/PVA injectable hydrogels for the controlled release of neomycin sulfate by using tetraethyl orthosilicate (TEOS) crosslinker. Swelling was inversely related to TEOS concentration [24].

6.5.2 Tissue Engineering

Tissue engineering is a procedure that is used for organ and tissue regeneration. The non-toxic, hemostatic, antimicrobial, biocompatible, and biodegradable nature of hydrogels enables them to be tremendous matrices in the field of tissue engineering. However, hydrogel-derived scaffolds have lesser mechanical and tensile strengths, which are dealt with by crosslinking and reinforcements. Tatiana et al. manufactured a thermo-sensitive collagen/poly(N-isopropyl acrylamide)/hydroxyapatite 3D hydrogel and determined that the properties of poly(N-isopropyl acrylamide) as an artificial extracellular matrix above 33°C can be a good substitute for the regeneration of bone tissues [25]. Likewise, Oguz and Ege prepared calcium phosphate reinforced methyl cellulose/gelatin 3D hydrogels. The quantity of inorganic filler is associated with the rise in transition temperature from 25 to 37°C, which enables injectable hydrogel frameworks to attain solid form in human bone tissue [26].

6.5.3 Wound Healing

It is preferred to keep wounds dry for quick healing. Biocompatibility and water-absorbing capacities of hydrogels improve re-epithelialization and cell migration [27]. Silk is added to PVA/PVP/κ-carrageenan hydrogel to improve mechanical strength; however, PVP/κ-carrageenan/PEG hydrogel dressings have proved to be superior in shelf life, mechanical performance, removability, and patient compliance [28]. Likewise, the reinforcement of k-carrageenan by nanosilicates increased stability, injectability, mechanical properties, absorption of proteins, cell adhesion, platelets binding, and the sustained release of therapeutic agents [29]. Hydrogels act as a barrier to microbial action, thus accelerating the wound-healing process.

6.5.4 Cancer Treatment

Cancer is a complex disease with symptoms such as lymph node enlargement, shortness of breath, cough, severe fatigue, bowel changes, nausea, vomiting, and eating problems. Genetic, reproductive, hormonal, alcohol intake, and physical inactivity are factors that can increase its risk. There are varieties of treatments for cancer like hormone therapy, radiotherapy, surgery, and chemotherapy. However, chemical agents affect cancerous cells as well as normal cells which leads to serious side effects and toxicities. Consequently, intelligent 3D hydrogel carriers loaded with anti-cancerous agents can overcome the limitations of conventional therapies that offer localized, targeted, controlled, and complementary systems for the delivery of drugs, radioisotopes, and genes. Hydrogel-enabled combined therapies are chemotherapy-hyperthermia, chemotherapy-gene therapy, and chemotherapy-radiotherapy. Furthermore, 3D hydrogel hybrids loaded with appropriate nanomagnetic materials have shown dependable, biocompatible, and biodegradable imaging platforms for tumor visualization and screening [30]. Injectable hydrogels have successfully activated anticancer immune actions and inhibit metastasis [31]. For instance, Akhlaq et al. prepared pH-sensitive gelatin/PVA hydrogel for the colon targeted release of methotrexate. Non-Fickian kinetics of release was reported in PBS solutions at pH 1.2, 6.8, and 7.4 [32]. It is expected that hydrogels will play a pivotal role in the development of clinically transformable next-generation cancer treatments.

6.5.5 Eye Treatments

Hydrogel 3D stimuli responsive hybrids are preferred in ocular eye drops due to their therapeutic effect, viscosity, and prolonger contact time. A study was reported by Fernández Ferreiro et al. for improvements in corneal contact time. The research group performed a series of experiments by using variable amounts of alginate, gellan gum, and carrageenan; an 80:20 gellan gum and k-carrageenan ratio considerably enhanced contact time at the cornea [33].

Mohammed et al. formulated a gentamicin and moxifloxacin containing antimicrobial hydrogel platform derived from chitosan and β-glycerophosphate. This 3D hybrid hydrogel efficiently retarded *Staphylococcus aureus* bacterial growth while thermo-sensitive gelation imparted the inside eye with β-glycerophosphate with the addition of drops [34].

6.5.6 Intelligent 3D Biosensors

The stimuli-responsive behavior of hydrogels can be explored for signal transduction in electrical, mechanical, and optical devices for the development of smart 3D optical, electrical, and mechanical sensors. Hydrogels can absorb fluids and host functional groups that recognize elements for higher sensitivities and selectivities.

6.5.6.1 Optical Sensors

Hydrogels act as optical sensors inherited from their transparent nature and optical quality. These sensors provide label-free recognition of biological moieties by surface plasmon resonance (SPR) and optical waveguide mode spectroscopy. Plasmonic particles demonstrate enhanced emission of plasmon signals stimulated by light. The sensitivity and selectivity of SPR-based sensors are considerably enhanced by coating hydrogel layers over sensors' surfaces. Therefore, swelling and de-swelling in the hydrogel matrix changes the refractive index, which regulates electrical signals accordingly [35]. Jabeen et al. fabricated chitosan/alginate/PVA hydrogels and their optical micrographs are displayed in Figure 6.6 [24].

6.5.6.2 Mechanical Sensors

Hydrogel based-polymeric matrices have the ability to expand and shrink, which is mediated by osmotic pressure fluctuations. For example, quartz crystal microbalance (QCM) was devised for virus, DNA, and biomarker detection by incorporating 3D hydrogels hybrids. QCM possesses piezoelectric elements built from quartz crystal that vibrate at characteristic frequency and amplitude. The hydrogel coated piezoelectric element changes its vibrating frequency upon interaction with analyte molecules. Subsequently, mass increases and mechanical mass detection is achieved [36].

6.5.6.3 Electrical Sensors

Electrical biosensors have higher sensitivities, however, they suffer from limited selectivity. Hence, electrical sensors are equipped with biological hybrid materials to achieve combined benefits for higher sensitivity, specificity,

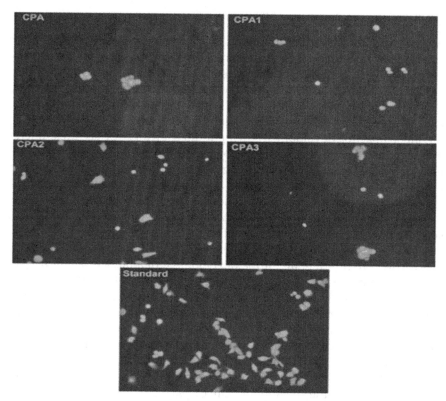

FIGURE 6.6
Optical micrographs for chitosan/alginate/PVA hydrogels after 24-hour incubation for cytotoxic analysis. (Adapted with permission [24], Copyright [2020], Elsevier.)

and selectivity for a variety of analytes. For example, Zhai et al. fabricated an electrode by doping a polyaniline hydrogel framework with platinum nanoparticles that not only immobilized glucose-specific enzymes but also presented a speedy response time for glucose detection [37].

6.5.7 Agriculture

Hydrogels act as a reservoir for water and aqueous solutions in their 3D architecture. These materials are not only superabsorbent but also serve as cargo carriers for agrochemicals, water-retentive agents, and soil conditioners in agriculture. Cheap and environmentally friendly polymeric hydrogels comprising cellulose, chitosan, pectin, starch, guar gum, Arabic gum, and alginate have been formulated for agriculture. A hydrogel platform could be ideal for the controlled and targeted release of fertilizer, nutrients, pesticides, and herbicides [38].

6.5.8 Supercapacitors

A supercapacitor is a device that stores energy by means of a Faradic process or ion adsorption. Many approaches have been explored to boost supercapacitor performance, including electrolyte stabilization, increments in the surface area of the electrode materials, and the development of numerous electrode materials. However, the energy storage capacities of supercapacitors remain low. Hydrogels are superior 3D materials that can serve as a potential choice for supercapacitor development owing to their greater surface area, microporous networks, and multidimensional pathways for electron transport. For example, Chen et al. developed a graphene-reinforced hydrogel-derived ultrafast supercapacitor as a result of a hydrothermal procedure doped via organic amine. The supercapacitor showed better performance and power density values of 113.8 F/g and 205 kW/kg, respectively [39]. Traditional supercapacitors present low values of capacitances due to the small thickness of electrodes in nanometers, which offers small mass loading. A study was carried out by Xu et al. for the development of a flexible 3D hydrogel/graphene-built supercapacitor containing H_2SO_4-PVA as the electrolyte with a much higher thickness of electrode at 120 μm. The stated supercapacitor exhibited exceptional area-specific capacitance and gravimetric-specific capacitance of 372 mF/cm^2 and 186 F/g at 1A/g, respectively. In addition, there was a tremendous capability rate with a 70% retention at 20 A/g [40]. Recently, a research team directed by Khan et al. reported a quasi-solid-state sodium ion-derived asymmetric supercapacitor device. The schematics for the development of the stated device are displayed in Figure 6.7 [41].

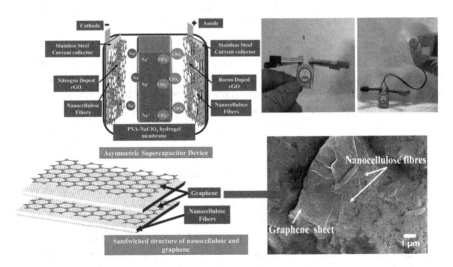

FIGURE 6.7
Schematic display for designing quasi-solid-state sodium ion derived asymmetric supercapacitor device. (Adapted with permission [41], Copyright [2023], Elsevier.)

6.5.9 Cosmetics

Hydrogels exhibit high charge density, biocompatibility, and non-toxic and mucoadhesive properties coupled with intrinsic bactericidal characteristics, a porous structure, minimal external body reactions, and film making and hair fixing properties in cosmeceutical use. In cosmetics, hydrogels are engaged in making body creams, hair conditioners, hair foams, and mascaras. These hydrogels are applied on skin surfaces to act as moisture promoters and humectants [42].

6.5.10 Wastewater Treatment

Water is known as the "blood of the Earth", therefore, it is very important to monitor water quality on a continuous basis. The selection of the treatment process for contaminated water is a difficult task involving the consideration of a number of factors including the available space for the construction of treatment facilities, process efficiency, waste disposal, and operating costs. A number of technologies are available, with varying degrees of success, to remove heavy metals from water. Hydrogels possess surface charge controlled by pH so swelling is directly related to electrostatic interaction among the same charges. Higher adsorptive capacities, faster kinetics, and recyclability make hydrogels special in the field of industrial wastewater treatment. For example, chitosan-based hydrogels adsorb metal ions, dyes, and organic molecules via hydrogen bonding, complexation, and electrostatic forces dependent on their structures. The length of the polymer chain, the degree of de-acetylation, and crystallinity are all factors that affect the formation of a metal-hydrogel complex. Table 6.1 describes some hydrogels reported for wastewater treatment applications.

6.6 Recent Developments

The occurrence of biopolymeric hydrogel is a turning point in modern analytical techniques with the performance of electrochemical sensors. Hydrogel-based selective and sensitive electrochemical biosensors have gained consideration in electrical, mechanical, and optical sensors [36]. Similarly, hydrogel beads derived from polymers are employed to remove heavy metals and harmful dyes from water. For instance, Azeem et al. prepared chitosan/acrylic acid hydrogel crosslinked by formaldehyde for the removal of toxic Pb^{+2} and Cd^{+2} ions from aqueous solutions [43]. Polymeric hydrogels were also developed for determining trace quantity of metals such as Hg^{+2}, Pb^{+2}, Cd^{+2}, Cu^{+2}, As^{+3}, and Cr^{+6}. Furthermore, hydrogels were explored for framing 3D hydrogel/MOF composites for electrode modifications and the

TABLE 6.1
Summary of Chitosan-based Hydrogels for Wastewater Applications

Hydrogels	Crosslinkers	Method	Removal study	pH	Kinetic Model	Reference
Chitosan-g-poly (sodium acrylate-co-acrylamide)/clay nanocomposite	N,N'-methylenebisacrylamide	Microwave-assisted grafting	Crystal violet, naphthol green, and sunset yellow	4	Langmuir and Freundlich	[45]
Poly(itaconic acid)-grafted crosslinked chitosan	Glutaraldehyde	Grafting polymerization	Hg^{2+} and Pb^{2+}	6 for Pb^{2+} 5 for Hg^{2+}	Langmuir and pseudo-second order	[46]
Chitosan-grafted sulfonate	Glutaraldehyde	Grafting polymerization	Pramipexole dihydrochloride	10	Langmuir and Freundlich	[47]
Chitosan/PAA-GLA	Glutaraldehyde	Simple one step	Cu^{2+}	4	Langmuir and pseudo-second order	[48]
Cellulose/sodium alginate/carrageenan	—	Ligand-exchange	Pb^{2+}	3	Langmuir and pseudo-second order	[49]
Agar/k-carrageenan	—	Free radical crosslinking	Methylene blue from water	7	Langmuir and pseudo-second order	[50]

detection of various pharmaceutical drugs [44]. In energetics, 3D hydrogel frameworks are used for water splitting, fuel cells, bioelectronics, electrode developments, supercapacitors, and oxygen evolution reactions. Likewise, conductive 3D hydrogels were fabricated by the incorporation of graphene and its derivatives. In the agriculture sector, hydrogel plays a pivotal role in the sustained release of fertilizers, water retention nutrient carriers, and soil conditioning. Moreover, the development of biopolymeric pH-sensitive hydrogels for biomedical applications is a vast area of research in the medical sector, which includes drug delivery, tissue engineering, wound healing, and ocular and dental treatments [2]. Hydrogel formulations are a significant breakthrough not only for the development of fourth-generation cancer treatments but also for the clinical diagnosis, warning, and detection of cancer.

6.7 Conclusions and Outlook

Among innumerable biopolymeric materials, 3D hydrogel frameworks have gained consideration owing to their versatility, biodegradability, cost-effectiveness, and biocompatibility, as well as their environmentally friendly and modifiable nature. Hydrogels are fabricated by natural-synthetic polymeric combinations because purely natural polymers produce crystalline hydrogel films with poor mechanical properties. Furthermore, 3D hydrogel films are reinforced with the addition of fillers such as graphene, metal oxide, nanoparticles, and natural clays, which not only improve the mechanical performance of 3D hydrogel matrices but also instruct responsiveness, stability, electrical conduction, and greater loading/release of adsorbents, fertilizers, and drugs. This chapter initially summarized 3D hydrogel hybrids, their properties, reinforcement materials, and polymeric matrices for the development of 3D hydrogel frameworks. Next, a brief summary of the medical and non-medical applications of 3D hydrogel was provided. In the current era, it is important to investigate the mechanical features, flexibility, and compatibleness of hydrogel frameworks. The addition of novel fillers in hydrogel frameworks has opened the door for the production of fascinating and attractive systems addressing several challenges in medico-biological, environmental, electrochemical, agricultural, and biotechnological applications.

Acknowledgment

The listed authors are highly thankful to their respective universities for providing literature services.

References

1. M. A.u.R. Qureshi, N. Arshad, A. Rasool, A. Islam, M. Rizwan, M. Haseeb, T. Rasheed, and M. Bilal, Chitosan and Carrageenan-based biocompatible hydrogel platforms for cosmeceutical, drug delivery, and biomedical applications, *Starch-Stärke*. (2022) 2200052.
2. A. Rasool, M. Rizwan, A. Islam, H. Abdullah, S.S. Shafqat, M.K. Azeem, T. Rasheed, and M. Bilal, Chitosan-based smart polymeric hydrogels and their prospective, applications in biomedicine, *Starch-Stärke*. (2021) 2100150.
3. J.M. Unagolla, and A.C. Jayasuriya, Hydrogel-based 3D bioprinting: A comprehensive review on cell-laden hydrogels, bioink formulations, and future perspectives, *Appl. Mater. Today*. 18 (2020) 100479.
4. R. Parenteau-Bareil, R. Gauvin, and F. Berthod, Collagen-based biomaterials for tissue engineering applications, Mater. 3 (2010) 1863–1887.
5. J.Y. Park, J.-C. Choi, J.-H. Shim, J.-S. Lee, H. Park, S.W. Kim, J. Doh, and D.-W. Cho, A comparative study on collagen type I and hyaluronic acid dependent cell behavior for osteochondral tissue bioprinting, *Biofabrication*. 6 (2014) 035004.
6. M. Petchsangsai, W. Sajomsang, P. Gonil, O. Nuchuchua, B. Sutapun, S. Puttipipatkhachorn, and U.R. Ruktanonchai, A water-soluble methylated N-(4-N, N-dimethylaminocinnamyl) chitosan chloride as novel mucoadhesive polymeric nanocomplex platform for sustained-release drug delivery, *Carbohydr. Polym.* 83 (2011) 1263–1273.
7. A. Rasool, S. Ata, and A. Islam, Stimuli responsive biopolymer (chitosan) based blend hydrogels for wound healing application, *Carbohydr. Polym.* 203 (2019) 423–429.
8. M. Naz, M. Rizwan, S. Jabeen, A. Ghaffar, A. Islam, N. Gull, A. Rasool, R.U. Khan, S.Z. Alshawwa, and M. Iqbal, Cephradine drug release using electrospun chitosan nanofibers incorporated with halloysite nanoclay, *Zeitschrift für Physikalische Chemie*. 236 (2022) 227–238.
9. S. Ata, A. Rasool, A. Islam, I. Bibi, M. Rizwan, M.K. Azeem, and M. Iqbal, Loading of Cefixime to pH sensitive chitosan based hydrogel and investigation of controlled release kinetic, *Int. J. Biol. Macromol.* 155 (2020) 1236–1244.
10. M. Naz, S. Jabeen, N. Gull, A. Ghaffar, A. Islam, M. Rizwan, H. Abdullah, A. Rasool, S. Khan, and R. Khan, Novel silane crosslinked chitosan based electrospun nanofiber for controlled release of benzocaine, *Front. Mater*. 9 (2022) 826251.
11. A. Rasool, S. Ata, A. Islam, and R.U. Khan, Fabrication of novel carrageenan based stimuli responsive injectable hydrogels for controlled release of cephradine, *RSC Adv*. 9 (2019) 12282–12290.
12. P. Sriamornsak, Chemistry of pectin and its pharmaceutical uses: A review. *Silpakorn University Int. J.* 3 (2003) 206–228.
13. G. Morello, A. Quarta, A. Gaballo, L. Moroni, G. Gigli, A. Polini, and F. Gervaso, A thermo-sensitive chitosan/pectin hydrogel for long-term tumor spheroid culture, *Carbohydr. Polym.* 274 (2021) 118633.
14. A. Siddiqua, N.M. Ranjha, S. Rehman, H. Shoukat, N. Ramzan, and H. Sultana, Preparation and characterization of methylene bisacrylamide crosslinked pectin/acrylamide hydrogels, *Polym. Bull.* 79 (2022) 7655–7677.

15. J. Venkatesan, I. Bhatnagar, P. Manivasagan, K.-H. Kang, and S.-K. Kim, Alginate composites for bone tissue engineering: A review, *Int. J. Biol. Macromol.* 72 (2015) 269–281.
16. A. Rasool, S. Ata, A. Islam, M. Rizwan, M.K. Azeem, A. Mehmood, R.U. Khan, and H.A. Mahmood, Kinetics and controlled release of lidocaine from novel carrageenan and alginate-based blend hydrogels, *Int. J. Biol. Macromol.* 147 (2020) 67–78.
17. J. Yi, G. Choe, J. Park, and J.Y. Lee, Graphene oxide-incorporated hydrogels for biomedical applications, *Polym. J.* 52 (2020) 823–837.
18. R.R. Palem, K.M. Rao, G. Shimoga, R.G. Saratale, S.K. Shinde, G.S. Ghodake, and S.-H. Lee, Physicochemical characterization, drug release, and biocompatibility evaluation of carboxymethyl cellulose-based hydrogels reinforced with sepiolite nanoclay, *Int. J. Biol. Macromol.* 178 (2021) 464–476.
19. A. Vicosa, A. Gomes, B. Soares, and C. Paranhos, Effect of sepiolite on the physical properties and swelling behavior of rifampicin-loaded nanocomposite hydrogels, *Express Polym. Lett.* 3 (2009) 518–524.
20. C.A. Holden, P. Tyagi, A. Thakur, R. Kadam, G. Jadhav, U.B. Kompella, and H. Yang, Polyamidoamine dendrimer hydrogel for enhanced delivery of antiglaucoma drugs. *Nanomed.: Nanotechnol. Biol. Med.* 8 (2012) 776–783.
21. M. El-Aassar, G. El Fawal, E.A. Kamoun, and M.M. Fouda, Controlled drug release from cross-linked κ-carrageenan/hyaluronic acid membranes, *Int. J. Biol. Macromol.* 77 (2015) 322–329.
22. J.S. Varghese, N. Chellappa, and N.N. Fathima, Gelatin–carrageenan hydrogels: Role of pore size distribution on drug delivery process, *Colloids Surf. B: Biointerfaces.* 113 (2014) 346–351.
23. N. Gull, S.M. Khan, O.M. Butt, A. Islam, A. Shah, S. Jabeen, S.U. Khan, A. Khan, R.U. Khan, and M.T.Z. Butt, Inflammation targeted chitosan-based hydrogel for controlled release of diclofenac sodium, *Int. J. Biol. Macromol.* 162 (2020) 175–187.
24. S. Jabeen, A. Islam, A. Ghaffar, N. Gull, A. Hameed, A. Bashir, T. Jamil, and T. Hussain, Development of a novel pH sensitive silane crosslinked injectable hydrogel for controlled release of neomycin sulfate, *Int. J. Biol. Macromol.* 97 (2017) 218–227.
25. N.M. Tatiana, V. Cornelia, R. Tatia, and C. Aurica, Hybrid collagen/pNIPAAM hydrogel nanocomposites for tissue engineering application, *Colloid Polym. Sci.* 296 (2018) 1555–1571.
26. Ö.D. Oğuz, and D. Ege, Rheological and mechanical properties of thermoresponsive methylcellulose/calcium phosphate-based injectable bone substitutes, Mater. 11 (2018) 604.
27. E.A. Kamoun, E.-R.S. Kenawy, and X. Chen, A review on polymeric hydrogel membranes for wound dressing applications: PVA-based hydrogel dressings, *J. Adv. Res.* 8 (2017) 217–233.
28. D.A. De-Silva, B.U. Hettiarachchi, L. Nayanajith, M.Y. Milani, and J. Motha, Development of a PVP/kappa-carrageenan/PEG hydrogel dressing for wound healing applications in Sri Lanka, *J. Natn. Sci. Foundation Sri Lanka.* 39 (2011) 25–33.
29. G. Lokhande, J.K. Carrow, T. Thakur, J.R. Xavier, M. Parani, K.J. Bayless, and A.K. Gaharwar, Nanoengineered injectable hydrogels for wound healing application, *Acta Biomater.* 70 (2018) 35–47.

30. M. Sepantafar, R. Maheronnaghsh, H. Mohammadi, F. Radmanesh, M.M. Hasani-Sadrabadi, M. Ebrahimi, and H. Baharvand, Engineered hydrogels in cancer therapy and diagnosis, *Trends Biotechnol.* 35 (2017) 1074–1087.
31. Y. Chao, Q. Chen, and Z. Liu, Smart injectable hydrogels for cancer immunotherapy, *Adv. Funct. Mater.* 30 (2020) 1902785.
32. M. Akhlaq, A.K. Azad, I. Ullah, A. Nawaz, M. Safdar, T. Bhattacharya, A.H. Uddin, S.A. Abbas, A. Mathews, and S.K. Kundu, Methotrexate-loaded gelatin and polyvinyl alcohol (Gel/PVA) hydrogel as a pH-sensitive matrix, *Polymers.* 13 (2021) 2300.
33. A. Fernández-Ferreiro, M.G. Barcia, M. Gil-Martínez, A. Vieites-Prado, I. Lema, B. Argibay, J.B. Méndez, M.J. Lamas, and F.J. Otero-Espinar, In vitro and in vivo ocular safety and eye surface permanence determination by direct and magnetic resonance imaging of ion-sensitive hydrogels based on gellan gum and kappa-carrageenan, *Eur. J. Pharm. Biopharm.* 94 (2015) 342–351.
34. S. Mohammed, G. Chouhan, O. Anuforom, M. Cooke, A. Walsh, P. Morgan-Warren, M. Jenkins, and F. De Cogan, Thermosensitive hydrogel as an in situ gelling antimicrobial ocular dressing, *Mater. Sci. Eng.* 78 (2017) 203–209.
35. X. Yin, L. Hesselink, Z. Liu, N. Fang, and X. Zhang, Large positive and negative lateral optical beam displacements due to surface plasmon resonance, *Appl. Phys. Lett.* 85 (2004) 372–374.
36. H.S. Song, O.S. Kwon, J.-H. Kim, J. Conde, and N. Artzi, 3D hydrogel scaffold doped with 2D graphene materials for biosensors and bioelectronics, *Biosens. Bioelectron.* 89 (2017) 187–200.
37. D. Zhai, B. Liu, Y. Shi, L. Pan, Y. Wang, W. Li, R. Zhang, and G. Yu, Highly sensitive glucose sensor based on Pt nanoparticle/polyaniline hydrogel heterostructures, *ACS Nano.* 7 (2013) 3540–3546.
38. M.R. Guilherme, F.A. Aouada, A.R. Fajardo, A.F. Martins, A.T. Paulino, M.F. Davi, A.F. Rubira, and E.C. Muniz, Superabsorbent hydrogels based on polysaccharides for application in agriculture as soil conditioner and nutrient carrier: A review, *Eur. Polym. J.* 72 (2015) 365–385.
39. P. Chen, J.-J. Yang, S.-S. Li, Z. Wang, T.-Y. Xiao, Y.-H. Qian, and S.-H. Yu, Hydrothermal synthesis of macroscopic nitrogen-doped graphene hydrogels for ultrafast supercapacitor, *Nano Energy.* 2 (2013) 249–256.
40. Y. Xu, Z. Lin, X. Huang, Y. Liu, Y. Huang, and X. Duan, Flexible solid-state supercapacitors based on three-dimensional graphene hydrogel films, *ACS Nano.* 7 (2013) 4042–4049.
41. M.S. Khan, B. SanthiBhushan, K.C. Bhamu, S.G. Kang, H.S. Kushwaha, A. Sharma, R. Dhiman, R. Gupta, M.K. Banerjee, and K. Sachdev, Polymer hydrogel based quasi-solid-state sodium-ion supercapacitor with 2.5 V wide operating potential window and high energy density, *Appl. Surf. Sci.* 607 (2023) 154990.
42. M. Rinaudo, Main properties and current applications of some polysaccharides as biomaterials, *Polym. Int.* 57 (2008) 397–430.
43. M.K. Azeem, M. Rizwan, A. Islam, A. Rasool, S.M. Khan, R.U. Khan, T. Rasheed, M. Bilal, and H.M. Iqbal, In-house fabrication of macro-porous biopolymeric hydrogel and its deployment for adsorptive remediation of lead and cadmium from water matrices, *Environ. Res.* 214 (2022) 113790.

44. M. Rizwan, V. Selvanathan, A. Rasool, D.N. Iqbal, Q. Kanwal, S.S. Shafqat, T. Rasheed, and M. Bilal, Metal–organic framework-based composites for the detection and monitoring of pharmaceutical compounds in biological and environmental matrices, *Water Air Soil Pollut.* 233 (2022) 1–29.
45. M. Nagarpita, P. Roy, S. Shruthi, and R. Sailaja, Synthesis and swelling characteristics of chitosan and CMC grafted sodium acrylate-co-acrylamide using modified nanoclay and examining its efficacy for removal of dyes, *Int. J. Biol. Macromol.* 102 (2017) 1226–1240.
46. H. Ge, T. Hua, and J. Wang, Preparation and characterization of poly (itaconic acid)-grafted crosslinked chitosan nanoadsorbent for high uptake of Hg^{2+} and Pb^{2+}, *Int. J. Biol. Macromol.* 95 (2017) 954–961.
47. G.Z. Kyzas, M. Kostoglou, N.K. Lazaridis, D.A. Lambropoulou, and D.N. Bikiaris, Environmental friendly technology for the removal of pharmaceutical contaminants from wastewaters using modified chitosan adsorbents, *Chem. Eng. J.* 222 (2013) 248–258.
48. J. Dai, H. Yan, H. Yang, and R. Cheng, Simple method for preparation of chitosan/poly (acrylic acid) blending hydrogel beads and adsorption of copper (II) from aqueous solutions, *Chem. Eng. J.* 165 (2010) 240–249.
49. G. Sharma, A. Khosla, A. Kumar, N. Kaushal, S. Sharma, M. Naushad, D.-V.N. Vo, J. Iqbal, and F.J. Stadler, A comprehensive review on the removal of noxious pollutants using carrageenan based advanced adsorbents, *Chemosphere.* 289 (2022) 133100.
50. O. Duman, T.G. Polat, C.Ö. Diker, and S. Tunç, Agar/κ-carrageenan composite hydrogel adsorbent for the removal of Methylene Blue from water, *Int. J. Biol. Macromol.* 160 (2020) 823–835.

7
Hydrogels for 3D Frameworks

Ismail Can Karaoglu, Esra Yalcin, Ayesha Gulzar,
Isilay Goktan, and Seda Kizilel

CONTENTS

7.1 Hydrogels for Engineered 3D Vascularized Tissue Models 128
    7.1.1 The Development of the Vascular System 128
        7.1.1.1 Vasculogenesis 128
        7.1.1.2 Angiogenesis 129
    7.1.2 Hydrogels for Vascularization 130
        7.1.2.1 Collagen and Fibrin 130
        7.1.2.2 Gelatin 131
7.2 PEG-based Hydrogels and Their Immunological Properties 132
    7.2.1 Modifications of PEG Hydrogels for Enhanced Properties
        of Biomimicry 132
    7.2.2 Crosslinking Strategies for PEG 133
7.3 An Overview of LMSC Therapy 135
    7.3.1 Treatment of LSCD 136
    7.3.2 The Potential Use of Hydrogels in LMSC 136
        7.3.2.1 Fibrin 137
        7.3.2.2 Collagen 137
        7.3.2.3 Gelatin 137
        7.3.2.4 Alginate 138
    7.3.3 New Therapies to Treat LSCD 138
7.4 Hydrogels in Pancreatic Islet Transplantation 139
    7.4.1 Strategies for Oxygenation 139
    7.4.2 Strategies for Vascularization 141
    7.4.3 Strategies for Immune Modulation 141
        7.4.3.1 Physical Isolation 142
        7.4.3.2 Functionalization of Hydrogels with Apoptotic
                Molecules and Immunosuppressive Drugs 142
7.5 Conclusion 143
Bibliography 143

7.1 Hydrogels for Engineered 3D Vascularized Tissue Models

Hydrogels can be used as biomaterials as they have the potential to be game-changing in the field of tissue engineering and regenerative medicine. Comprising a water content of 70–90%, hydrogels are like tissue in terms of physical properties, hence having the potential to achieve excellent biocompatibility. The delivery of oxygen and nutrients and conveying waste products through a distributed vascular network to tissues and organs are essential for survival. Many diseases occur due to the dysregulation of the vascular network, such as diabetes and coronary artery disease, which cause insufficient vascular perfusion of the heart. These have been the main causes of morbidity and mortality. On the other hand, hyper-vascularization could occur in cancer or diabetic retinopathy. In recent years, using hydrogels in vascularization for tissues has increased for disease modeling, drug screening, and tissue regeneration. In this section, hydrogels from naturally derived and synthetic scaffolds supporting cell proliferation and cell-specific functions are examined for fabricating vascularized tissues. Transplanted biomaterial that cannot be integrated with the host vascular perfusion or lacks vascularization for the local area cannot hold functional and structural integrity for a long time. However, the functionalization of the hydrogel, such as releasing the signaling molecules like growth factors to the transplanted area, could provide vascularized three-dimensional (3D) models that resemble complex human tissues [1]. Therefore, sophisticated applications relevant to the vascularization of hydrogels are highlighted in relevant biomedicine applications.

7.1.1 The Development of the Vascular System

The vasculature is the component of the circulatory system that branches into vessels ranging from several micrometers to several millimeters in diameter. Arteries, capillary beds, and veins have blood-carrying lumens to distribute blood throughout the body. However, all vasculatures have a different composition in terms of mechanical strength, wall thickness, cell types, and extracellular matrix (ECM) distribution. To mimic fluid flow and pressure, single-lumen structures have been engineered for vascularized tissue construction. Khademhosseini and his colleagues encapsulated human umbilical vein endothelial cells (HUVECs) into hydrogel with a hollow structure, and the early maturation of the vessel tissue was observed [2].

7.1.1.1 Vasculogenesis

Vasculogenesis is the process of the differentiation of precursor cells (angioblasts) into endothelial cells, which then form a primitive vascular network, as seen in Figure 7.1. In this strategy, the incorporation of endothelial cells

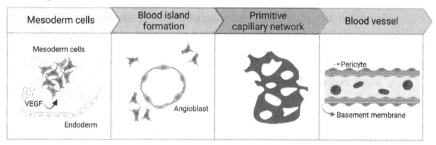
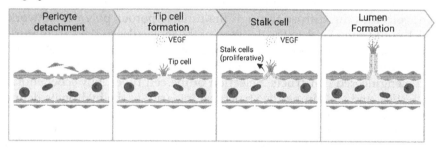

FIGURE 7.1
Schematics of vasculogenesis and angiogenesis.

(ECs) with targeted cells occurs into hydrogel scaffolds in vitro to constitute pre-vascularization. Encapsulated or seeded ECs trigger the ingrowth of the host vessels. Consequently, anastomosis formation of the host vasculature develops, and in vitro endothelial networks can be expedited by pre-existing endothelial networks in the engineered scaffolds.

7.1.1.2 Angiogenesis

Angiogenesis is the growth of microvessels, which sprout to the size of capillary blood vessels. Initially, degradation of the ECM occurs, pericytes are detached, and tip formation develops. The tip endothelial cells are non-proliferative, so they can be directed to proangiogenic factors with dividing stalk cells. Further, the smooth muscle cells (SMCs) are recruited and blood vessels become mature. Considering the importance of growth factors in the mechanisms of angiogenesis, some strategies are based on the loading or conjugating growth factors into the hydrogel scaffold to stimulate capillary ingrowth. The combination of the capillary network and the host vasculature proceeds slowly. Since time is crucial for capillary growth and its integration, the delicate engineered designs often lead to regeneration failure or necrosis.

7.1.2 Hydrogels for Vascularization

The development of an engineered vascularized tissue has become interesting for researchers to mimic a disease model in vitro. In 1980, for the first time, the formation of vascular networks in vitro was mentioned by Folkman et al. Formation of capillary tubes was achieved when the capillary ECs were cultured in a tumor-conditioned medium [3]. The characterization of tissue and an efficient 3D framework for vessels requires 3D matrices that support biochemical and biophysical dynamic interactions similar to the natural ECM. Notably, artificial ECM can be adjusted by the incorporation of growth factors and bioactive peptides by immobilization or diffusion. Distributed hydrophilic polymer chains of hydrogels keep the environment hydrated for entrapped or seeded cells. Additionally, bioactive molecules can diffuse through the tunable porous polymer network, which could contribute to the delivery of the molecules to the vasculature. Furthermore, the molecules driven to angiogenesis could be integrated into the hydrogel to initiate signaling pathways and develop vascular growth in the engineered constructs. Selection of the cell types, physicochemical properties, and sustaining the structural integrity of the hydrogels define the quality of the engineered vascularized tissue scaffold [4]. Pore sizes and interconnectivity of the pores could be adjusted by the polymer concentration and molecular weight. Subsequently, the space maintaining the diffusion of oxygen, nutrients, and waste exchange is required for most cells in the body, which are found no more than 100–200 µm from the adjacent capillary [5].

7.1.2.1 Collagen and Fibrin

ECM-inspired protein polymers are the most attractive class among the polymer scaffolds. The reason is that containing cell-instructive bioactivity features provides cell adhesion and proteolytic degradable sites. Nicosia et al. observed that fibrin and collagen type I hydrogels could present microcapillary-like structures ex vivo in the 1980s [6]. Based on these studies, ECM protein-based hydrogels have been the core of vascular engineering. Fibrin hydrogels enable micro-capillary development in vitro while co-culturing the endothelial and mesenchymal stem cells. Liu et al. incorporated peripheral blood-derived mesenchymal stem cells (PBMSCs) and endothelial colony-forming cells (ECFCs) into fibrin gel, then mixed with poly(lactic-co-glycolic acid) (PLGA) microspheres. They achieved pre-vascularization in the fibrin scaffold, and after implantation, the capillary-like structures formed by ECFCs were not only anastomosed with the host vasculature but also stimulated the rapid growth of natural capillaries into the scaffold from the surrounding tissue [7]. Another example of the importance of fibrin is in Jiang et al., who used porous poly(ethylene glycol) (PEG) hydrogels as scaffolds providing mechanical and structural support, and fibrin was loaded within the pores to induce vascularized tissue formation. They showed that PEG hydrogels with fibrin loading induced more vascularization, with a significantly higher vascular density [8].

Collagen-based hydrogels are popular scaffolds considering ECM contains an abundance of collagen, and especially type-1 collagen is extensively used in the tissue engineering field. Davis and co-workers used collagen hydrogels resulting in 3D micro-capillary networks by endothelial and perivascular cells [9]. Salamone et al. utilized a 3D co-culture system for cross-talking between islets of Langerhans and adipose-derived microvascular fragments providing efficient vascularization. While islets were growing on type I collagen hydrogel, islet vascularization had been observed by adipose-derived microvascular fragments [10]. Although low collagen concentrations show the 3D culture of micro-capillary networks, scaffolds exhibit mechanically weak structures. To overcome the weakness, collagen hydrogels can be chemically modified to improve structural integrity. Hybrid materials like photo-crosslinkable collagen-PEG acquire improved mechanical properties related to material density [11].

7.1.2.2 Gelatin

Gelatin, a partially hydrolyzed product of collagen, is composed of proteins such as glycine, proline, and 4-hydroxy proline residues. As gelatin contains matrix metalloproteinase (MMP) protein sequences, gelatin-based constructs are degradable and allow the cells to deposit new ECM on the degraded sites. The weak mechanical properties of gelatin and its sol-gel transition at physiological conditions make the structural integrity difficult to control. However, methacrylic anhydride modification represents gelatin into photocrosslinkable methacrylated gelatin (GelMA) hydrogels for a broad range of tissue engineering applications. Khademhosseini et al. showed the suitability of photo-crosslinkable GelMA hydrogel to reinforce the human-progenitor-cell-based formation of 3D vascular networks in vitro and in vivo. The researchers observed that 3D co-cultures of human-blood-derived endothelial colony-forming cells and bone marrow-derived mesenchymal stem cells provoke pervasive capillary-like networks combining functional anastomoses with the existing vasculature upon implantation into immunodeficient mice. Moreover, they demonstrated that GelMA hydrogels did not evoke any host inflammatory responses [12]. One specific example of musculoskeletal tissue regeneration was performed by Zhou et al. by performing dual release drug and growth factor. They designed a composite hydrogel including GelMA-basic fibroblast growth factor (bFGF) and mineral and microparticle (MCM)-bone morphogenetic proteins (BMP-2). The bFGF molecules were rapidly released, while the BMP-2 molecules were released over time from composite gels. According to their results, a significantly enhanced vascularized bone regeneration was observed compared with the single-release system. The authors showed the efficiency of a burst release of bFGF from

GelMA-initiated vascularization mediated by endothelial cells, while the sustained release of BMP-2 from MCM enhanced ossification [13]. In addition, microfluidic chips have been used to guide the location of endothelial cells. For instance, Baker et al. produced temporally and spatially defined vascular endothelial growth factor (VEGF) gradients within 3D gelatin/collagen gels on microfluidic chips, providing endothelial sprouting from the channels [14].

7.2 PEG-based Hydrogels and Their Immunological Properties

PEG is a highly hydrophilic synthetic polymer that has been used for therapeutic applications since the 1970s. Owing to its hydrophilic nature, it facilitates water uptake into the hydrogel, which makes PEG an advantageous candidate for tissue engineering applications due to its ECM-like properties. Additionally, it is known to be non-cytotoxic and non-immunogenic while also having protein-repellent properties, which allow additional masking of the hydrogel from the host's immune system [15]. Bal et al. showed that PEG hydrogels can be used as immunologically acceptable composite material in cellular therapy with the addition of acrylate modified cholesterol-bearing pullulan (CHPOA) nanogels [16].

Another advantage of PEG is that, as it has many hydroxyl groups, it can be functionalized with various functional groups such as amines, thiols, vinyl sulfones, maleimides, acrylates, and methacrylates. This provides endless alternatives for the researcher as the route of crosslinking can be easily varied depending on the aim of the application and the available materials for the synthesis. Owing to multiple available functionalization sites, the crosslinking of PEG into hydrogels can be carried out using a step-growth polymerization approach that facilitates the formation of a highly organized hydrogel structure [15].

7.2.1 Modifications of PEG Hydrogels for Enhanced Properties of Biomimicry

Despite its considerable advantages of non-toxicity, non-immunogenicity, and hydrophilicity, the properties of PEG-based hydrogels can be further improved through various modifications, which will allow a broader area of application.

One strategy for the advancement of PEG-based hydrogels is the conjugation of the polymer with peptides of cell adhesive properties during crosslinking. These peptides can be derived from various ECM proteins that lie under the categories of fibronectin, laminin, collagen, and elastin. In the body, these peptides function as ligands to a major family of receptors called integrins, which are responsible for cell-cell and cell-ECM interactions. The

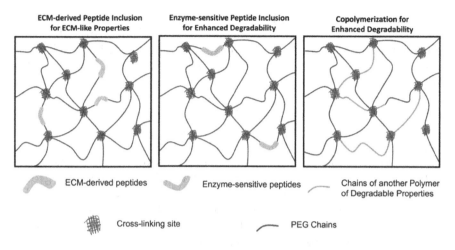

FIGURE 7.2
Schematic of the strategies for modifications on PEG hydrogels for enhanced properties.

most commonly used peptide for this purpose is a tripeptide RGD (arginine, glycine, aspartate). Hence, the integration of RGD peptide into (or onto) the hydrogel structure facilitates increased affinity toward the integrins and enhances biological mimicry through increased cell adhesion (Figure 7.2). In a study by Bal et al., in 2014, a PEG-based hydrogel was designed for enhanced viability of mesenchymal stem cells and increased insulin secretion. In this work, ECM-derived peptides, namely fibronectin-derived RGDS and laminin-derived IKVAV, were incorporated in the PEG-based hydrogel structure to promote cell adhesion and growth, in combination with insulinotropic glucagon-like peptide (GLP-1), which stimulated insulin secretion. With the incorporation of these peptides, the produced hydrogel had a promising improvement in pancreatic islet functionality concerning alterations in glucose levels [17].

Photo-crosslinked PEG-diacrylate hydrogels contain sterically protected acrylates as their end groups. Due to being sterically very stable, the in vivo degradation rate of the hydrogel is innately very slow, which may lead to the inhibition of tissue repair or regeneration. To eliminate this problem, an alternative is to incorporate other hydrolytically degradable polymers, such as PLA and PGA, during crosslinking reactions to conduct a co-polymerization. Another strategy can be the incorporation of enzyme-sensitive peptides into the crosslinked network to allow the proteolytic degradation of the hydrogel through MMP enzymes (Figure 7.2).

7.2.2 Crosslinking Strategies for PEG

Even though PEG hydrogel formation can be carried out via both physical and chemical crosslinking, chemical crosslinking provides a more stable structure owing to covalent bond formation. In addition, chemical crosslinking allows for tunability of the degree of crosslinking, which allows the researcher to modify the physicochemical properties of the hydrogel, which may include permeability, water content (swelling), viscoelastic modulus, and degradation rates [18].

The covalent crosslinking of PEG polymers can be mainly carried out by using one of the following methods: chain-growth polymerization (free-radical polymerization) and condensation (step-growth polymerization) (Figure 7.3).

In step-growth polymerization, the reaction requires using PEG molecules that contain bi- or multifunctional groups in which these functional groups are complementary. In these reactions, the presence of an initiator such as radicals is not required, as the functional groups are innately reactive, which also results in no termination reactions for polymerization. In free-radical polymerization, the reaction commences through the radicals that can be formed by thermal energy, redox reactions, or the photo-cleavage of the initiator molecules. Following initiation, the mechanism moves to the propagation phase, in which the free radicals continue to form carbon-carbon double bonds through the unsaturated vinyl bonds of the PEG molecules. The reaction terminates when two free radicals react with each other, which results in no available energy for further polymerization of the PEG molecules. Given these reaction mechanisms, chain-growth polymerization occurs through a series of reactions that are caused by subsequent intermolecular interactions between a PEG molecule and the functional site of the other. This concept

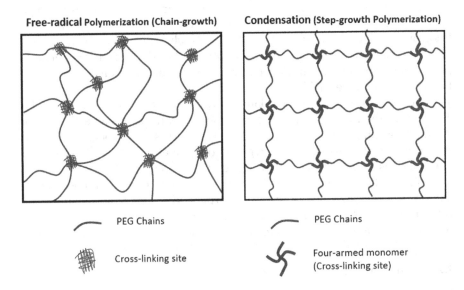

FIGURE 7.3
Schematic of structural comparison of different crosslinking strategies.

provides a controllable molecular weight of the end product and hence lower molecular weight distributions.

As a subcategory of free-radical polymerization, photopolymerization is one of the most used approaches for PEG-based hydrogels, as it greatly decreases the overall reaction time, requires less energy, and is carried out more easily in situ to obtain a spatially specific hydrogel formation. Additionally, in comparison with the thermally activated crosslinking mechanisms, photo-crosslinking can be preferable for uses that involve the delivery of encapsulated cells or other biological reagents such as peptides, proteins, and nucleic acids. For photopolymerization, PEGs with functional groups of acrylate derivatives are utilized, such as PEGDA, PEG-dimethacrylate (PEGDMA), and multi-armed versions of PEG-acrylates. In a study by Bal et al., a PEG hydrogel was synthesized through surface-initiated photopolymerization, aiming to discover the effects of the crosslinking density of the hydrogel on the delivery of GLP-1 and a model protein BSA. The study successfully found that a gradient in crosslinking density achieved by surface-initiated photopolymerization yielded a promising scaffold for the controlled release of bioactive molecules and therefore suitable for the covalent incorporation of ligands, which further supported cell viability [19].

7.3 An Overview of LMSC Therapy

Limbal stem cell deficiency (LSCD) is a pathology that causes vision loss. It is characterized by difficulty in corneal epithelialization, permanent epithelial defect, corneal vascularization, and compromised corneal vision. There are many approaches in the treatment of LSCD, such as surface reconstruction with the amniotic membrane, autologous or allogeneic limbal tissue transplantation, and keratoplasty. Among these options, the most successful result so far is obtained with LSC transplantation [20].

The first condition for the realization of the visual function is that the cornea should be transparent. For the cornea to be transparent, it is imperative that the corneal epithelial cell layer, which has completed its differentiation, renews regularly, and maintains its avascular structure [21]. The source of the regenerated epithelial cells is the LSCs located in the limbus region of the eye as seen in Figure 7.4. However, if LSCs are damaged and absent, corneal reconstruction is delayed, resulting in serious inflammatory reactions. Polymorphonuclear leukocytes secrete large amounts of proteases that disrupt extracellular matrix components, causing corneal ulcers, revascularization, opacification, and eventual loss of vision.

Loss of function of LSCs occurs due to primary or secondary reasons such as chemical/heat burns, Stevens–Johnson syndrome, congenital/genetic

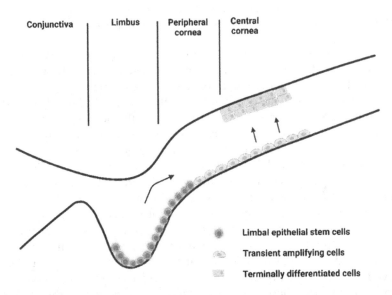

FIGURE 7.4
Schematic view of the limbal stem cell niche.

anomalies causing conjunctival epithelium to migrate over the cornea, and severe vision loss with the development of scar-vascularization [22].

7.3.1 Treatment of LSCD

For the treatment of damaged and non-healable corneas, surgical interventions such as corneal transplantation or amniotic membrane transplantation can be performed. However, there are limitations in corneal transplantation, such as donor deficiency and tissue rejection [23]. Today, effective treatment options for this pathology are very limited. For healthy corneal regeneration, there is a need to develop treatment options based on existing LSCs in terms of clinical predictability, feasibility, and economic feasibility. For these reasons, attempts have been made to propagate LMSCs as explant culture on a biocompatible material and surgical transplantation of this tissue. However, a biocompatible environment that mimics the niche environment of LSCs has not yet been established in vitro. This deficiency is the most important factor affecting the success rate of LSC transplantation [24].

7.3.2 The Potential Use of Hydrogels in LMSC

Based on clinical and pre-clinical studies, hydrogels such as fibrin, hyaluronic acid, collagen, gelatin, and alginate are promising candidates for the therapy of LSCD.

7.3.2.1 Fibrin

Fibrin gels are regenerated from the combination of fibrinogen and thrombin. These hydrogels have been extensively used in LSCD. James et al. mixed fibrinogen with purified extracellular vesicles (EVs) to reduce corneal fibrosis and inflammation. In mice models, the corneal epithelium and a thin layer of anterior stroma were mechanically removed. The wound was then covered with fibrinogen mixed with purified extracellular vesicles. The wound bed was treated with thrombin to form fibrin gel. After 24 hours, the corneal epithelial layer regenerated over the injured tissue, and corneal scarring was avoided in treated eyes after 2 weeks [25].

In another study by Rama et al., cultured limbal epithelial cells on fibrin gels were used to treat patients with these limbal epithelial cells; 14 out of 18 patients treated with these cells regained useful visual acuity. Within the first week, their corneas underwent re-epithelialization, and within the first to four weeks, reduced inflammation and vascularization were observed. In a 12–27-month follow-up, their corneal surfaces were clinically and cytologically stable [26].

7.3.2.2 Collagen

Collagen is the primary extracellular matrix protein of native cornea and has been widely used as a biomaterial for corneal tissue engineering applications. Collagen-based hydrogels are biocompatible, biodegradable, and inert. They have been shown to encapsulate cells and biomolecules. In an experiment, Griffith et al. synthesized collagen-based composites by combining collagen with poly(N-isopropylacrylamide) (pNIPAAm), which is a synthetic acrylic-based polymer. They further tested these composites in ocular surface reconstruction in rabbits. The collagen-pNIPAAm scaffolds were implanted through a 3 mm circular incision in the central cornea of Japanese white rabbits. The results indicated that 87.5% of the rabbits had regrown epithelium after three days. While the epithelium was renewed on the composite scaffold, histological analysis revealed that the differentiation of epithelial cells on the polymer was abnormal, which was indicated by disorganized cell stratification on the scaffold [27].

Li et al. synthesized a composite collagen hydrogel by blending collagen and a copolymer poly(N-isopropylacrylamide-coacrylicacid-coacryloxysuccinimide) (TERP) and grafted a laminin adhesion pentapeptide motif (YIGSR) onto the hydrogel (TERP 5). Implantation of TERP 5 in pig corneas resulted in rapid recruitment of host corneal epithelial and stromal cells, as well as functional innervation inside the implant [28].

7.3.2.3 Gelatin

Gelatin is composed of peptides and proteins produced by partial hydrolysis of collagen extracted from bone, skin, and connective tissue. Gelatin gels have been successfully used to culture corneal epithelial cells and stromal cells, which indicates that such scaffolds may support limbal epithelial cells as well. Porous gelatin was shown to be suitable for the engineering of the corneal stroma [29].

Gelatin, either raw or chemically modified, has been used as soluble ocular drug inserts and has shown potential for limbal epithelial culture [30]. Wright et al. incorporated epidermal growth factor (EGF) into cationized gelatin film and tested it in rabbit corneas with an epithelial defect. The gelatin-EDF films not only reduced the defect but there was enhanced epithelial proliferation. Gelatin and polyvinyl alcohol (PVA) polymeric blends also show promising results as ocular inserts for the prolonged release of antibiotics in the eye [29].

7.3.2.4 Alginate

Alginate is a non-toxic, degradable, linear copolymer consisting of unbranched binary co-polymers of 1–4 linked β-d-mannuronic acid and α-l-guluronic acid. It is widely used as a biocompatible material. However, the application of alginate hydrogels for ocular treatment is relatively new. Alginate microspheres embedded in collagen hydrogels have previously been shown to be a feasible composite construct for controlled drug delivery and growth of human corneal epithelial cells. In another study, alginate membranes were coated with chitosan and were used as a base matrix for limbal epithelial cell cultivation. The cells remained attached and remained viable by spreading and growing on the alginate matrices [31]. Another composite hydrogel made of sodium alginate dialdehyde and hydroxypropyl chitosan was used to encapsulate corneal endothelial cells for their transplantation onto Descemet's membrane. The encapsulated corneal endothelial cells not only remained viable but also retained their morphology [32].

7.3.3 New Therapies to Treat LSCD

Surgical procedures for LSCD are classified according to the source of stem cells, the kind of stem cell transplant, and the use of ex vivo stem cell growth in cell culture. It is now obvious that any stem cell-based treatment method will fail if the right stem cell niche is not restored. In the future, treatments should be focused on enhanced graft survival and function, potentially with topical growth factors or innovative biomaterial platforms. The creation of HLA-matched iPSCs and reprogrammed differentiated cell lineages are of great interest. The use of mesenchymal stem cells and other stem cell sources in clinical practice is an active field of research with clinical trials underway. Secretomes derived from ex vivo cultures of mesenchymal stem cells contain many growth factors. These secretomes can be applied topically and enhance the viability and regeneration of the limbal stem cell niche [24].

7.4 Hydrogels in Pancreatic Islet Transplantation

Type 1 diabetes is an autoimmune disease in which a patient's immune cells destroy insulin-secreting beta cells in the pancreas, leading to a lack of insulin in the body. This causes an increase in blood sugar levels, which can damage the organs and tissues and lead to severe complications such as heart disease, stroke, nerve damage, and kidney failure. Type 1 diabetes usually develops in childhood or adolescence but can also occur in adults. It is treated with insulin replacement therapy, regular blood sugar monitoring, and a healthy lifestyle [33]. However, each of these therapies has its downsides, such as multiple daily injections, finger prick tests, and restrictions on diet. Hence, researchers have come up with new solutions like hydrogels to improve patient compliance.

Pancreatic islets are clusters of endocrine cells that produce insulin and other hormones involved in glucose metabolism. Islet transplantation is a promising treatment for type 1 diabetes, in which the islets are transplanted into the recipient's liver to restore normal blood sugar levels. However, this approach has several challenges, such as the need for immunosuppressive drugs to prevent rejection, the limited availability of donor islets, and the risk of islet loss and dysfunction due to the harsh environment of the transplantation site. Besides the downsides of current treatment methods, hydrogels have been utilized as a potential solution to the challenges of islet transplantation. By encapsulating the islets in a hydrogel, the cells can be protected from the immune system and the external environment while exchanging nutrients and waste with the surrounding tissue. Moreover, hydrogels can be designed to provide mechanical support and release bioactive molecules that can support islet survival and function [34]. For islet transplantation, isolation of pancreatic islets interrupts their vascular and nervous connections. This causes the islets to depend on their environment for survival with diffusion [35]. Accordingly, this part will discuss strategies like oxygenation of the implant, vascularization, and local immune modulation, which have been conducted over the last decade to make islet transplantation possible in clinics (Figure 7.5).

7.4.1 Strategies for Oxygenation

Oxygenation is a vital part of the transplantation of pancreatic islets. Since pancreatic islets contain highly metabolic cells that depend on oxygen consumption and glucose oxidation for their energy source production and insulin secretion, they have higher consumption rates than other cell types. After isolation, islets are entirely dependent on diffusion from the environment. Thus, oxygenation ensures that islet cells receive the oxygen they need to survive and function properly. Without essential oxygen, the islet cells may not be able to regenerate and integrate into

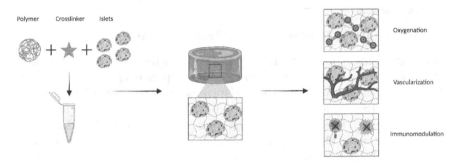

FIGURE 7.5
Strategies for oxygenation, vascularization, and immunomodulation of hydrogels with pancreatic islets.

the recipient's body, which can result in the failure of transplantation. In addition, adequate oxygenation can help to reduce the risk of complications, such as tissue damage or infection. Overall, ensuring that islet cells are adequately oxygenated during transplantation is critical for the procedure's success.

Several strategies can be used to promote oxygenation in islet transplantation, such as oxygen-generating biomaterials, scaffold-type devices for optimal islet spacing, or oxygen flushing to the transplantation site before and after the operation. Among these, oxygen-generating biomaterials draw attention. Pedraza et al. designed PDMS-based materials where they encapsulated solid calcium peroxide to activate oxygen generation hydrolytically. They showed that oxygen is generated at an average rate of 0.026 mM per day for more than six weeks, which improves metabolic function and glucose-dependent insulin secretion in transplanted islets by eliminating hypoxia-induced cell dysfunction and death [36]. In another study, Barkai et al. designed a bioartificial pancreas (BAP) that can supply oxygen to the transplanted islets. Their subcutaneous implant was made of alginate hydrogel and consisted of a gas chamber, a gas-permeable membrane, an external membrane, and a mechanical support. With this device, they achieved normoglycemia in streptozotocin-induced diabetic rats for up to 6 months by supplying oxygen to grafts when needed [37]. Moreover, Toftdal et al. showed that hydrogel composed of thiolated hyaluronic acid (tHA), 8-arm-poly(ethylene glycol)-acrylate (PEGA), and calcium peroxide (CPO) could be used as an oxygen-releasing system. By optimizing the CPO amount, they found that 30% CPO-containing hydrogels could release oxygen for at least 30 hours, which results in improved islet viability [38]. Finally, strategies that promote vascularization, which will be discussed in detail, such as using growth factors or stem cells, can also help improve oxygenation by increasing the formation of new blood vessels to support the transplanted islet.

7.4.2 Strategies for Vascularization

Pancreatic islets are highly vascularized tissues, leading them to respond and secrete quickly in blood glucose changes. However, the isolation of pancreatic islets loses the islet vasculature because of isolation procedures. Revascularization, or the re-establishment of the blood vessel in the transplanted islets, starts on the first three days of transplantation and can be completed around day 14. Before vasculature forms, transplanted islets meet the need for essential molecules like oxygen, nutrients, and growth factors by simple diffusion. Without an adequate blood supply, the transplanted islets will not receive the oxygen and nutrients they need to survive, which might refer to being incapable of functioning as regulating the blood sugar levels of the patient and the death of significant numbers of islets.

Several strategies can be used to promote vascularization in pancreatic islet transplantation. These include decellularization, pre-vascularized hydrogels, delivery of growth factors, and co-culture with endothelial cells [39]. Jiang et al. generated 3D ECM-hydrogels using decellularized porcine bladder, lung, and pancreas tissues. These hydrogels enabled long-term islet survival and function [40]. Weaver et al. designed a macro device system consisting of two components: a hydrogel core crosslinked with a non-degradable PEG dithiol and a vasculogenic outer layer crosslinked with a proteolytically sensitive peptide to promote degradation and enhance localized vascularization. They showed that islets encapsulated with this system exhibit high viability in vitro. The degradable hydrogel layer at the external interface of the implant enhanced vascularization within the rat omentum, resulting in improved encapsulated islet viability in vivo [41]. In another study, Yin et al. synthesized alginate hydrogel functionalized with VEGF to promote angiogenesis. With the optimum concentration of VEGF (100 ng/ml), vascularization is triggered, and encapsulated islets' functionality and survival are enhanced, leading to long-term therapeutic effects of up to 50 days in vivo [42]. As mentioned earlier, vascularization is necessary for the survival and function of the transplanted islets, but it can also make them more susceptible to immune rejection. This is because the formation of new blood vessels in the transplanted tissue can expose the islets to immune cells, which can attack the transplanted tissue. In the next section, strategies for immune modulation will be discussed to overcome this issue.

7.4.3 Strategies for Immune Modulation

When islets from a donor are transplanted into a recipient, the recipient's immune system may recognize them as foreign and attack them. This is known as the immune rejection of transplanted islets. The immune system uses various mechanisms to identify and attack foreign cells, including antibodies and specialized immune cells such as T and B cells. There are several ways to prevent the immune rejection of transplanted islets, including immunosuppressive drugs and other immune-modulating therapies. Immunosuppressive drugs work by suppressing the recipient's immune system, which can reduce the risk of rejection. These drugs must be carefully monitored and adjusted to ensure that they are effective without causing significant side effects. Hydrogels can be utilized for the islet encapsulation design since they can prevent the immune rejection of transplanted islets by either creating a physical barrier to immune cells or delivering immunosuppressive drugs directly to the transplant site, which can further reduce the risk of rejection.

7.4.3.1 Physical Isolation

Physical isolation of transplanted islets by hydrogels can help to prevent immune rejection. The hydrogels can physically separate the transplanted islets from the recipient's immune cells, which can prevent the immune cells from recognizing and attacking the islets. This can reduce the risk of rejection and improve the success of the transplant. Previously, our group showed that PEG hydrogels can be used as a semipermeable physical barrier to improve immune protection from cytokines. By co-culturing islets with mesenchymal stem cells (MSCs), islets are further protected from immune attack by TGF-β1 secretion from MSCs [43]. Weber et al. generated a two-layered PEG hydrogel consisting of PEG-laminin as the core and an exterior inert PEG layer. The islets survived 28 days in culture in the biologically active core, while the outer shell provided a protective barrier against a host immune response [44]. In another study, Kim et al. designed a novel enzymatic crosslinking-based hydrogel nanofilm caging system using monophenol-modified glycol chitosan and hyaluronic acid. They showed that the hydrogel system protects cells from high shear stress and reduces immune response by interfering with cell-cell interaction [45].

Although the encapsulation strategy protects islets from infiltration of cytotoxic T cells and reduces host-cell interactions, cytokines and antibodies, which are the smaller compounds of the immune system, can still potentially diffuse through the passive barrier. This may lead to islet destruction and reduce the longevity of the transplant. Therefore, researchers have developed more advanced strategies like conjugating apoptotic agents and immunosuppressive drugs for hydrogels.

7.4.3.2 Functionalization of Hydrogels with Apoptotic Molecules and Immunosuppressive Drugs

Hydrogels can be designed to release drugs over a period of time, which can provide sustained levels of immunosuppression at the site of the transplant. This can help to reduce the risk of rejection and improve the success of the transplant by minimizing the side effects of the drugs, as they are delivered only to the specific area where they are needed. Similarly, hydrogels can be functionalized with apoptotic molecules to prevent immune rejection. By functionalizing hydrogels with apoptotic molecules, the hydrogels could potentially induce apoptosis, or programmed cell death, in immune cells that encounter them. This could help prevent the immune cells from attacking the transplanted islets and improve the chances of successful transplantation. Medina et al. synthesized PEG microgel functionalized with an analog of IL-2 (IL-2D) and Fas ligand (FasL) to achieve localized immunotolerance. The hydrogel significantly changes T cell subpopulations to regulatory T cells (Treg), but long-term allograft function was not achieved [46]. Here, it is important to note that functionalizing hydrogels with apoptotic molecules does not guarantee against immune rejection. The immune system is complex and robust, and it can sometimes reject transplanted tissue even when hydrogels and other measures protect it. In these cases, additional immunosuppressive drugs may be necessary to prevent the body from rejecting the transplanted islets. In that sense, Headen et al. combined the functionalization of FasL to hydrogel with a short course of an immunosuppressant called rapamycin. They showed that the islets survived up to 31 days with FasL-presenting microgels in vivo, while the survival of islets increased to over 200 days when rapamycin was injected into recipient animals [47].

7.5 Conclusion

In conclusion, hydrogels are a versatile and promising material for tissue engineering, particularly in treating eye-related diseases, islet transplantation, vascularization, and immunomodulation. By providing a supportive environment for the growth of cells and tissues, hydrogels can help improve the success of these therapies, offering new hope for patients with various conditions.

Bibliography

1. U. Blache, M. Ehrbar, Inspired by nature: Hydrogels as versatile tools for vascular engineering, *Adv Wound Care* 7(7) (2018) 232–46.
2. Y. Wang, R.K. Kankala, C. Ou, A. Chen, Z. Yang, Advances in hydrogel-based vascularized tissues for tissue repair and drug screening, *Bioact Mater* 9 (2022) 198–220.
3. J. Folkman, C. Haudenschild, Angiogenesis in vitro, *Nature* 288(5791) (1980) 551–6.

4. Y.J. Blinder, A. Freiman, N. Raindel, D.J. Mooney, S. Levenberg, Vasculogenic dynamics in 3D engineered tissue constructs, *Sci Rep* 5 (2015) 17840.
5. R.K. Jain, P. Au, J. Tam, D.G. Duda, D. Fukumura, Engineering vascularized tissue, *Nat Biotechnol* 23(7) (2005) 821–3.
6. R.F. Nicosia, J.A. Madri, The microvascular extracellular matrix. Developmental changes during angiogenesis in the aortic ring-plasma clot model, *Am J Pathol* 128(1) (1987) 78–90.
7. J. Liu, G. Chen, H. Xu, K. Hu, J. Sun, M. Liu, F. Zhang, N. Gu, Pre-vascularization in fibrin Gel/PLGA microsphere scaffolds designed for bone regeneration, *NPG Asia Mater* 10(8) (2018) 827–39.
8. B. Jiang, T.M. Waller, J.C. Larson, A.A. Appel, E.M. Brey, Fibrin-loaded porous poly(ethylene glycol) hydrogels as scaffold materials for vascularized tissue formation, *Tissue Eng Part A* 19(1–2) (2013) 224–34.
9. G.E.D. Kayla J. Bayless, The Cdc42 and Rac1 GTPases are required for capillary lumen formation in three dimensional extracellular matrices, *J Cell Sci* 115(6) (2002) 1123–36.
10. M. Salamone, S. Rigogliuso, A. Nicosia, S. Campora, C.M. Bruno, G. Ghersi, 3D Collagen hydrogel promotes in vitro langerhans islets vascularization through ad-MVFs angiogenic activity, *Biomedicines* 9(7) (2021) 739.
11. R.K. Singh, D. Seliktar, A.J. Putnam, Capillary morphogenesis in PEG-collagen hydrogels, *Biomaterials* 34(37) (2013) 9331–40.
12. Y.C. Chen, R.Z. Lin, H. Qi, Y. Yang, H. Bae, J.M. Melero-Martin, A. Khademhosseini, Functional human vascular network generated in photo-crosslinkable gelatin methacrylate hydrogels, *Adv Funct Mater* 22(10) (2012) 2027–39.
13. X. Zhou, J. Chen, H. Sun, F. Wang, Y. Wang, Z. Zhang, W. Teng, Y. Ye, D. Huang, W. Zhang, X. Mo, A. Liu, P. Lin, Y. Wu, H. Tao, X. Yu, Z. Ye, Spatiotemporal regulation of angiogenesis/osteogenesis emulating natural bone healing cascade for vascularized bone formation, *J Nanobiotechnol* 19(1) (2021) 420.
14. B.M. Baker, B. Trappmann, S.C. Stapleton, E. Toro, C.S. Chen, Microfluidics embedded within extracellular matrix to define vascular architectures and pattern diffusive gradients, *Lab Chip* 13(16) (2013) 3246–52.
15. C.-C. Lin, K.S. Anseth, PEG hydrogels for the controlled release of biomolecules in regenerative medicine, *Pharm Res* 26(3) (2009) 631–43.
16. T. Bal, I.C. Karaoglu, F.S. Murat, E. Yalcin, Y. Sasaki, K. Akiyoshi, S. Kizilel, Immunological response of polysaccharide nanogel-incorporating PEG hydrogels in an in vivo diabetic model, *J Biomater Sci Polym Ed.* (just-accepted) 1794–1810 (2022) 1–16.
17. T. Bal, C. Nazli, A. Okcu, G. Duruksu, E. Karaöz, S. Kizilel, Mesenchymal stem cells and ligand incorporation in biomimetic poly(ethylene glycol) hydrogels significantly improve insulin secretion from pancreatic islets, *J Tissue Eng Regen Med* 11(3) (2017) 694–703.
18. C.C. Lin, A.T. Metters, Hydrogels in controlled release formulations: Network design and mathematical modeling, *Adv Drug Deliv Rev* 58(12–13) (2006) 1379–408.
19. T. Bal, B. Kepsutlu, S. Kizilel, Characterization of protein release from poly(ethylene glycol) hydrogels with crosslink density gradients, *J Biomed Mater Res Part A* 102(2) (2014) 487–95.

20. K. Sejpal, P. Bakhtiari, S.X. Deng, Presentation, diagnosis and management of limbal stem cell deficiency, *Middle East Afr J Ophthalmol* 20(1) (2013) 5–10.
21. A. Gulzar, E. Yıldız, H.N. Kaleli, M.A. Nazeer, N. Zibandeh, A.N. Malik, A.Y. Taş, I. Lazoğlu, A. Şahin, S. Kizilel, Ruthenium-induced corneal collagen cross-linking under visible light, *Acta Biomater* 147 (2022) 198–208.
22. D. Robaei, S. Watson, Corneal blindness: A global problem, *Clin Exp Ophthalmol* 42(3) (2014) 213–4.
23. D.H.K. Ma, J.K. Chen, F. Zhang, K.Y. Lin, J.Y. Yao, J.S. Yu, Regulation of corneal angiogenesis in limbal stem cell deficiency, *Prog Retin Eye Res* 25(6) (2006) 563–90.
24. A.M. Elhusseiny, M. Soleimani, T.K. Eleiwa, R.H. ElSheikh, C.R. Frank, M. Naderan, G. Yazdanpanah, M.I. Rosenblatt, A.R. Djalilian, Current and emerging therapies for limbal stem cell deficiency, *Stem Cell Transl Med* 11(3) (2022) 259–68.
25. G. Shojaati, I. Khandaker, M.L. Funderburgh, M.M. Mann, R. Basu, D.B. Stolz, M.L. Geary, A. Dos Santos, S.X. Deng, J.L. Funderburgh, Mesenchymal stem cells reduce corneal fibrosis and inflammation via extracellular vesicle-mediated delivery of miRNA, *Stem Cell Transl Med* 8(11) (2019) 1192–201.
26. P. Rama, S. Bonini, A. Lambiase, O. Golisano, P. Paterna, M. De Luca, G. Pellegrini, Autologous fibrin-cultured limbal stem cells permanently restore the corneal surface of patients with total limbal stem cell deficiency, *Transplantation* 72(9) (2001) 1478–85.
27. S. Shimmura, C.J. Doillon, M. Griffith, M. Nakamura, E. Gagnon, A. Usui, N. Shinozaki, K. Tsubota, Collagen-poly(N-isopropylacrylamide)-based membranes for corneal stroma scaffolds, *Cornea* 22(7) Supplement (2003) S81–8.
28. F. Li, M. Griffith, Z. Li, S. Tanodekaew, H. Sheardown, M. Hakim, D.J. Carlsson, Recruitment of multiple cell lines by collagen-synthetic copolymer matrices in corneal regeneration, *Biomaterials* 26(16) (2005) 3093–104.
29. B. Wright, S.L. Mi, C.J. Connon, Towards the use of hydrogels in the Treatment of limbal stem cell deficiency, *Drug Discov Today* 18(1–2) (2013) 79–86.
30. M. Mathurm, R.M. Gilhotra, Glycerogelatin-based ocular inserts of aceclofenac: Physicochemical, drug release studies and efficacy against prostaglandin E-2-induced ocular inflammation, *Drug Deliv* 18(1) (2011) 54–64.
31. W.G. Liu, M. Griffith, F.F. Li, Alginate microsphere-collagen composite hydrogel for ocular drug delivery and implantation, *J Mater Sci-Mater M* 19(11) (2008) 3365–71.
32. Y. Liang, W.S. Liu, B.Q. Han, C.Z. Yang, Q. Ma, F.L. Song, Q.Q. Bi, An in situ formed biodegradable hydrogel for reconstruction of the corneal endothelium, *Colloid Surf B* 82(1) (2011) 1–7.
33. M.A. Atkinson, G.S. Eisenbarth, A.W. Michels, Type 1 diabetes, *Lancet* 383(9911) (2014) 69–82.
34. L. Xu, Y. Guo, Y. Huang, Y. Xu, Y. Lu, Z. Wang, Hydrogel materials for the application of islet transplantation, *J Biomater Appl* 33(9) (2019) 1252–64.
35. J. Lau, J. Henriksnas, J. Svensson, P.O. Carlsson, Oxygenation of islets and its role in transplantation, *Curr Opin Organ Transplant* 14(6) (2009) 688–93.
36. E. Pedraza, M.M. Coronel, C.A. Fraker, C. Ricordi, C.L. Stabler, Preventing hypoxia-induced cell death in beta cells and islets via hydrolytically activated, oxygen-generating biomaterials, *Proc Natl Acad Sci USA* 109(11) (2012) 4245–50.

37. U. Barkai, G.C. Weir, C.K. Colton, B. Ludwig, S.R. Bornstein, M.D. Brendel, T. Neufeld, C. Bremer, A. Leon, Y. Evron, K. Yavriyants, D. Azarov, B. Zimermann, S. Maimon, N. Shabtay, M. Balyura, T. Rozenshtein, P. Vardi, K. Bloch, P. de Vos, A. Rotem, Enhanced oxygen supply improves islet viability in a new bioartificial pancreas, *Cell Transplant* 22(8) (2013) 1463–76.
38. M.S. Toftdal, N. Taebnia, F.B. Kadumudi, T.L. Andresen, T. Frogne, L. Winkel, L.G. Grunnet, A. Dolatshahi-Pirouz, Oxygen releasing hydrogels for beta cell assisted therapy, *Int J Pharm* 602 (2021) 120595.
39. M.A. Nazeer, I.C. Karaoglu, O. Ozer, C. Albayrak, S. Kizilel, Neovascularization of engineered tissues for clinical translation: Where we are, where we should be?, *APL Bioeng* 5(2) (2021) 021503.
40. K. Jiang, D. Chaimov, S.N. Patel, J.P. Liang, S.C. Wiggins, M.M. Samojlik, A. Rubiano, C.S. Simmons, C.L. Stabler, 3-D physiomimetic extracellular matrix hydrogels provide a supportive microenvironment for rodent and human islet culture, *Biomaterials* 198 (2019) 37–48.
41. J.D. Weaver, D.M. Headen, M.D. Hunckler, M.M. Coronel, C.L. Stabler, A.J. Garcia, Design of a vascularized synthetic poly(ethylene glycol) macroencapsulation device for islet transplantation, *Biomaterials* 172 (2018) 54–65.
42. N. Yin, Y. Han, H. Xu, Y. Gao, T. Yi, J. Yao, L. Dong, D. Cheng, Z. Chen, VEGF-conjugated alginate hydrogel prompt angiogenesis and improve pancreatic islet engraftment and function in type 1 diabetes, *Mater Sci Eng C Mater Biol Appl* 59 (2016) 958–64.
43. T. Bal, Y. Inceoglu, E. Karaoz, S. Kizilel, Sensitivity study for the key parameters in heterospheroid preparation with insulin-secreting β-cells and mesenchymal stem cells, *ACS Biomater Sci Eng* 5(10) (2019) 5229–39.
44. L.M. Weber, C.Y. Cheung, K.S. Anseth, Multifunctional pancreatic islet encapsulation barriers achieved via multilayer PEG hydrogels, *Cell Transplant* 16(10) (2007) 1049–57.
45. M. Kim, H. Kim, Y.S. Lee, S. Lee, S.E. Kim, U.J. Lee, S. Jung, C.G. Park, J. Hong, J. Doh, D.Y. Lee, B.G. Kim, N.S. Hwang, Novel enzymatic cross-linking-based hydrogel nanofilm caging system on pancreatic beta cell spheroid for long-term blood glucose regulation, *Sci Adv* 7(26) (2021).
46. J.D. Medina, G.F. Barber, M.M. Coronel, M.D. Hunckler, S.W. Linderman, M.J. Quizon, V. Ulker, E.S. Yolcu, H. Shirwan, A.J. Garcia, A hydrogel platform for co-delivery of immunomodulatory proteins for pancreatic islet allografts, *J Biomed Mater Res A* 110(11) (2022) 1728–37.
47. D.M. Headen, K.B. Woodward, M.M. Coronel, P. Shrestha, J.D. Weaver, H. Zhao, M. Tan, M.D. Hunckler, W.S. Bowen, C.T. Johnson, L. Shea, E.S. Yolcu, A.J. Garcia, H. Shirwan, Local immunomodulation Fas ligand-engineered biomaterials achieves allogeneic islet graft acceptance, *Nat Mater* 17(8) (2018) 732–9.

8
3D Printing of Hydrogels

Janet de los Angeles Chinellato Díaz, Santiago P. Fernandez Bordín, Ramses S. Meleán, and Marcelo R. Romero

CONTENTS

8.1 Introduction 147
8.2 Types of 3D Printing 148
    8.2.1 FDM 149
    8.2.2 SLA 149
    8.2.3 SLS 151
    8.2.4 Direct Ink Writing-Liquid Direct Modeling 151
8.3 Type of Polymers Printed 151
    8.3.1 Natural Hydrogels 152
        8.3.1.1 Polysaccharide Hydrogels 152
        8.3.1.2 Protein Hydrogels 155
    8.3.2 Hydrogels from Synthetic Polymers 155
        8.3.2.1 PEG 156
        8.3.2.2 Poly(vinyl alcohol) 156
        8.3.2.3 Pluronics 156
    8.3.3 Hydrogel Composites 156
8.4 3D Printing of Hydrogel 157
    8.4.1 Methods 157
    8.4.2 Experimental Procedure 157
    8.4.3 Quality Parameters 158
8.5 Applications: Literature Review 160
8.6 Prospects 164
References 165

8.1 Introduction

In 1980, Hideo Kodama was the first to register a three-dimensional (3D) method while working for the Public Research Institute in the city of Nagoya, Japan. During the period 1980–1981, he began the process of patenting the "rapid prototyping" (RP) method, although he did not complete the requirements to finish the process. This researcher proposed the RP method using a laser for resin curing and published his findings [1]. Later, in 1984,

Jean-Claude André, Alain le Méhauté, and Olivier de Witte coined the term stereolithography (SLA) and proposed a patent application to the French General Electric Company. However, it was rejected by the directors "for a lack of business perspective". That same year, the inventor Chuck Hull filed a patent for the creation of objects layer by layer using UV lasers for curing and defined this system as a "system for generating three-dimensional objects by creating a cross-sectional pattern of the object to be formed" [2]. Carl Deckard, a few years later in 1988, proposed the selective laser sintering (SLS) method since he worked for several years with traditional methods based on material removal and wanted to find a way to develop prototypes using a simpler procedure. With this instrument, he used a laser to fuse the particles and obtain the pieces layer by layer of molten material. Years later, Scott Crump developed the fused deposition method (FDM), which is currently the most widely used due to its low cost and simplicity of printer construction. Scott patented his method in 1989 and two years later introduced the world's first printer model [3]. In that same period, the German Hans Langer developed the method of metal laser sintering based on SLS, becoming a market leader. Other technological milestones were subsequently carried out in 1996 with direct ink writing (DIW), and analogous robo printing where the material in paste form (originally for ceramic compositions) is deposited onto the created structure. Also, electron beam melting technology was created in 1997 by Arcam, and two years later, Wake Forrest Institute for Regenerative Medicine printed the first human tissue of a synthetic human bladder. In the 2000s, the projects advanced enormously, but the most important milestone of the decade was the massification of printers and free software projects after patent expiration.

Currently, printers are widespread in areas that involve the creation of biostructures and cutting-edge technological developments thanks to a large number of materials available, among which polymers are remarkable. In particular, hydrogels are formidable materials that have high biocompatibility and are adaptable to 3D printing. This chapter details the fundamental aspects of 3D printing, the main concepts of hydrogels, their uses in 3D printing, and their relevant applications.

8.2 Types of 3D Printing

A 3D printer is a device that builds objects from polymers or other types of materials in an additive manufacturing process. During the 40 years of history of this technology, different printing techniques have been developed and improved. Each type of methodology has been positioned in many fields according to its advantages and limitations. In Figure 8.1, prominent 3D printing techniques are outlined, which will also be detailed in

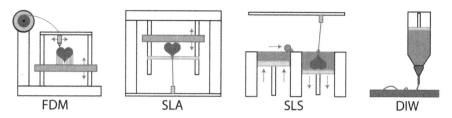

FIGURE 8.1
Schematic representation of the main printing techniques.

the following paragraphs, and their main characteristics are summarized in Table 8.1.

The process of printing an object is performed in three main stages [5]:

- Pre-processing: the object to be printed is 3D designed in a computer-aided design (CAD) file and then imported into specific software that allows configuration of the printing options and generating the cuts (layers) in the design.
- Production: the piece is built either by adding material layer by layer or synthesizing each layer from a reservoir of ink.
- Post-processing: the final object and the support material that is loosely attached are removed from the printer.

8.2.1 FDM

FDM has become the most widely used method for manufacturing polymeric composite parts due to its simplicity, the low maintenance of the printers, and the wide variety of materials suitable for printing [4, 6]. FDM builds 3D parts by extruding a thermoplastic polymer. The deposition of the material is carried out layer by layer through a computer-controlled head. The layers are melted and then solidified, producing objects ready for use. Some control parameters during printing are layer thickness, print direction, screen width and angle, and air gaps, among others. The variation of these parameters can influence the manufacturing time, the cost of the part, and its quality [7]. With FDM, objects with complex geometries can be made quickly, which allows for making several prototypes [5]. Accordingly, the possible polymers must meet specific characteristics like adequate viscosity and melting point [4].

8.2.2 SLA

SLA is a widely used 3D printing process due to its versatility, high accuracy, and precision compared with other 3D printing techniques. SLA uses liquid photopolymers that react under exposition to UV lasers producing a solid

TABLE 8.1
Summary of the Characteristics of the Main 3D Printing Techniques

Technique	State of starting materials	Working principle	Resolution (Z direction, μm)	Advantages	Disadvantages
FDM	Filament	Extrusion and deposition	50–200	Low cost, good strength, multi-material capability	Anisotropy, nozzle clogging
SLA	Liquid photopolymer	Laser scanning and UV-induced curing	10	High printing resolution	Material limitation, cytotoxicity, high cost
SLS	Powder	Laser scanning and heat-induced sintering	80	Good strength, easy removal of support powder	High cost, powdery surface
DIW	Liquid or paste	Pressurized syringe extrusion, and heat or UV-assisted curing	5–200	High printing resolution, soft materials capability	Low mechanical strength, slow

Reprinted from [4], Copyright (2017), with permission from Elsevier.

3D Printing of Hydrogels

product. Through the use of the specific software, the movement of the UV laser is controlled, illuminating the surface of the resin to form the first solid layer on a build platform. Once this layer is completed, the platform is moved down, and the curing process is repeated [8]. In this method, it is necessary to know the exposure time, the scanning speed, and the laser power, among others [9]. Among the advantages of SLA are high printing resolution, a good surface finish, and printing times are short without heat [8].

On the other hand, the main disadvantages are the limited number of usable resins, the high cost, and the toxicity of the residual reagents [4].

8.2.3 SLS

SLS is a 3D printing process based on the solidification of successive layers of powdered material on top of each other. To perform this process, a high-powered laser beam heats the powder in the print zone, and the partially molten particles are joined together and form the first layer of an object. Then, the base of the print area moves down, and a new layer of powder is deposited. One of the advantages of this technique is that there is a wide variety of materials, such as polymers, metals, and ceramics [10]. SLS makes it possible to print complex geometries with interlocking designs without support [11]. However, it often requires a polishing process before obtaining the final piece [12]. Another disadvantage of this technique compared with other 3D printing processes is the high cost due to the use of a high-power laser [13]. The finishing quality of the printed object and its resolution are affected by the size of the powder particles, the power of the laser, the step width and the speed of the scan, the thickness of the layer, the position, and manufacturing orientation [14].

8.2.4 Direct Ink Writing-Liquid Direct Modeling

The DIW printing technique is very similar to FDM since both use an analogous procedure to print 3D objects. But DIW is performed by adding an ink layer by layer [11]. Here, the ink is usually a viscous solution that passes through a pressurized syringe [4], and once the printing finishes, the solvent is removed from the structure. Then, the curing of the material can be done by ultraviolet light or heat. Another method is to inject two reactive inks through a mixing nozzle [4]. There are a lot of materials suitable for inks from these printers, such as colloidal suspensions, ceramics, and polymers [15].

8.3 Type of Polymers Printed

The advantage of polymers compared with other materials lies in their versatility. They can be prepared in different physical forms like adjustable mechanical properties, and are biodegradable, biocompatible, and even bioabsorbable [16]. In the bioprinting field, polymeric hydrogels are ideal systems for simulating extracellular matrix (ECM) microenvironments, biomolecular structures, and growth factors. Therefore, hydrogels are usually selected for bioprinting technologies, 3D printing for tissue engineering, and regenerative medicine [17]. There are different classifications for hydrogels according to synthesis methods, physical properties, crosslinking type, origin, and so on [18]. In this sense, the hydrogel source is the most common classification into natural polymer hydrogels, synthetic polymer hydrogels, and hydrogel composites.

8.3.1 Natural Hydrogels

Natural polymers are characterized by high biocompatibility, biodegradability, and highly hydrated networks that allow good oxygen permeability and provide conditions for cell adhesion and proliferation, transport of nutrients, and the generation of water-soluble metabolites [19].

8.3.1.1 Polysaccharide Hydrogels

8.3.1.1.1 Alginate

Alginate is a naturally occurring hydrophilic anionic copolymer based on different ratios between the monomers β-D-mannuronic acid (M) and α-L-guluronic acid (G) (Figure 8.2a). The M and G monomer ratio of the copolymer, the structure of the sequence, and the length of the chains define their physical properties. Alginate hydrogels are obtained in a slightly acidic medium, and they self-assemble by hydrogen bonds giving gels. On the other hand, when divalent or trivalent cations interact with G monomers, ionic bonds are formed between alginate chains. Also, they can be produced using poly(ethylene glycol) (PEG) to establish a covalent crosslink with the alginate polymer chains or, using carbodiimide chemistry [20]. Alginate has a similar structure to ECM that aids cell adhesion, thus showing potential characteristics as a scaffold material for tissue engineering [21].

8.3.1.1.2 κ-Carrageenan

Carrageenan is a hydrophilic linear polymer composed of carrageenan isomers (k-, l-, and λ-carrageenan), k-carrageenan (k-Ca), with alternating sulfated D-galactose, and 3,6-anhydro-D-galactose (Figure 8.2b), and characterized by forming strong, rigid, and thermoreversible hydrogels. At high temperatures, the k-Ca in an aqueous solution disperses in random spirals, but when the temperature drops, the spirals associate into double helices, resulting in gelation. Also, aqueous solutions containing cations can be cooled, binding spirals to these cations [22]. The following order of cations is preferred for stronger gelation: $K^+ > Ca^{2+} \gg Na^+$ and a lower concentration

3D Printing of Hydrogels

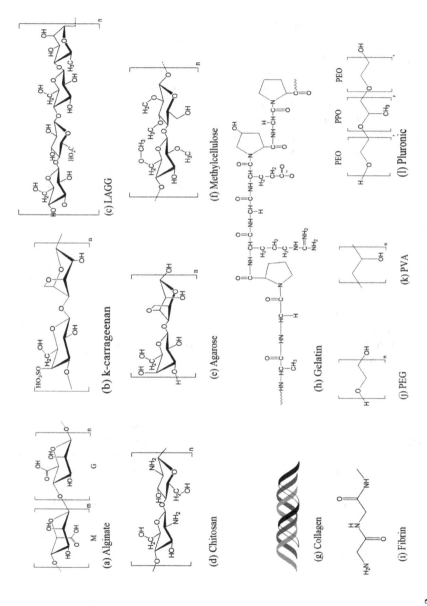

FIGURE 8.2
Classification of hydrogel polymers according to their source.

limit of ~0.2 M to avoid destabilization of the hydrogels [18]. The thermoreversible gelation characteristic of these k-Ca hydrogels has allowed their study as 3D printing patterns [23].

8.3.1.1.3 Gellan Gum

Gellan gum (GG) is a highly water-soluble anionic linear polysaccharide that consists of 1,4-a-L-rhamnose, 1,3-b-D-glucose, 1,4-b-D-glucose, and 1,4-b-D-glucuronic acid. GG can be found in two variants in its original acetylated GG form (HAGG), a GG with high acyl content, and deacetylated GG (LAGG). LAGG is the product of alkaline hydrolysis of HAGG, known as GG of low acyl content (Figure 8.2c). Both GGs can form thermoreversible gels, going from a coiled structure at elevated temperatures (>90°C) to a double helix shape at lower temperatures. They also have good compatibility with other polymers [18]. This range of properties has been of interest to the field of 3D printing, where some researchers have been able to fabricate GG-based scaffolds and brain-like 3D biostructures, encapsulating neuronal cells in peptide-modified GG hydrogels [24].

8.3.1.1.4 Chitosan

Chitosan is a cationic polysaccharide composed of β-(1→4)-linked D-glucosamine and N-acetyl-D-glucosamine (Figure 8.2d). This polymer is produced by the partial alkaline deacetylation of chitin (~50%) from marine organisms [25]. The formation of chitosan hydrogels can be carried out by different chemical or ionic crosslinking pathways [18]. The gelation and solubilization characteristics are a function of the pH, which allows the selection of the conditions of the medium to form specific hydrogels. Furthermore, by controlling the properties of the chitosan ink (rheology and hydrogel formation stages) and the printing parameters, improvements have been achieved in these systems [26].

8.3.1.1.5 Agarose

Agarose is a linear polysaccharide composed of D-galactose and 3,6-anhydro-L-galactose (Figure 8.2e) soluble in water at temperatures higher than 65°C and capable of getting gel between 17°C and 40°C, depending on the chemical structure of the polymer [18]. Agarose hydrogels are formed by physical crosslinking, are thermally reversible, and have high viscosity and shape precision after gelation. They are commonly used as a support material in 3D printing because they have limited bioactivity. However, Nadernezhard et al. used nanosilicate as an additive to adapt the flow behavior during extrusion, achieving 3D structures with high shape fidelity and structural integrity, likewise, an improvement in the bioactivity of encapsulated cells was found [27].

8.3.1.1.6 Methylcellulose

Methylcellulose (MC) is the main ether derived from cellulose by partial substitution of hydroxyl groups by methoxy groups (Figure 8.2f). In bioprinting processes, the printing bed is adjusted to a temperature of ~37°C so that the

MC bioink undergoes the sol-gel process when it comes into contact with the bed. Furthermore, Negrini et al. designed a bioink based on MC hydrogels for the creation of cell sheets with complex shapes, which showed stability in water after 7 days of creation [28].

8.3.1.2 Protein Hydrogels

8.3.1.2.1 Collagen

Collagen can be found in bones, tendons, ligaments, blood vessels, skin, cornea, and ECM. The most common collagen structure consists of a triple helix that tends to clump into microfibrils (type I) (Figure 8.2g). The formation of type I collagen hydrogels consists of three stages, starting with the generation of fibrils from the triple helices, the linear growth of the fibrils, and finally, the formation of a network structure of linear fibrils [29]. Furthermore, they are an ideal material to mimic aspects of the ECM microenvironment [30]. Studies have been reported on the preparation of collagen-alginate and collagen-genipin inks for the conformation of 3D printed scaffolds, with excellent osteogenic activity in the case of collagen-genipin-based inks [31].

8.3.1.2.2 Gelatin

Gelatin is a heterogeneous mixture of proteins derived from the original collagen protein (see Figure 8.2h). Accordingly, some chemical modification is usually carried out that allows its use [18]. The biocompatibility of gelatin hydrogels as well as their high elasticity have extended their use to the 3D printing field as an ink for printing scaffolds for the generation of artificial tissues and organs [32].

8.3.1.2.3 Fibrin

Fibrin is a natural polymer (Figure 8.2i) derived from the protein monomer fibrinogen. Fibrin hydrogels can be obtained by thermal, chemical, or enzymatic gelling processes [17]. The ability of these hydrogels to promote angiogenesis and increase the propagation of neurites has been successfully verified [33]. Fibrin has low mechanical resistance and rapid degradation, however, these disadvantages are compensated by adding other polymers, for example, poly-L-lysine prevents shrinkage of the gel [34], TPU or PCL can help to increase mechanical rigidity [35], and PEG can be added to improve the stability of the structure of fibrin hydrogels [36].

8.3.2 Hydrogels from Synthetic Polymers

Hydrogels obtained from synthetic polymers usually have better mechanical properties compared with natural sources, since the synthesis processes are more controlled and can be adjusted to obtain higher mechanical performance. However, the biocompatibility and biodegradability properties are limited or reduced.

8.3.2.1 PEG

PEG (Figure 8.2j) is a hydrophilic linear polyester with high water solubility and is widely used for its high biocompatibility and performance in protein adsorption and cell adhesion. PEG is not hydrolytically degradable, limiting its use to macromolecules with molecular weights below 50 kDa [18]. In addition, for 3D printing systems, hydrogel inks made from PEG-derived polymers, such as PEG monoacrylate, PEG diacrylate (PEGDA), PEG methacrylate, and PEG dimethacrylate, have shown better results [29].

8.3.2.2 Poly(vinyl alcohol)

Poly(vinyl alcohol) or PVA (Figure 8.2k) is a polymer commonly synthesized from the polymerization of vinyl acetate. Physically crosslinked hydrogels are often preferred for tissue engineering due to their high mechanical strength [18], low manufacturing costs, and absence of toxicity. Also, alginate-PVA-hydroxyapatite hydrogels were used as bioink for the creation of scaffolds with high cell viability [37]. Subsequently, alginate/PEGDA/PVA showed favorable 3D printability and biocompatibility [38].

8.3.2.3 Pluronics

Pluronics or poloxamers are triblock copolymers made up of a central block of poly(propylene oxide) surrounded by blocks of poly(ethylene oxide) (Figure 8.2l). Pluronic F127 is one of the most widely used in the medical field because it gels below body temperature and exhibits shear thinning behavior and good shear recovery, making it suitable for precise 3D extrusion printing. Studies have shown the possibility of encapsulating cells in these hydrogels for a short period without any detrimental effect on their viability [39]. This unsatisfactory feasibility is improved blending with hydrogels of natural origin [29].

8.3.3 Hydrogel Composites

In the search to obtain hydrogels with high resistance and high recovery capacity, there are hydrogels with interpenetrating polymeric networks. On the other hand, nanocomposite hydrogels are characterized by introducing an inorganic material that reinforces the mechanical and biological properties of the pure hydrogel. Another way to improve the mechanical behavior of polymeric hydrogels is by incorporating fibers. However, it is difficult to obtain a uniform distribution and a predicted alignment of the fibers within the matrix, which limits the quality of materials printed by 3D printing methods [40].

8.4 3D Printing of Hydrogel

Three-dimensional printing technology can use hydrogels to transform into personalized objects with high spatial resolution, enabling the creation of complex and highly customizable scaffold structures that support cell adhesion, as well as the proliferation of tissue engineering in the obtention of programmable biomimetic actuators [41]. The main printing methods used for hydrogel polymers or hydrogel bioinks are detailed in the following sections.

8.4.1 Methods

Within the medical field, a widely used technique is extrusion or nozzle printing using viscous hydrogels as raw material. In addition to the traditional extrusion method, FDM, this technique has been innovated by creating variants such as micro-extrusion, extrusion-based gradient, and the use of a static mixer [19]. The hydrogel extrusion technique is considered successful when the hydrogel ink flows through the needle during printing, depositing the material on the printing bed without changing, and hydrogels must have good structural integrity for adequate layer-by-layer adhesion. For its part, the inkjet printing method can be performed using a thermal or piezoelectric inkjet. Thermal inkjet printing uses a heater to generate rapidly expanding steam bubbles expelled from the print head. Thus, the piezoelectric action supplies pulses that expel the ink from the printer's chamber [18, 29]. Laser printing methods include other light and photopolymerization techniques such as hybrid laser printing, digital light processing, and optical projection lithography [19]. Laser printing methods usually generate high-resolution droplets, between 50 and 60 µm, and offer a suitable medium for bioprinting processes, since the absence of mechanical forces during printing is ideal for obtaining systems with high cell viability (95%). Furthermore, the absence of direct contact with the inks or the dispenser reduces the risks of contamination [29].

8.4.2 Experimental Procedure

In any 3D printing process, the starting point is the preparation of the 3D design from a CAD and the configuration of the relevant printing parameters, which depend on the selected printing method and the type and characteristics of the employed hydrogel ink. In FDM-based methods, it is necessary to optimize the printing temperature to guarantee good adhesion between the layers of the material while maintaining shape fidelity, and cell viability must also be guaranteed when working with bioprinting systems. As a general rule, it must be ensured that the material is not so dense, to avoid clogging the nozzle, nor so fluid as to collapse the strands of the printed material [28].

Frequently, the hydrogels need viscosities between 300 and 30,000 cps. In an extrusion study carried out by He et al. for different viscosities and surface tensions, it was concluded that the pressure must be greater than 5 kPa to come out from the nozzle but inferior to 35 kPa [42] for a controllable extrusion, with good fidelity of shape and print quality. In addition, the speed of the nozzle directly influences the width of the printed line. Sometimes, when this value is high, the line tends to stretch, becoming thinner and thinner, promoting material breakage. In a study with a mixture of sodium alginate and gelatin, the impression of hydrogel lines with right angles was observed for a printing distance of less than 0.4 mm and the curvature of the corners when this value was >0.9 mm [42]. In some cases, 3D hydrogels undergo a subsequent drying stage, depending on the final use of the material [43].

8.4.3 Quality Parameters

One of the main challenges for 3D printing of hydrogels is precision and reproducibility due to the high water content. For this, certain parameters are defined to quantify the 3D printing quality of hydrogels [44]. These parameters include printability, extrudability, precision, fidelity, and accuracy of printing. Printability is defined as the ability of bioink to be extruded and maintain high resolution, quality, and fidelity of the printed object [45]. Printability depends on the material and is affected by its rheological, thermal, and mechanical properties [46]. Printability can be determined by the following Equation (8.1) [44, 46]:

$$P_r = \frac{L^2}{16A} \qquad (8.1)$$

Where P_r is the printability index, L is the perimeter of the object, and A the geometrical area.

Another way to quantify printability is through injection speed (nozzle speed), a parameter strongly dependent on the hydrogel fluidity, according to Equation (8.2):

$$Nozzle\,speed = \frac{4Q}{\pi(D_s)^2} \qquad (8.2)$$

Where Q is the flow rate and D_s is the diameter of the extruded wire.

The flow rate is associated with strand printability, defined as the ability of the bioink to flow at a constant speed after being deformed and maintaining its rheological properties. This deformation affects the viscosity, surface tension, and contact angle of the extruded material [44, 46]. Taking into consideration the injection speed, the strand printability is defined according to Equation (8.3):

$$Strand\,printability = 1 - \frac{D_s - D_{exp}}{D_{exp}} \qquad (8.3)$$

Where D_{exp} is the diameter of the experimentally extruded thread.

An important factor that interferes with printability is irregularity. This parameter takes into account the printing layers of the piece that is made and compares the differences between the experimental length and the CAD design. This irregularity is defined by the following Equation (8.4) [44, 46]:

$$Irregularity = \frac{Experimental\ length_{x,y,z}}{Design\ length_{x,y,z}} \times 100 \qquad (8.4)$$

Another parameter in the 3D printing quality of hydrogels is the precision of the impression that can be quantified through the optimization index (*POI*). This parameter depends on the nozzle used in the printer and the mechanical properties of the bioink and influences the construction of the printed object, affecting its size and shape [44]. *POI* can be defined through Equation (8.5):

$$POI = \frac{1}{tline * DG * p} \qquad (8.5)$$

Where DG is the nozzle gauge; p is the pressure, and $tline$ is the thickness of the printed line.

On the other hand, to analyze the degree of coincidence between the original design and the printed one, fidelity is defined. This parameter takes into account the dimensions and is dependent on the bioink ability to maintain its shape after being deposited on the printing surface, without collapsing as the height of the printed object increases [44]. For fidelity determination, the relationship between experimental and designed length [44, 46] (Equation 8.6) is calculated:

$$Fidelity = \frac{Length\ experimental}{Length\ designed} \times 100 \qquad (8.6)$$

For its part, to determine the integrity of the structure, the diffusion rate or deflection ratio (*Dfr* (%)) is used. *Dfr* (%) is calculated as the ratio of the areas of the experimental and CAD-printed pores. This parameter is related to the fidelity and quality of the print (see Equation (8.7)):

$$Dfr(\%) = \frac{A_{experimental}}{A_{designed}} \times 100 \qquad (8.7)$$

Where $A_{experimental}$ is the area of the printed pores and $A_{designed}$ is the area of the designed pores.

Finally, to quantify the accuracy by determining the reproducibility of the printed parts based on geometry, size, and spatial location, dimensional stability (Δd (%)) is used. This parameter is measured based on the length, width, and height of the printed part based on the original according to Equation (8.8):

$$\Delta d(\%) = \frac{d_r - d_0}{d_0} \times 100 \tag{8.8}$$

Where, d_r is the height, width, or length of the printed part, and d_o, is the width, length, and height of the initial piece [44].

8.5 Applications: Literature Review

Currently, 3D printing of hydrogels is present in a wide range of applications: medicine, the food industry, fashion, architecture, energy, and the manufacturing of actuators and sensors, among others. Three-dimensional printing has been selected in applications such as cancer treatment, bone, tissue, and vascular engineering, in the area of neurology, and in pharmaceutical research [19]. In the field of tissue engineering and regenerative medicine, the ability to deposit biomaterials with micrometric precision, in conditions suitable for cells, gives it an advantage over other techniques to create scaffolds.

Markstedt et al. [47] proposed a bioink that combines the rheological properties of nanofibrillated cellulose with the crosslinking ability of alginate to print living soft tissue with cells. From magnetic resonance imaging (MRI) and computed tomography (CT), CAD files were made to print anatomically shaped cartilage structures, such as a human ear and a sheep meniscus (Figure 8.3). Human chondrocytes bioprinted in the non-cytotoxic nanocellulose-based bioink showed ~80% cell viability. In this way, the potential use of nanocellulose for the 3D bioprinting of living tissues and organs can be demonstrated.

With the progress of 3D printing techniques and the appearance of new inks, the complexity of the applications has increased, focusing goals on the possibility of creating functional organs. Hinton et al. [48] printed complex 3D biological structures from soft protein and polysaccharide hydrogels. To perform this task, they implemented a method of building printed hydrogel structures, such as femurs, hearts, coronary arteries, and brains, with robust mechanics and complex anatomical structures (Figure 8.4). Design models were obtained from 3D optical data, CT, and MRI.

Wei et al. [49] developed a local therapy strategy in which they used hydrogels to deliver drugs for diseases such as cancer. In their studies, they prepared hydrogel fibers/scaffolds with a specialized core for controlled drug delivery. To do this, they coaxially printed fibers using ink mixtures

3D Printing of Hydrogels

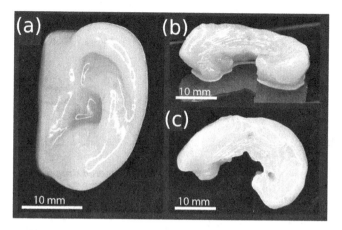

FIGURE 8.3
(a) A 3D printed human ear, (b) sheep meniscus side view, and (c) sheep meniscus (top view). (Reprinted with permission from [47], Copyright 2015, American Chemical Society.)

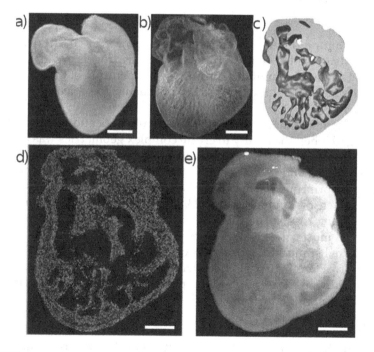

FIGURE 8.4
Three-dimensional imaging data and model design for an embryonic chicken heart: (a) A dark-field image. (b) A 3D image was stained for nuclei, fibronectin, and F-actin, and imaged with a confocal microscope. (c) A cross-section of the 3D CAD model based on the confocal imaging data. (d) A cross-section of the 3D print showing a recreation of the internal trabecular structure. (e) A dark-field image of the 3D-printed heart. (Reprinted with permission [48]. Copyright The Authors, some rights reserved; exclusive licensee Science. Distributed under a Creative Commons Attribution License 4.0 (CC BY) https://creativecommons.org/licenses/by/4.0/)

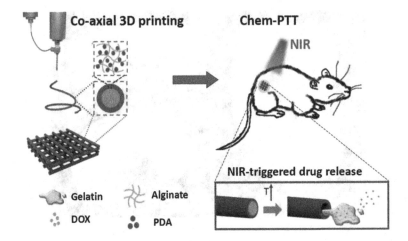

FIGURE 8.5
Schematic diagram of the localized drug delivery system developed [49]. (Reprinted with permission [49], Copyright 2020, Elsevier.)

of polydopamine (PDA) and concentrated alginate to coat the fibers; the core was drug-loaded temperature-sensitive hydrogels. With near-infrared, the PDA can be stimulated, raising the temperature of the nucleus with the subsequent gel-solution transition producing the release of the drug (Figure 8.5).

A significant advance in the development of applications from 3D printing emerged when the ability to create complex geometries was combined with the conformational changes caused by external stimuli typical of smart hydrogels and shape memory polymers. This type of combination is called 4D printing. In this way, multifunctional devices can be created that act as sensors and actuators [33]. Most of the devices manufactured by 4D printing respond to the electromagnetic field, the concentration of ions, and in most cases, temperature, and transform the energy extracted from external stimuli to perform specific tasks [50]. Han et al. [51] worked with electroactive hydrogels that deform when subjected to an external electric field. This material has the potential for use in soft robots and artificial muscles (Figure 8.6). Using the 3D printing technique, they can create parts capable of manipulating microstructures with smooth robotic locomotion.

Yao et al. [52] inspired by nature sought to develop materials and systems that simultaneously produce a change in color and a transformation caused by certain stimuli. Starting from a supramolecular hydrogel coordinated with lanthanide ions, they printed by DIW a material that can dynamically control opacity and luminescence in response to humidity variation or the hydration/dehydration process (Figure 8.7).

3D Printing of Hydrogels

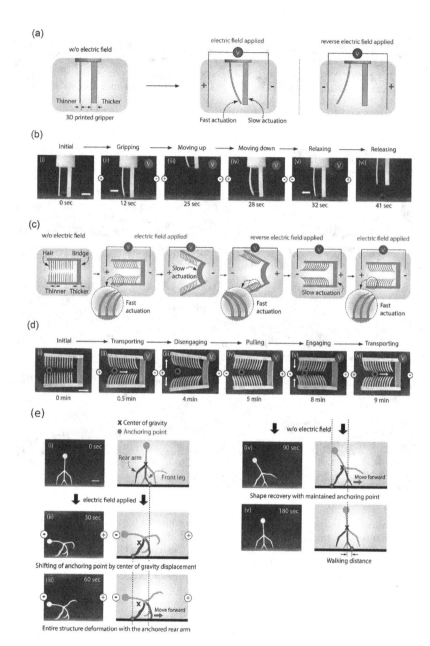

FIGURE 8.6
Soft robotic manipulation with 3D printing through the application of electric fields. (a) Diagrams of a clamp and (b) applied to grab an object. (c) Diagrams of a transporter using hairs and (d) applied to transport an object. All scale bars indicate 5 mm. (e) Walking driven by an electric field and variation of the center of gravity. Scale bars = 5 mm. (Reprinted with permission [51], Copyright 2018, American Chemical Society.)

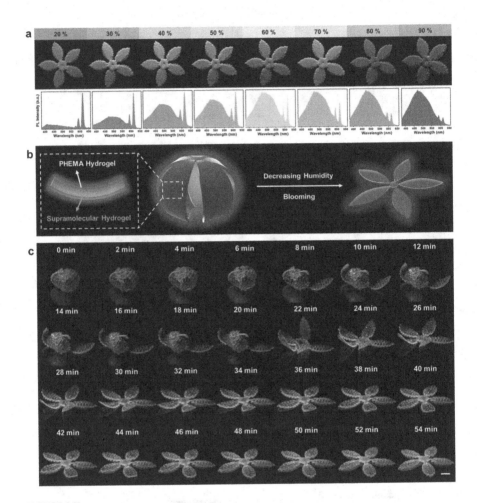

FIGURE 8.7
Four-dimensional printing of objects with a simultaneous change of morphology and full-color luminance adjustment. (a) Printed flowers and their respective photoluminescence spectra at different relative humidity percentages. (b) Design of a bilayer hydrogel flower. (c) temporal evolution of the shape and color of the bilayer hydrogel flower with increasing relative humidity. Scale bar = 10 mm. (Adapted with permission [52], Copyright 2020, American Chemical Society.)

8.6 Prospects

Currently, 3D printing with hydrogels has proven to be successful for the fabrication of organ regeneration scaffolds, drug delivery systems, and devices capable of specific and complex torsion, bending, and conformational changes. The implementation of intelligent hydrogels led to the definition of the concept of 4D printing, capable of responding to external stimuli such as electric, magnetic or acoustic fields, temperature, and pH variations. Furthermore, we envision that the joint efforts of researchers and companies will make it possible to improve both techniques and inks, overcoming the challenges and indicating a promising future for the 3D printing of hydrogels.

References

1. H. Kodama, 'Automatic method for fabricating a three-dimensional plastic model with photo-hardening polymer', *Review of Scientific Instruments*, vol. 52, no. 11, pp. 1770–1773, 1981.
2. I. 3D Systems, '3D systems', 2022. https://www.3dsystems.com/ (accessed Jul. 07, 2022).
3. I. Stratasys, 'Stratasys', 2022. www.stratasys.com (accessed Jul. 07, 2022).
4. X. Wang, M. Jiang, Z. Zhou, J. Gou, and D. Hui, '3D printing of polymer matrix composites: A review and prospective', *Compos B Eng*, vol. 110, pp. 442–458, 2017.
5. V. Surange and P. Gharat, '3D printing process using fused deposition modelling (FDM)', *IRJET*, Vol. 3, pp. 1403–1406, 2016.
6. K. B. Mustapha and K. M. Metwalli, 'A review of fused deposition modelling for 3D printing of smart polymeric materials and composites', *Eur Polym J*, vol. 156, p. 110591, 2021.
7. A. K. Sood, R. K. Ohdar, and S. S. Mahapatra, 'Parametric appraisal of mechanical property of fused deposition modelling processed parts', *Mater Des*, vol. 31, no. 1, pp. 287–295, 2010.
8. I. Karakurt, A. Aydoğdu, S. Çıkrıkcı, J. Orozco, and L. Lin, 'Stereolithography (SLA) 3D printing of ascorbic acid loaded hydrogels: A controlled release study', *Int J Pharm*, vol. 584, p. 119428, 2020.
9. Y. H. Cho, I. H. Lee, and D.-W. Cho, 'Laser scanning path generation considering photopolymer solidification in micro-stereolithography', *Microsyst Technol*, vol. 11, no. 2, pp. 158–167, 2005.
10. J. P. Kruth, X. Wang, T. Laoui, and L. Froyen, 'Lasers and materials in selective laser sintering', *Assem Autom*, vol. 23, no. 4, pp. 357–371, 2003.
11. A. D. Valino, J. R. C. Dizon, A. H. Espera, Q. Chen, J. Messman, and R. C. Advincula, 'Advances in 3D printing of thermoplastic polymer composites and nanocomposites', *Prog Polym Sci*, vol. 98, p. 101162, 2019.
12. A. Lamikiz, J. A. Sánchez, L. N. López de Lacalle, and J. L. Arana, 'Laser polishing of parts built up by selective laser sintering', *Int J Mach Tools Manuf*, vol. 47, no. 12, pp. 2040–2050, 2007.
13. G. Kim and Y. T. Oh, 'A benchmark study on rapid prototyping processes and machines: Quantitative comparisons of mechanical properties, accuracy, roughness, speed, and material cost', *Proc Inst Mech Eng B-J Eng Manuf*, vol. 222, pp. 201–215, 2008.
14. I. Gibson and D. Shi, 'Material properties and fabrication parameters in selective laser sintering process', *Rapid Prototyp J*, vol. 3, no. 4, pp. 129–136, 1997.

15. A. Zhang et al., '3D printing hydrogels for actuators: A review', *Chin Chem Lett*, vol. 32, no. 10, pp. 2923–2932, 2021.
16. T. D. Ngo, A. Kashani, G. Imbalzano, K. T. Q. Nguyen, and D. Hui, 'Additive manufacturing (3D printing): A review of materials, methods, applications and challenges', *Compos B Eng*, vol. 143, pp. 172–196, 2018.
17. R. Pugliese, B. Beltrami, S. Regondi, and C. Lunetta, 'Polymeric biomaterials for 3D printing in medicine: An overview', *Annals 3D Print Med*, vol. 2, p. 100011, 2021.
18. J. Li, C. Wu, P. K. Chu, and M. Gelinsky, '3D printing of hydrogels: Rational design strategies and emerging biomedical applications', *Mater Sci Eng: R: Rep*, vol. 140, p. 100543, 2020.
19. C. Liu, N. Xu, Q. Zong, J. Yu, and P. Zhang, 'Hydrogel prepared by 3D printing technology and its applications in the medical field', *Colloid Interface Sci Commun*, vol. 44, p. 100498, 2021.
20. M. Farokhi, F. Jonidi Shariatzadeh, A. Solouk, and H. Mirzadeh, 'Alginate based scaffolds for cartilage tissue engineering: A review', *Int J Polym Mater Polym Biomater*, vol. 69, 2019.
21. J. L. Drury, R. G. Dennis, and D. J. Mooney, 'The tensile properties of alginate hydrogels', *Biomaterials*, vol. 25, no. 16, pp. 3187–3199, 2004.
22. W Lim, G.J. Kim, H.W. Kim, J. Lee, X. Zhang, M.G. Kang, J.W. Seo, J.M. Cha, H.J. Park, M.Y. Lee, S.R. Shin, S.Y. Shin, and H. Bae, 'Kappa-Carrageenan-based dual crosslinkable bioink for extrusion type bioprinting', *Polymers*, vol. 12, p. 2377, 2020.
23. S. Liu and L. Li, 'Ultrastretchable and self-healing double-network hydrogel for 3D printing and strain sensor', *ACS Appl Mater Interfaces*, vol. 9, no. 31, pp. 26429–26437, 2017.
24. R. Lozano, L. Stevens, B.C. Thompson, K.J. Gilmore, R. 3rd Gorkin, E.M. Stewart, M. inhet Panhuis, M. Romero-Ortega, and G.G. Wallace, '3D printing of layered brain-like structures using peptide modified gellan gum substrates', *Biomaterials*, vol. 67, pp. 264–273, 2015.
25. M. Rinaudo, 'Chitin and chitosan: Properties and applications', *Prog Polym Sci*, vol. 31, no. 7, pp. 603–632, 2006.
26. Q. Wu, D. Therriault, and M.-C. Heuzey, 'Processing and properties of chitosan inks for 3D printing of hydrogel microstructures', *ACS Biomater Sci Eng*, vol. 4, no. 7, pp. 2643–2652, 2018.
27. A. Nadernezhad, O. S. Caliskan, F. Topuz, F. Afghah, B. Erman, and B. Koc, 'Nanocomposite bioinks based on agarose and 2D nanosilicates with tunable flow properties and bioactivity for 3D bioprinting', *ACS Appl Bio Mater*, vol. 2, no. 2, pp. 796–806, 2019.
28. N. Contessi Negrini, L. Bonetti, L. Contili, and S. Farè, '3D printing of methylcellulose-based hydrogels', *Bioprinting*, vol. 10, p. e00024, 2018.
29. S. Liu, X. Chen, and Y. Zhang, 'Chapter 14: Hydrogels and hydrogel composites for 3D and 4D printing applications', in *3D and 4D Printing of Polymer Nanocomposite Materials*, K. K. Sadasivuni, K. Deshmukh, and M. A. Almaadeed, Eds. Elsevier, 2020, pp. 427–465.
30. M. McCoy, B. R. Seo, S. Choi, and C. Fischbach, 'Collagen I hydrogel microstructure and composition conjointly regulate vascular network formation', *Acta Biomater*, vol. 44, pp. 200–208, 2016.

31. Y. B. Kim, H. Lee, and G. H. Kim, 'Strategy to achieve highly porous/biocompatible macroscale cell blocks, using a collagen/genipin-bioink and an optimal 3D printing process', *ACS Appl Mater Interfaces*, vol. 8, no. 47, pp. 32230–32240, 2016.
32. X. Wang, 'Bioartificial organ manufacturing technologies', *Cell Transplant*, vol. 28, p. 096368971880991, 2018.
33. T. Ahmed, E. Dare, and M. Hincke, 'Fibrin: A versatile scaffold for tissue engineering applications', *Tissue Eng Part B Rev*, vol. 14, pp. 199–215, 2008.
34. S. Jockenhoevel, G. Zund, S.P. Hoerstrup, K. Chalabi, J.S. Sachweh, L. Demircan, B.J. Messmer, and M. Turina, 'Fibrin gel: Advantages of a new scaffold in cardiovascular tissue engineering', *Eur J Cardiothorac Surg*, vol. 19, pp. 424–430, 2001.
35. C. Lee, S. Grad, K. Gorna, S. Gogolewski, A. Goessl, and M. Alini, 'Fibrin–polyurethane composites for articular cartilage tissue engineering: A preliminary analysis', *Tissue Eng*, vol. 11, pp. 1562–1573, 2005.
36. D. Dikovsky, H. Bianco-Peled, and D. Seliktar, 'The effect of structural alterations of PEG-fibrinogen hydrogel scaffolds on 3-D cellular morphology and cellular migration', *Biomaterials*, vol. 27, no. 8, pp. 1496–1506, 2006.
37. S. Bendtsen, S. Quinnell, and M. Wei, 'Development of a novel alginate-polyvinyl alcohol-hydroxyapatite hydrogel for 3D bioprinting bone tissue engineered scaffolds', *J Biomed Mater Res A*, vol. 105, 2017.
38. F. Yu, X. Han, K. Zhang, B. Dai, S. Shen, X. Gao, H. Teng, X. Wang, L. Li, H. Ju, W. Wang, J. Zhang, Q. Jiang, 'Evaluation of a polyvinyl alcohol-alginate based hydrogel for precise 3D bioprinting', *J Biomed Mater Res A*, vol. 106, no. 11, pp. 2944–2954, 2018.
39. M. Müller, J. Becher, M. Schnabelrauch, and M. Zenobi, 'Nanostructured pluronic hydrogels as bioinks for 3D bioprinting', *Biofabrication*, vol. 7, p. 035006, 2015.
40. D. Myung, D. Waters, M. Wiseman, P.E. Duhamel, J. Noolandi, C.N. Ta, C.W. Frank, 'Progress in the development of interpenetrating polymer network hydrogels', *Polym Adv Technol*, vol. 19, pp. 647–657, 2008.
41. F. Puza and K. Lienkamp, '3D printing of polymer hydrogels: From basic techniques to programmable actuation', *Adv Funct Mater*, vol. 32, no. 39, p. 2205345, 2022.
42. Y. He, F. Yang, H. Zhao, Q. Gao, B. Xia, and J. Fu, 'Research on the printability of hydrogels in 3D bioprinting', *Sci Rep*, vol. 6, no. 1, p. 29977, 2016.
43. T. Tagami, M. Ando, N. Nagata, E. Goto, N. Yoshimura, T. Takeuchi, T. Noda, T. Ozeki, 'Fabrication of naftopidil-loaded tablets using a semisolid extrusion-type 3D printer and the characteristics of the printed hydrogel and resulting tablets', *J Pharm Sci*, vol. 108, no. 2, pp. 907–913, 2019.
44. S. Bom, R. Ribeiro, H. M. Ribeiro, C. Santos, and J. Marto, 'On the progress of hydrogel-based 3D printing: Correlating rheological properties with printing behaviour', *Int J Pharm*, vol. 615, p. 121506, 2022.
45. E. Pulatsu and M. Lin, 'A review on customizing edible food materials into 3D printable inks: Approaches and strategies', *Trends Food Sci Technol*, vol. 107, pp. 68–77, 2021.
46. Y.-W. Ding, X.-W. Zhang, C.-H. Mi, X.-Y. Qi, J. Zhou, and D.-X. Wei, 'Recent advances in hyaluronic acid-based hydrogels for 3D bioprinting in tissue engineering applications', *Smart Mater Med*, vol. 4, pp. 59–68, 2023.

47. K. Markstedt, A. Mantas, I. Tournier, H. Martínez Ávila, D. Hägg, and P. Gatenholm, '3D bioprinting human chondrocytes with nanocellulose–alginate bioink for cartilage tissue engineering applications', *Biomacromolecules*, vol. 16, pp. 1489–1496, 2015.
48. T.J. Hinton, Q. Jallerat, R.N. Palchesko, J.H. Park, M.S. Grodzicki, H.J. Shue, M.H. Ramadan, A.R. Hudson, A.W. Feinberg, 'Three-dimensional printing of complex biological structures by freeform reversible embedding of suspended hydrogels', *Sci Adv*, vol. 1, no. 9, p. e1500758, 2022.
49. X. Wei, C. Liu, Z. Wang, and Y. Luo, '3D printed core-shell hydrogel fiber scaffolds with NIR-triggered drug release for localized therapy of breast cancer', *Int J Pharm*, vol. 580, p. 119219, 2020.
50. M. Champeau, D. A. Heinze, T. N. Viana, E. R. de Souza, A. C. Chinellato, and S. Titotto, '4D printing of hydrogels: A review', *Adv Funct Mater*, vol. 30, no. 31, p. 1910606, 2020.
51. D. Han, C. Farino, C. Yang, T. Scott, D. Browe, W. Choi, J.W. Freeman, H. Lee, 'Soft robotic manipulation and locomotion with a 3D printed electroactive hydrogel', *ACS Appl Mater Interfaces*, vol. 10, pp. 17512–17518, 2018.
52. Y. Yao, C. Yin, S. Hong, H. Chen, Q. Shi, J. Wang, X. Lu, and N. Zhou, 'Lanthanide-ion-coordinated supramolecular hydrogel inks for 3D printed full-color luminescence and opacity-tuning soft actuators', *Chem Mater*, vol. 32, pp. 8868–8876, 2020.

9
Hydrogels for CO_2 Reduction

Jesús A. Claudio-Rizo, Martín Caldera-Villalobos,
Denis A. Cabrera-Munguía, and María I. León-Campos

CONTENTS

9.1 Introduction 169
9.2 Hydrogels Based on Synthetic Polymers for CO_2 Reduction 171
9.3 Hydrogels Based on Natural Polymers for CO_2 Reduction 175
9.4 Hydrogels Based on Inorganic Components for CO_2 Reduction 179
    9.4.1 General Synthesis of Aerogels 179
        9.4.1.1 Gelation Stage 179
        9.4.1.2 Drying 181
        9.4.1.3 Carbonization 182
    9.4.2 Aerogels for CO_2 Capture 182
    9.4.3 Aerogels for CO_2 Reduction 183
9.5 Conclusion 184
References 185

9.1 Introduction

Carbon dioxide (CO_2) is one of the main gases associated with the greenhouse effect, so innovating strategies to reduce its presence in the atmosphere is of vital importance in order not to promote harmful effects on the climate. Within the strategies for the reduction of this gas, the generation of materials in the hydrogel state has been used, highlighting the high absorption/adsorption capacities of the gas according to the composition and structure of the 3D matrix. Hydrogels can be synthesized from synthetic and natural polymers, and coupled with inorganic components, can perform well in the transformation of CO_2 into value-added products or components that are not harmful to the environment through its reduction (Figure 9.1).

Various synthetic polymers such as poly(vinyl alcohol) (PVA), polyacrylic acid derivatives (PMMA, PAA), and polyethyleneimine have been used for this purpose. If there are nucleophilic atoms or coordination complexes that have a high reduction potential to reduce CO_2, they can perform the process of chemical transformation of this gas. Transition metals such as gold, silver, and copper show the redox potential, however, amino group atoms

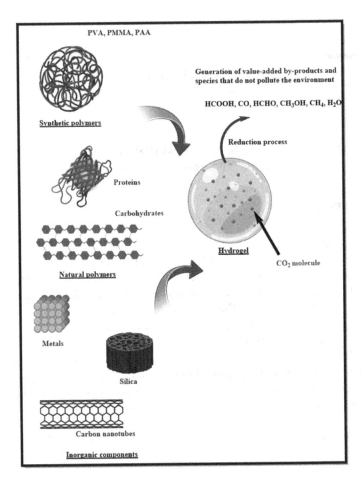

FIGURE 9.1
Representative outline for the generation of hydrogels for the adsorption/sorption and reduction of CO_2.

have also been shown to exhibit this property. CO_2 reduction products using this type of material are carbamic acids, formic acid, CO, hydrogen, and water, among others. The use of biopolymers has attracted the attention of researchers, since carbohydrates and proteins present in nature, such as chitosan, cellulose, starch, and arginine, have shown potential for the removal and reduction of this gas, taking advantage of the fact that hydrogels made up of biopolymers show biodegradation, thus avoiding the accumulation of synthetic materials that aggravate environmental contamination.

Finally, systems with a high selective capacity for CO_2 reduction have also been studied, where inorganic components such as silica, metal-organic frameworks, and carbon derivatives can be used to generate hydrogels and/or aerogels with the specific surface modification that enhances the chemical transformation of carbon dioxide. This chapter aims to present the current state of studies for CO_2 reduction using materials in a hydrogel state.

9.2 Hydrogels Based on Synthetic Polymers for CO_2 Reduction

In modern society, reduction processes to produce sustainable fuels or chemicals using renewable energy are expected to decouple economic growth from CO_2 release. CO_2 reduction processes can be classified into three categories according to the obtained product: (1) CO, (2) formic acid, and (3) hydrocarbons and alcohols. These reactions require the use of metal catalysts: the reduction of CO_2 to CO is catalyzed commonly by Au and Ag, the reduction to formic acid is catalyzed by Sn and Pb, and the reduction to hydrocarbons, such as methane and ethylene, is catalyzed by Cu, although they have also been used as catalysts of Mn and Fe [1].

CO_2 is kinetically inert, so electrochemical reduction with acceptable reaction yields requires highly active catalysts that allow energy barriers to be lowered. These catalysts can be homogeneous, heterogeneous, or immobilized on surfaces. In homogeneous catalysis, the electronic transference happens to take after the dissemination of the catalyst toward an anode where proton-coupled electron exchange happens. Heterogeneous catalysts have been explored in greater detail and offer the advantages of their ease of synthesis and their high activity associated with a high surface area. Finally, today, catalysts immobilized on a surface are relevant only for investigations that seek to elucidate reaction mechanisms [2].

Converting CO_2 into high-value chemicals requires CO_2 capture first and then electrochemical reduction. In both processes, polymer-based materials can be implemented because, depending on the structure, they can function as absorbents and/or catalysts or they can have a different function. Dixit and Majumder performed electrochemical CO_2 reduction in a continuous multi-compartment reactor using KOH as an electrolyte and adsorbent. By using a stainless-steel cathode in a batch system, formic acid was obtained as the main product. While using a copper cathode in a continuous system, a mixture of methanol and methane was obtained as products. In this work, a PVA hydrogel was also used, whose function is to bind the catalyst that is the particulate, so it does not participate in the reduction process [3].

Multiple investigations have shown that polymers with basic nitrogen atoms and their hydrogels can chemically absorb and retain CO_2 molecules. For example, polyethyleneimine is a polymer with secondary nitrogen atoms with a high affinity for CO_2 molecules. Figure 9.2a shows the reaction between a secondary amine and CO_2 to form a carbamate. From the point of view of adsorption, this process is chemisorption. It is important to highlight the reversible nature of this reaction since it allows the regeneration

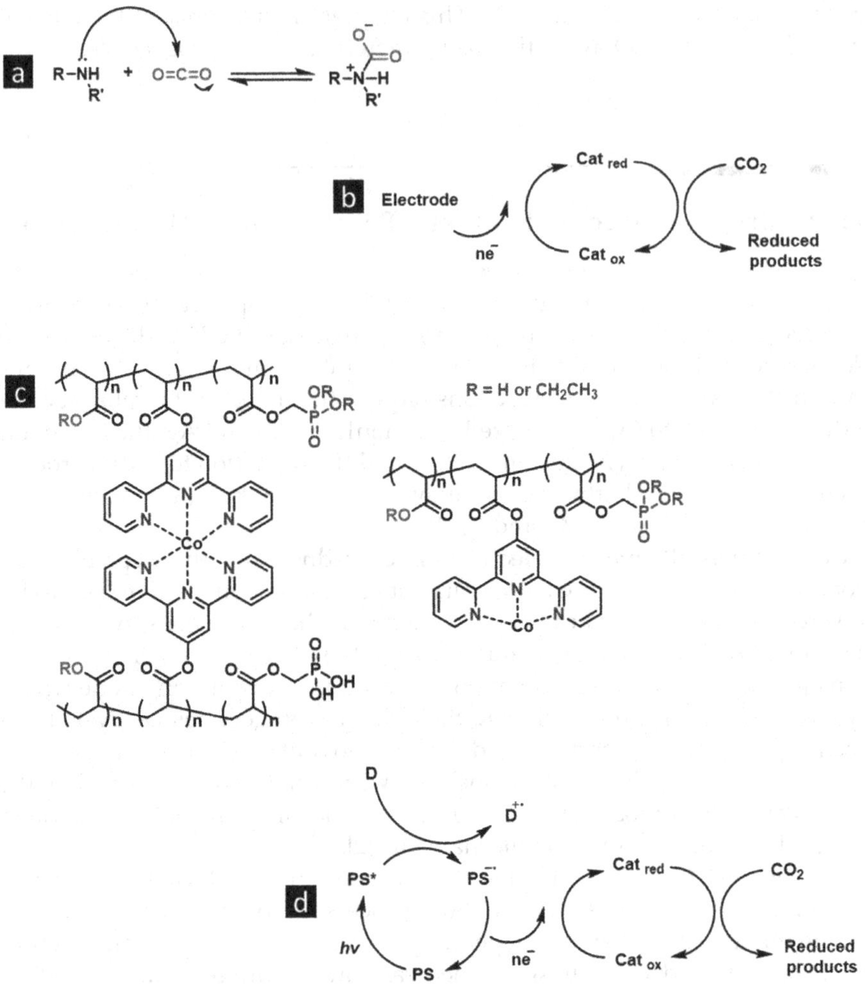

FIGURE 9.2
(a) Reaction mechanism for the nucleophilic addition of amines to CO_2, (b) catalytic cyclic for the electrochemical reduction of CO_2, (c) structure of polymer complex with terpyridine ligands, and (d) catalytic cycles for the photochemical reduction of CO_2.

of the material. Polyethyleneimine is a liquid with low volatility and low cost, which increases its attractiveness as a CO_2 adsorbent. However, the performance of this polymer with a branched structure can be significantly improved if it is crosslinked with epichlorohydrin, obtaining a material capable of forming hydrogels and with a high surface area, which increases its absorption capacity [4].

However, these hydrogels are not capable of completely eliminating the CO_2 that they have captured, since chemisorption is reversible. Heating the hydrogel with captured CO_2 causes carbamate cleavage regenerating the CO_2 molecule and restoring its basic character to nitrogen. This behavior has also been observed in polyvinylamine hydrogels that are used as absorbent films [5], in copolymers of poly(N-cyclopropyl acrylamide-co-L-arginine methyl ester acrylamide) [6], and polyamidoamine [7]. Therefore, these materials do not have useful catalytic properties to carry out CO_2 reduction and they are only used as absorbents. Therefore, it is necessary to use a metal or a metal oxide as a catalyst.

Metal oxides are easy to prepare and stable in water, so they could be good catalysts. However, their hydrophilic nature favors the evolution of H_2 over the reduction of CO_2. The general mechanism of electrocatalysis CO_2 reduction is illustrated in Figure 9.2b. It consists of the transfer of electrons from the electrode to the catalyst to produce the catalyst in its reduced form [8]. This species is oxidized in the presence of CO_2 to produce an oxidized form of the catalyst and CO_2 reduction products. This mechanism can also occur in molecular catalysts – not only in metal oxides – and polymers have been designed for CO_2 reduction under this principle.

The rational design of applicable polymers to CO_2 reduction must consider the formation of porous structures and guarantee a certain hydrophobic environment around the catalytic centers where the reduction occurs to avoid the evolution of CO_2. For this, Leung et al. designed terpolymers synthesized from (1) a monomer with terpyridine groups (tpy) capable of complexing with cobalt, (2) an easily hydrolyzable diethylphosphonate monomer to confer a hydrophilic character to the material, and (3) a methacrylate with alkyl substituents of length variable that provide hydrophobic character. The hydrogels resulting from this terpolymer have catalytic activity capable of carrying out CO_2 reduction through the formation of mono(tpy) and bis(tpy) complexes resulting from the complexation of cobalt with 1 or 2 tpy monomeric units, respectively. The structure of these complexes is illustrated in Figure 9.2c. In this material, an equilibrium is established between mono(tpy) and bis(tpy) complexes that favor the electrocatalytic evolution of CO instead of the evolution of hydrogen [9].

The molecular perspective in the design of catalysts for CO_2 reduction has led to the obtaining of a wide variety of complexes with interesting electrocatalytic properties. In addition, some of these complexes also have photocatalytic properties applicable to CO_2 reduction; some examples are shown in Table 9.1. These complexes are soluble and stable in water and can be directly incorporated into matrices in the hydrogel state and applied to the processes of reduction. However, based on Leung's work, the complexes shown could be converted into vinyl derivatives capable of copolymerizing with other hydrophilic monomers to obtain hydrogels with the necessary catalytic properties to reduce CO_2. Hydrogels with catalytic properties can be remarkably interesting materials for CO_2 reduction. Due to their nature, they combine characteristics of homogeneous and heterogeneous systems since the hydrogel forms a homogeneous phase where CO_2 can diffuse and facilitates electron transport. At the same time, due to their reticulated structure,

TABLE 9.1
Complexes for Electrochemical CO_2 Reduction

Complex	Type	Medium	Products	Ref.
Mn(R-bpy)(CO)$_3$Br, R = H or CH_3	Electrocatalyst	H_2O/MeCN/H^+	CO	[10]
Mn(HOOC-bpy)(CO)$_3$Br	Electrocatalyst	H_2O at pH 9	CO	[11]
Ni pyridine-diimine complex $^{2+}$	Electrocatalyst	MeCN/H_2O (1:2)	CO	[12]
Ni cyclam $^{2+}$ (ClO$_4^-$)$_2$	Electrocatalyst	H_2O at pH 4.1	CO	[12]
Ni cyclam $^{2+}$	Photocatalyst	H_2O at pH 5	CO/HCOO$^-$	[12]
Cu(phen)$_2$	Electrocatalyst	$KHCO_3$ 0.1 M	CO, HCOOH, C_2H_4	[13]
Re(HOCH$_2$-bpy)(CO)$_3$(OH$_2$)	Electrocatalyst	H_2O at pH 6.9	CO	[14]

hydrogels form matrices with a porous structure with a high surface area that increases the number of reactive sites available to conduct the reduction.

The careful selection of the monomers used for the synthesis of the hydrogel can help to obtain the necessary pH conditions to conduct the reaction. As mentioned earlier, some complexes suitable for the electrocatalytic reduction of CO_2 are also suitable for photocatalyzed reduction. The mechanism by which photocatalysis occurs is shown in Figure 9.2d and requires the presence of two additional chemical species, the sensitizer, and the electron donor. The reaction begins when the sensitizer absorbs photons to produce an active species that, when interacting with the electron donor, generates ionized radical species. This species relaxes by transferring electrons to the catalyst to form a reduced structure of the catalyst, which is oxidized in the presence of CO_2, returning the catalyst to its oxidized form and generating CO_2 reduction products [8]. If the hydrogel where the catalyst is immobilized is transparent, it will not prevent the passage of light into the reaction medium, and therefore will not interfere with the development of the reaction.

The development of technologies for the reduction of CO_2 using hydrogels obtained from synthetic polymers is still in its initial stages and much remains to be investigated. However, the incorporation of said catalysts in polymeric materials in a hydrogel state offers a window of opportunity for the development of innovative technologies that help to mitigate the effects of CO_2 on the environment

9.3 Hydrogels Based on Natural Polymers for CO_2 Reduction

The increase in nursery gas emanations is evidence of genuine natural issues. Diminishing CO_2 outflows is a major challenge [15]. It has been assessed that fossil power will still constitute the major source of human activities for a long time. In this manner, it is imperative to capture CO_2 from the combustion gas of fossil power materials and diminish it, maintaining a strategic distance from its discharge into the air [16].

There are numerous strategies accessible for CO_2 capture, including partition into fluid assimilation, strong adsorption, layer division, and hydrogels for CO_2 decrease. To capture CO_2 from the combustion gas of fossil power, chemical assimilation is required because the substance of CO_2 within combustion gas is ordinarily low (5 vol.%~20 vol.%) [17]. The outflow of CO_2, which is one of the foremost reasons for worldwide warming when fossil fuel is burned, cannot be maintained at a strategic distance. Fossil-fueled power-production innovation plays a critical part in contributing to the outflow of nursery gas into the air. By lessening the outflow of CO_2 into the environment, and by changing to an elective control era with zero emissions, it is conceivable to anticipate future impacts. Carbon capture utilization and capacity (CCUS) strategies and advances are among the numerous ways to decrease CO_2 emanations [18].

The capture of CO_2 from blended streams is fundamental to numerous businesses, such as nursery gas outflow control (CO_2/N_2) and common gas overhauling (CO/CH_4), therefore, research has been conducted on CO_2 separation technology [19]. It is vitally important to create more energy-effective and cost-effective innovations for CO_2 capture and chemical reduction. Film gas partition has demonstrated successful partition innovation due to its low facile utilization, simple operation, and cheap upkeep. In expansion to layer composition, on the off chance that catalytic locales exist, CO_2 reduction can be accomplished [20]. For a long time, a parcel of common polymers has been utilized to manufacture CO_2-responsive shape memory hydrogels. Compared with conventional shape memory hydrogels, CO_2-responsive shape memory hydrogels have numerous focal points. For instance, after a few reversible cycles, the execution does not blur and there is no massing, which shows an extraordinary potential application [21].

The utilization of biopolymers can essentially diminish CO_2 emanations by up to 90% in comparison with the generation of customary plastics [22]. There are ordinary forms (e.g., in the cement industry) that are the source of around 5% of worldwide nursery gases. In this case, soils treated with hydrogels based on polysaccharides such as xanthan gum and gellan gum have made strides in their mechanical resistance and capacity for the adsorption of CO_2 and other nursery gases [23]. This sort of elective can be utilized by joining a catalyst that diminishes CO_2 to anticipate its outflow into the climate. For the most part, biopolymers are recognized as routinely manufactured polymers because of savings, low environmental effects, non-toxicity, and non-secondary contamination. Figure 9.3 shows a schematic outline of CO_2 decrease using hydrogels based on biopolymers.

The high presence of basic nitrogen in proteins, amino acids, and polysaccharides such as chitosan can be used to stimulate CO_2 reduction once it is absorbed by the hydrogel under specific radiation conditions and catalytic sites of action. On the other hand, the abundance of hydroxyl groups in polysaccharide chains could induce a reduction potential that stimulates the transformation of CO_2 to other species that do not represent environmental consequences. Creating productive and reasonable CO_2 capture may be a noteworthy way to decrease carbon emissions [24]. In this approach, the plan of biopolymer-based hydrogels speaks to a promising line of inquiry about creating maintainable materials. A few studies have detailed the utilization of hydrogels for CO_2 diminishment, for example, chitosan can be utilized as amino-functionalized structures for CO_2 capture and diminishment. Numerous mechanical forms might diminish their emissions using these frameworks. Moreover, there are numerous other choices where chitosan can be utilized to decrease nursery gas outflows [25].

Primo et al. detailed the era of carbon circles from normal biopolymers (alginate and chitosan), which were effectively synthesized by warm treatment between 400 and 800°C in dormant air. All the tests, counting the untreated common biopolymers, as well as the carbon materials, displayed an exceptional

Hydrogels for CO_2 Reduction

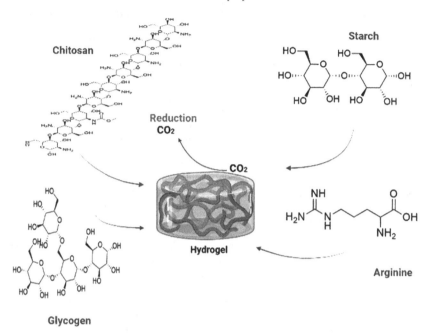

FIGURE 9.3
Schematic illustration of CO_2 adsorption and/or reduction using hydrogel-based natural polymers.

CO_2-adsorption capacity. The test that showed the most noteworthy adsorption capacity was the carbonization of alginate at 800°C and ensuing treatment with KOH at 800°C. This material showed a particular surface range of 765 m²/g, a particular micropore volume of 0.367 cm³/g, an ultra-micropore volume of 0.185 cm³/g, a normal ultra-micropore estimate of 0.7 nm, and a CO_2-adsorption capacity of 5 mmol/g [26]. This sort of carbon circle may be coupled inside hydrogels based on common polymers to amplify their adsorption execution and following CO_2 diminishing, on the off chance that a metallic catalyst can be utilized. Wang and S. Okubayashi showed how a novel polyethyleneimine-crosslinked cellulose (PCC) hydrogel sorbent was arranged by sol-gel preparation, hydrolysis response, and crosslinking response. The particular surface range of permeable PCC hydrogel was still held at 234.2 m²/g when the substance of nitrogen was 17.4 wt.%. The CO_2 adsorption capacity of the PCC matrix was 2.31 mmol/g at 25°C beneath the immaculate dry CO_2 environment. Pseudo-second arrangement demonstrated that it is appropriate to anticipate the CO_2 adsorption behaviors of PCC hydrogels at diverse temperatures [27]. Aerogels (hydrogels dried by the activity of a particular gas) might be optimized for CO_2 reduction by consolidating a catalytic dynamic location. Table 9.2 shows hydrogels for the adsorption and/or lessening of CO_2.

TABLE 9.2

Hydrogels Prepared with Natural Polymers for the Adsorption and/or Reduction of CO_2

Chemical composition	Type of hydrogel	Main characteristics	Ref.
Chitosan	A novel chitosan/lithium sulfonate hydrogel synthesized by electron radiation.	As a result of the permeable structure and nearness of CO_2-philic amino and lithium sulfonate, the hydrogel can capture CO_2 effectively (67.9 mg g^{-1} at 298 K, 0.1 MPa). Since the material has physical and chemical steadiness, down-to-earth applications in capturing CO_2 are conceivable in a vent gas environment.	[24]
Chitosan	Hydrogel layer with carboxymethyl chitosan and carbon nanotubes.	Great CO_2 selectivity and penetrability with the potential diminishment of adsorbed gas.	[28]
Chitosan	An arrangement of hydrogel films from chitosan (CS)/poly ether-block-amide (Pebax) mixes were arranged and utilized for CO_2 division.	These materials have great CO_2 porousness and moderate soundness. CO_2 porousness essentially expands with the increase of Pebax substance. When the mass proportion of CS to Pebax was 1:1, the film showed the most elevated CO_2 porousness (2884 Barrer) and a direct division factor (23.2 for CO_2/CH_4 and 65.3 for CO_2/N_2). These hydrogels might be combined with complexes for the lessening of adsorbed CO_2.	[29]
Chitosan	Chitosan hydrogels; this centers on the improvement of a modern adsorption technique by CO_2-activated chitosan.	High-temperature adsorption gives an upgrade of adsorption capacities and expulsion efficiencies of the color by the carbamate crosslinking of chitosan with CO_2. Considered a novel elective for the chemical change of CO_2.	[30]
Arginine-chitosan-graphene oxide	A chitosan-based hydrogel utilized as a CO_2 adsorbent and its CO_2 gas adsorption capacity was measured utilizing the TGA strategy on its strong surface.	CO_2 gas adsorption rose to 24.15 wt.% (5.48 mmol g^{-1}) at 35°C for 35 min. Chitosan-graphene oxide and arginine-based aerogels showed a much higher CO_2 capture capacity than that of unadulterated chitosan, as well as a decreased capacity. Chitosan-graphene oxide and arginine-based materials were utilized as catalysts for the inclusion of CO_2 into natural particles, showing the arrangement of unused natural particles related to decreasing CO_2 processes.	[31]
Starch	Starch graft copolymer (SGC) hydrogel.	For treating channels and helping with CO_2 huff-n-puff preparations and diminishing CO_2 emanations.	[32]

9.4 Hydrogels Based on Inorganic Components for CO_2 Reduction

There are many options to reduce CO_2 concentration in the environment – the choice depends on the physicochemical characteristics of every material. However, in the case of hydrogels based on inorganic components devoted to CO_2 reduction, aerogels (dry hydrogels) by physical adsorption or chemical reduction of CO_2 with electro-photocatalytic materials are an option. In this section, the general synthesis of hydrogels based on inorganic components, the adsorbent materials applied for CO_2 capture, and the inorganic materials used for CO_2 reduction into less toxic molecules will be explained [33]. Table 9.3 details some hydrogels based on inorganic components for CO_2 capture and reduction, emphasizing the type of reduction mechanism, material composition, surface functionalization to achieve this performance, and the main characteristics of the systems.

9.4.1 General Synthesis of Aerogels

The characteristics of CO_2 sorbents are great adsorption capacity, high selectivity for CO_2 adsorption, gentle conditions for recovery, versatile mechanical solidness and resistance, and fast adsorption energy. Within the case of aerogels, those constituted by carbon and silica particles can be valuable materials for CO_2 sequestration or lessening, particularly when they are altered with amines, carbon nitride (C_3N_4), and metal particles and/or metal oxides. By and large, the preparation for most silica and carbon aerogels incorporates three stages: gelation (arrangement of hydrogel), drying, and carbonization, where the conditions of every stage have an impact on the ultimate properties (Figure 9.4) [33].

9.4.1.1 Gelation Stage

Carbon aerogels were first arranged by polymerizing resorcinol with formaldehyde in soluble conditions. At that point, other monomers such as melamine formaldehyde, phenol-formaldehyde, cresol formaldehyde, phenol-furfural, and polymers like polyacrylonitriles, polystyrenes, and polyurethanes, were utilized as carbon feedstock. These days, other carbon aerogels based on graphene, graphene oxide, carbon nanotubes, and biomass have been developed [33].

The gelation step of carbon hydrogels incorporates hydrogel arrangement by polymerizing and/or crosslinking single antecedent particles. Resorcinol-formaldehyde hydrogel includes four steps: (1) in soluble conditions (e.g., Na_2CO_3, K_2CO_3, $Ca(OH)_2$), resorcinol misplaces protons and effortlessly responds with formaldehyde to make hydroxymethylen bunches (–CH_2OH);

TABLE 9.3

Hydrogels Based on Inorganic Components for the Capture and Reduction of CO_2

Type of reduction	Hydrogel composition	Surface functionalization	Adsorption of CO2 or reduced species	Ref.
Adsorption	Resorcinol/formaldehyde carbon	Activated by KOH	3.0 mmol/g	[33]
	Resorcinol/formaldehyde/silsesquioxane carbon	Amine	2.57 mmol/g	
	Carbon	Activated by KOH	4.90 mmol/g	
	Graphene oxide	Polyethylenimine	2.55 mmol/g	
	Graphene oxide	Carbon nitride	0.43 mmol/g	
	Silica (TEOS, precursor)	(3-aminopropyl) triethoxysilane	6.97 mmol/g humid 1.95 mmol/g dry	[34]
	Silsesquioxane (TEOS precursor)	Polyethylenimine	3.30 mmol/g dry	
	Silica (TEOS, polystyrene latex)	Tetraethylenepentamine	7.90 mmol/g humid	
	Sodium-based (TEOS, Na_2CO_3)	(3-aminopropyl) triethoxysilane	2.51 mmol/g humid	
Electro-photocatalysis	Graphitic carbon nitride (g-C_3N_4)	Amine	CH_4, CH_3OH	[35]
		UIO66	CO	
		N-TiO_2	CO	
		$Bi_4O_5I_2$	CO, CH_4, H_2, O_2	
		ZnO	CO, CH_4, CH_3OH, C_2H_5OH	
	Graphene	Iron-porphyrin	CO	[36]
	Reduced graphene oxide	Nanostructurerod-likeTiO_2	CH_4, CH_3OH	[37]
	Reduced graphene	Carbon dot-coupled $BiVO_4$	CO, CH_4	[38]

TEOS, tetraethyl orthosilicate.

(2) the polymerization of hydroxymethylen resorcinol through methylene (–CH_2) and ether–methylene (–CH_2OCH_2) bridge to make the clusters; (3) arrangement of 3D hydrogel through the accumulation and crosslinking of the clusters; and (4) maturing, which incorporates the development and fortifying of the hydrogel with the unfinished chemical responses [33].

Carbon aerogels based on graphene organize by physically or chemically associating graphene sheets with each other with crosslinking operators. Whereas carbon aerogels based on carbon nanotubes are crosslinked by van der Waals force, this complies with the characteristic electrical conductivity

Hydrogels for CO_2 Reduction

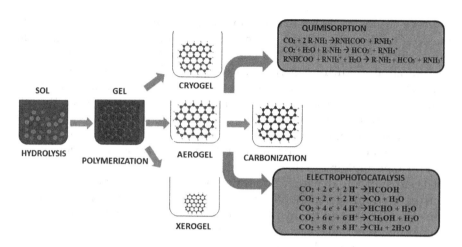

FIGURE 9.4
General synthesis and applications of hydrogels based on inorganic components for CO_2 reduction.

of carbon nanotubes. However, carbon nanotube aerogels have a destitute 3D system and poor mechanical quality and versatility. To overcome these impediments, a surfactant to improve the surface action and mechanical versatility of carbon nanotubes is included as a polymer to fortify the system [33].

In the case of silica aerogels, to begin with, the hydrogel is shaped by blending a metal alkoxide, as an antecedent fabric, with a dissolvable (water/alcohol) and an acidic and/or essential catalyst at room temperature. Within the sol-gel preparation, the forerunner is broken up and changed into a colloidal gel structure by a set of hydrolysis and polycondensation responses [39]. Sol–gel antecedents for silica aerogels incorporate tetramethoxysilane (TMOS), tetraethoxysilane (TEOS), monomeric alkylalkoxy silanes, bis trialkoxy alkyl or aryl silanes, and polyhedral silsesquioxanes, among others [40].

9.4.1.2 Drying

In this step, the solvent trapped in the hydrogel is removed to get a solid network (aerogel). In this case, drying is a critical stage that affects the properties and final structure of aerogels. The three most common methods for drying are supercritical drying, freeze drying, and ambient drying.

In supercritical drying, the solvent is removed under supercritical conditions (temperature and pressure), eliminating the surface tension, and the shrinkage and collapse of the structure are avoided, obtaining an aerogel with high porosity and pore volume. Some disadvantages are the high pressure required and the time, resulting in a high-cost technique that cannot be applied at a large scale [33].

Freeze drying includes the freezing and sublimation of solvent at low pressure to obtain a cryogel that maintains a similar porous structure to supercritical drying. The factors to control are solidifying rate, antecedent concentration, maturing time, and dissolvability. Thus, a long maturing time improves the mechanical quality of the aerogel, whereas a dissolvable with a MoO extinction coefficient and tall sublimation weight can decrease the collapse and shrinkage of the 3D structure. A fast-freezing rate leads to aerogels with a small pore measure and large surface region, driving structure collapse, at that point, solidifying drying is favored to arrange hydrogels with a visible shape applying to slow freezing [33].

The surrounding drying includes the dissolvable dissipation beneath the climatic environment and it is a low-cost method. Be that as it may, the structure can collapse, and the hydrogel tends to shrink which can lead to xerogels. At that point, to create aerogels with low thicknesses and a great pore structure, it is vital to diminish the dissolvable surface pressure by dissolvable substitution or surfactant expansion and fortifying the system structure to stand up to capillary pressure [33].

9.4.1.3 Carbonization

Carbonization is significant for aerogels based on polymer and biomass carbon but superfluous for aerogels based on graphene and carbon nanotubes. In this step, aerogels are warmed to temperatures over 600°C beneath an inactive climate (N_2 or Ar), where oxygen and hydrogen moieties are killed within the frame of gases, to create the carbon structure. In this step, macropores are diminished due to the shrinkage of the structure, whereas the mesopores and micropores expand, leading to an increment within the particular surface zone. The pore volume diminishes with temperature increments when it is within 700 to 900°C [33].

9.4.2 Aerogels for CO_2 Capture

The parameters that impact both the rate and extent/capacity of CO_2 adsorption are temperature, the fractional weight of CO_2, sorbent textural properties, thickness of the sorbent, and so on. Promising aerogels for CO_2 capture require low crude materials, low warming capacity, quick energy, high CO_2 adsorption capacity, and excellent CO_2 selectivity, as well as chemical and mechanical stability beneath impressive cycling (Figure 9.4) [41].

Inorganic aerogels are for the most part based on metal oxides as the antecedent, such as SiO_2, TiO_2, SnO_2, ZrO_2, V_2O_5, Y_2O_3, Al_2O_3, and Nb_2O_5 [41]. In expansion, lime-based sorbents and soluble base metal oxides have been connected in CO_2 adsorption. Calcium oxide and hydroxides respond with CO_2 to make thermodynamically steady carbonates, whereas K_2CO_3 and Na_2CO_3 interact with CO_2 and water to make hydrogen carbonates [42]. Among these materials, silica aerogels have an expansive surface range, great porosity volume, and low thickness [39]. Within the case of nanotubes, the adsorption happens in the tube, on the external surface, within the interstitial channels between carbon nanotubes, and in a groove between adjoining nanotubes. CO_2 adsorption on graphene, graphene derivates, and fullerenes has not been broadly examined [42]. The chemical surface of silica and carbon aerogels can be effectively functionalized by the sol-gel strategy. In this sense, amine-functionalized carbon and silica aerogels are exceptional materials for CO_2 capture. This complies with the chemical liking of amine moiety uncommonly within dampness, hence the amine response with CO_2 particles to deliver carbamic corrosive. This reaction is reversible, and immersed carbon aerogels can be recovered by warming up to 100°C [33, 43].

Physically sorbed amines are bigger and add sorption capacities for CO_2 and higher working temperatures, but they have destitute cyclic solidness due to dissipation of the amine amid warm recovery. Covalently fastened amines to silica (e.g., mono-, di-, and trialkylaminotrimethoxysilane) have cyclic soundness whereas the mono and di amino moieties offer excellent sorption/desorption energy because they use more pore space for gas trade [44]. Subsequently, atoms such as polyethyleneimine, alkanolamines, diethanolamine, monoethanolamine, and methyldiethanolamine are valuable to extend selectivity toward CO_2 adsorption [33, 43]. The amine forerunners diffuse into the hydrogel with capillary strength and reach the gel surface with the help of covalent bonds, Van der Waals divider bonds, π bonds, or hydrogen bonds [39]. Commercially, sorbents ought to minimize water sorption while maximizing CO_2 sorption due to warm recovery. Joining of amino-functionalized hydrophobic silica aerogels into pellets for fluidized bed CO_2-capture frameworks proposes that these sorbents will be used for commercial application in the future [44].

9.4.3 Aerogels for CO_2 Reduction

Electro- and photochemistry are appealing choices to diminish air CO_2 concentration, create important chemicals and fluids such as formic corrosive, and as utilization as a feedstock. The materials applied are 3D hydrogels with the shape of cathodes with tunable porosity, a large electrochemical surface region, and a 3D conductive pathway with high wettability [33].

Electron-hole sets over semiconductor materials are the initiator of photocatalytic CO_2 decrease. The semiconductor retains vitality upon illumination, in the event that this ingested vitality is bigger than its band crevices, electrons of the valence band are advanced to the conduction band, clearing out gaps within the valence band. The electron-hole sets can move along the fabric to reach the dynamic locales on the surface of the catalyst to perform oxidation responses, where electrons within the conduction band take a portion within the CO_2 decrease response, whereas the gaps within the valence band are included in water oxidation. The high number of electrons (two to eight) required for CO_2 reduction alongside a large recombination rate of electron-hole sets are the foremost overwhelming issues in photocatalytic CO_2 reduction (Figure 9.4) [45].

Metal oxide aerogels and carbon aerogels have been broadly examined and connected in electrocatalysts. However, carbon erosion and MoO conductivity of oxides are the foremost issues confronted by these materials in electrocatalytic applications. As 3D self-supported hydrogels with remarkable electronic conductivity, metal aerogels can maintain a strategic distance from catalyst structure annihilation caused by the decay of carbon at great potential and are broadly utilized as electrocatalyst materials [46].

Another approach is the potential of chemicals to change CO_2 without extra electric input into chemicals such as formic corrosive, formaldehyde, methane, and methanol [33]. In expansion, metal organic frameworks (MOFs) carry on as semiconductors due to the natural ligand that acts as a radio wire to capture photons with electrons being energized from the valence band to the conductive band and are exchanged to the metal hubs through a ligand-to-metal cluster charge move instrument, which is encouraged by the MOF's crystalline structure. Hence, MOF-based aerogels are promising materials that might show catalytic action due to their unsaturated metal particles or catalytically dynamic natural ligands [45].

On the other hand, carbon-based materials (e.g., graphite, graphene oxide, graphene, carbon nanotubes) can be chemically or physically combined with hydrogel lattices. Carbon-based materials exhibit high thickness, great quality, and high hardness. Their applications are CO_2 decrease, H_2 photoelectron era, batteries, the generation of clean vitality from the oxygen reduction response, the improvement of sensors, and materials for the treatment of wastewater and water [33]. Graphene oxide presents hydrophilic moieties (–Goodness, –C–O–C–, and –COOH) that advance the dispersibility of the graphene oxide in polymer frameworks. The gathering of graphene oxide sheets and polymers is made by electrostatic strength. Its toughness and solidness are clarified by coulombic, hydrogen holding, hydrophobic strengths, and p-p stacking of graphene oxide sheets. The anionic surface of graphene oxides permits its functionalization with cationic polymers with amine moieties such as polyethyleneimine [46]. In addition, carbon nanotubes possess thermal, mechanical, and electronic properties, and their tubular shape generates a high surface area that promotes the adsorption of pollutants such as CO_2 [46].

9.5 Conclusion

The use of hydrogels based on synthetic, natural polymers and inorganic components such as coordination complexes, metals, carbon derivatives, and silica have shown excellent performance in the removal and reduction capacity of carbon dioxide, generating value-added by-products and/or less environmentally harmful species. Currently, there is still the challenge of designing new hydrogel compositions that could be used on a large scale for the elimination of this greenhouse gas. Research in this field requires further growth to understand the interfacial phenomena that contribute to the reduction of CO_2 according to the chemical composition of hydrogels.

References

1. A. Bagger, W. Ju, A. S. Varela, P. Strasser and J. Rossmeisl, "Electrochemical CO_2 reduction: A classification problem," *ChemPhysChem*, 18(22), 3266–3273, 2017.
2. S. Zhang, Q. Fan, R. Xia and T. J. Meyer, "CO_2 reduction: From homogeneous to heterogeneous electrocatalysis," *Acc. Chem. Res.*, 53(1), 255–264, 2020.
3. R. J. Dixit and C. B. Majumder, "CO_2 capture and electro-conversion into valuable organic products: A batch and continuous study," *J. CO2 Util.*, 26, 80–92, 2018.
4. X. Xu, B. Pejcic, C. Heath and C. D. Wood, "Carbon capture with polyethylenimine hydrogel beads (PEI HBs)," *J. Mater. Chem. A*, 6(43), 21468–21474, 2018.
5. M. Yue, K. Imai, C. Yamashita, Y. Miura and Y. Hoshino, "Effects of hydrophobic modifications and phase transitions of polyvinylamine hydrogel films on reversible CO_2 capture behavior: Comparison between copolymer films and blend films for temperature responsive CO_2 absorption," *Macromol. Chem. Phys.*, 218(8), 1600570, 2017.
6. S. Lin, P. Schattling and P. Theato, "Thermo-and CO_2-responsive linear polymers and hydrogels as CO_2 capturing materials," *Sci. Adv. Mater.*, 7(5), 948–955, 2015.
7. H. Choi, S. Lee, S. U. Jeong, Y. K. Hong and S. Y. Kim, "Synthesis and CO_2 capture of porous hydrogel particles consisting of hyperbranched poly(amidoamine)s," *Gels*, 8(8), 1–9, 2022.
8. Y. Kuramochi, O. Ishitani and H. Ishida, "Reaction mechanisms of catalytic photochemical CO_2 reduction using Re(I) and Ru(II) complexes," *Coord. Chem. Rev.*, 373, 333–356, 2018.
9. J. J. Leung, J. A. Vigil, J. Warnan, E. Edwardes Moore and E. Reisner, "Rational design of polymers for selective CO_2 reduction catalysis," *Angew. Chemie*, 58(23), 7697–7701, 2019.
10. A. Sinopoli, N. T. La Porte, J. F. Martinez, M. R. Wasielewski and M. Sohail, "Manganese carbonyl complexes for CO_2 reduction," *Coord. Chem. Rev.*, 365, 60–74, 2018.
11. J. J. Walsh, G. Neri, C. L. Smith and A. J. Cowan, "Water-soluble manganese complex for selective electrocatalytic CO_2 reduction to CO," *Organometallics*, 38(6), 1224–1229, 2019.
12. J. W. Wang, W. J. Liu, D. C. Zhong and T. B. Lu, "Nickel complexes as molecular catalysts for water splitting and CO_2 reduction," *Coord. Chem. Rev.*, 378, 237–261, 2019.

13. J. Wang et al., "A water-soluble cu complex as molecular catalyst for electrocatalytic CO_2 reduction on graphene-based electrodes," *Adv. Energy Mater.*, 9(3), 1803151, 2019.
14. A. Nakada and O. Ishitani, "Selective electrocatalysis of a water-soluble rhenium(I) complex for CO_2 reduction using water as an electron donor," *ACS Catal.*, 8(1), 354–363, 2018.
15. M. E. Boot, J. C. Abanades, E. Anthony, M. Blunt, S. Brandani, N. Mac Dowell and J. Fernández, "Carbon capture and storage update," *Energ. Environ. Sci.* 7, 130–189, 2014.
16. W. L. Xu, H. J. Chen, C. Y. Wang, S. Liu, X. Y. Wan, H. L. Peng and K. Huang, "Chitin-derived fibrous carbon microspheres as support of polyamine for remarkable CO_2 capture," *GreenChE.* 3(3), 267–279, 2022.
17. F. Russo, F. Galiano, A. Lulianelli, A. Brasile and A. Figoli, "Biopolymers for sustainable membranes in CO_2 separation: A review," *Fuel Process. Technol.*, 213, 106643, 2021.
18. P. Madejsk, K. Chmiel, N. Subramanian and T. Ku's, "Methods and techniques for CO_2 capture: Review of potential solutions and applications in modern energy technologies," *Energies*, 15, 887, 2021.
19. N. Macdowell, N. Florin, A. Buchard, J. Hallett, A. Galindo, G. Jackson, C. Adjiman, C. Williams, N. Shah and P. Fennell, "An overview of CO_2 capture technologies," *Energy Environ. Sci.* 3, 1645–1669, 2010.
20. P. Luis and B. Van der Bruggen, "The role of membranes post combustion CO_2 capture," *Sci. Technol.* 3, 318–337, 2013.
21. N. He, X. Chen, J. Wen, Q. Cao, Y. Li and L. Wang, "Carbon dioxide and nitrogen-modulated shape transformation of chitosan-based composite nanogels," *ACS Omega*, 4(25), 21018–21026, 2019.
22. F. Barrillari and F. Chini, "Biopolymers: Sustainability for the automotive value-added chain," *ATZ Worldw* 122(11), 36–39, 2020.
23. I. Chang, T. P. Tran, J. Im and G. C. Choo, "Soil–Water characteristics of xanthan gum biopolymer containing soils," The 19th International Conference on Soil Mechanics and Geotechnical Engineering, 2017.
24. Z. Liu, R. Ma, W. Du, G. Yang and T. Chen, "Radiation-initiated high strength chitosan/lithium sulfonate double network hydrogel/aerogel with porosity and stability for efficient CO_2 capture," *RSC Adv.* 11(33), 20486–20497, 2021.
25. S. Maliki, G. Sharma, A. Kumar, M. M. Zamoramo, O. Moradi, J. Baselga, F. Staldler and A. Garcia, "Chitosan as a tool for sustainable development: A mini review," *Polymers* 14, 1475, 2022.
26. A. Primo, A. Forneli, A. Corma and H. García, "From biomass wastes to highly efficient CO_2 adsorbents: Graphitisation of chitosan and alginate biopolymers," *ChemSusChem* 5(11), 2207–2214, 2012.
27. C. Wang and S. Okubayashi, "Polyethyleneimine-crosslinked cellulose aerogel for combustion CO_2 capture," *Carbohydr. Polym.* 225, 115248, 2019.
28. R. Bargohain, N. Jain, B. Prasad, B. Mandal and B. Su, "Carboxymethyl chitosan/carbon nanotubes mixed matrix membranes for CO_2 separation," *React. Funct. Polym.* 104431, 2019.
29. Y. Liu, S. Yu, H. Wu, Y. Li, S. Wang, Z. Tian and Z. Jiang, "High permeability hydrogel membranes of chitosan/poly ether-block-amide blends for CO_2 separation," *J. Membr. Sci.* 1(8), 469, 2014.

30. H. Q. Le, Y. Sekiguchi, D. Ardiyanta and Y. Shimoyama, "CO_2 activated adsorption: A new approach to dye removal by chitosan hydrogel," *ACS Omega* 3, 14103–14110, 2018.
31. N. Hsan, P. Dutta, S. Kumar and J. Koh, "Arginine containing chitosan-graphene oxide aerogels for highly efficient carbon capture and fixation," *J. CO2 Util.* 59(8), 101958, 2022.
32. H. Hao, D. Yuan, J. Hou, W. Guo and H. Liu, "Using starch graft copolymer gel to assist the CO2 huff-n-puff process for enhanced oil recovery in a water channeling reservoir," *RSC Adv.* 12(31), 19990–20003, 2022.
33. G. Gan, X. Li, S. Fan, L. Wang, M. Qin, Z. Yin and G. Chen, "Carbon aerogels for environmental clean-up," *Eur. J. Inorg. Chem.* 3126–3141, 2019.
34. L. Keshavarz, M. R. Ghaani, J. M. Don MacElroy and N. J. English, "A comprehensive review on the application of aerogels in CO_2-adsorption: Materials and characterisation," *Chem. Eng. J.* 412, 128604, 2021.
35. Z. Sun, H. Wang, Z. Wu and L. Wang, "g-C3N4 based composite photocatalysts for photocatalytic CO_2 reduction," *Catal. Today* 300, 160–172, 2018.
36. J. Choi, J. Kim, P. Wagner, S. Gambhir, R. Jalili, S. Byun, S. Sayyar, Y. M. Lee, D. R. MacFarlane, G. G. Wallace and D. L. Officer, "Energy efficient electrochemical reduction of CO_2 to CO using a three-dimensional porphyrin/graphene hydrogel," *Energy Environ. Sci.* 12, 747–755, 2019.
37. S. Liu, T. Jiang, M. Fan, G. Tan, S. Cui and X. Shen, "Nanostructure rod-like TiO2-reduced graphene oxide composite aerogels for highly-efficient visible-light photocatalytic CO_2 reduction," *J. Alloys Compd.* 861, 158598, 2021.
38. C. Ma, Z. Xie, W. C. Seo, S. T. U. Din, J. Lee, Y. Kim, H. Jung and W. Yang, "Carbon dot-coupled BiVO4/reduced graphene hydrogel for significant enhancement of photocatalytic activity: Antibiotic degradation and CO_2 reduction," *Appl. Surf. Sci.* 565, 150564, 2021.
39. B. Yay and N. Gizli, "A review on silica aerogels for CO2 capture applications," *Pamukkale Univ. Muh. Bilim. Derg.* 25(7), 907–913, 2019.
40. N. A. D. Ho and C. P. Leo, "A review on the emerging applications of cellulose, cellulose derivatives and nanocellulose in carbon capture," *Environ. Res.* 197, 111100, 2021.
41. A.Verma, S. Thakur, G. Goel, J. Raj, V. K. Gupta, D. Roberts and V. K. Thakur, "Bio-based sustainable aerogels: New sensation in CO_2 capture," *Curr. Opin. Green Sustain. Chem.* 3, 100027, 2020.
42. J. E. Amonette and J. Matyáš, "Functionalized silica aerogels for gas-phase purification, sensing, and catalysis: A review," *Micropor. Mesopor. Mater.* 250, 100–119, 2017.
43. Y. Guo, J. Bae, Z. Fang, P. Li, F. Zhao and G. Yu, "Hydrogels and hydrogel-derived materials for energy and water sustainability," *Chem. Rev.* 120(15), 7642–7707, 2020.
44. I. I. Alkhatib, C. Garlisi, M. Pagliaro, K. Al-Ali and G. Palmisano, "Metal-organic frameworks for photocatalytic CO_2 reduction under visible radiation: A review of strategies and applications," *Catal. Today.* 340, 209–224, 2020.
45. R. Zhang and Y. Zhao, "Preparation and electrocatalysis application of pure metallic aerogel: A review" *Catalysts* 10, 1376, 2020.
46. A. G. B. Pereira, F. H. A. Rodrigues, A. T. Paulino, A. F. Martins and A. R. Fajardo, "Recent advances on composite hydrogels designed for the remediation of dyecontaminated water and wastewater: A review," *J. Clean. Prod.* 284, 124703, 2021.

10

Recent Advancements in Hydrogels for Electrocatalytic Activities

Allen Davis, Teddy Mageto, Felipe M. de Souza,
Anuj Kumar, and Ram K. Gupta

CONTENTS

10.1 Introduction 189
10.2 Synthesis of Hydrogels 192
10.3 Factors Affecting the Properties of Hydrogel 193
10.4 Hydrogel-based Electrocatalysts 197
    10.4.1 Hydrogels for Oxygen Reduction Reactions 197
    10.4.2 Hydrogels for Oxygen Evolution Reactions 200
    10.4.3 Hydrogels for Hydrogen Evolution Reactions 202
    10.4.4 Hydrogels for CO_2 Reductions Reactions 203
    10.4.5 Hydrogels as Multifunctional Electrocatalysts 205
10.5 Conclusion and Perspectives 207
References 208

10.1 Introduction

As the global population rapidly surpasses 8 billion, energy demands have risen to match. It is estimated that global energy use will rise by 50% in less than 30 years from the time of writing this chapter. As such, diversifying global energy infrastructure is critical to society's further advancement. After all, what good would a cell phone or tablet be without the electricity to power it? As it stands, there are currently three major categories of energy production. The first and most intensively used category is that of fossil fuels. Coal has been burned since before the industrial revolution. Oil and natural gas represent most of the world's current energy infrastructure. This is due to the high energy density of hydrocarbon materials. However, this ancient source of energy comes at a price. Rising oil and gas prices result from the growing scarcity of both the source material and refinery capacity. Meanwhile, accidents in oil refinement and extraction have taken a toll on both human lives and the environment. The burning of fossil fuels has accelerated global climate change toward a dangerous tipping point.

Consequently, weather extremes have increased in magnitude, requiring more energy to stabilize climate control in our cities and towns. As such, it falls to the two remaining categories of energy to return society from the brink of climate collapse. Nuclear energy was once touted as the energy source of the future, able to power cities with just mere kilograms of fissile material. However, social stigma has pushed this power source to the wayside. While it is currently responsible for 20% of America's energy production, widespread implementation of nuclear power is in limbo until public opinion changes. Thus, renewable energy remains, the modern power source of the future and today. Renewable energy represents a solid chunk of the market share, in no small part due to its flexibility and positive reputation. If the climate is sunny then install solar panels; if it is windy then erect a few wind turbines. While these methods are flexible in some ways, these sources quickly fail when access to sun and wind is restricted.

Electrochemical energy, on the other hand, can be used anywhere it is needed. Many forklifts use electrochemical fuel cells as a power source, due in large part to increased availability [1]. Meanwhile, research has been done on using electrical grids to electrochemically produce hydrogen with excess power [2]. You as the reader likely have an electrochemical energy source in your pocket, as the most common use of electrochemical energy storage is in the form of batteries. While batteries are the most well-known electrochemical device, many other forms of electrochemical energy sources exist. Pseudocapacitors, for example, act as a middle ground between chemical batteries and purely electric capacitors, storing some charge electrically and the rest chemically [3]. Other devices have been designed to convert water into hydrogen and oxygen for later use as electrochemical reagents [4]. Electrochemistry is a field of chemistry that specializes in the electrochemical properties of certain chemicals. More specifically, this field deals with the oxidation and reduction potentials of materials, known colloquially as redox potentials. The advantage of these materials lies in their propensity for energy transfer, converting electrical energy to chemical energy and vice versa. While batteries are useful electrochemical devices, they can only store energy, not generate it. To produce energy through electrochemical reactions, one would need to employ the use of a fuel cell. Fuel cells encapsulate the concept of green energy, converting the chemical energy stored in hydrogen and oxygen to water and electrical energy. There are many varieties of fuel cells available, however, their basic concept is universal.

At the anode, a fuel source is oxidized, releasing electrons in the process. These electrons flow outward to power a device, then flow back into the cathode of the fuel cell. At the cathode, the ions released during oxidation are combined with oxygen in a reduction reaction, wherein the byproduct is most often water [5]. The entirety of the process and its reverse, electrolysis, are shown in Figure 10.1. To facilitate the redox reactions required for electrochemical energy generation, electrocatalysts are required. Electrocatalysts often vary in both form and function depending on their intended use.

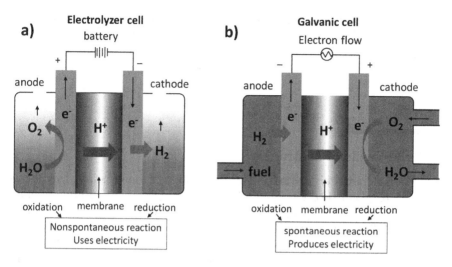

FIGURE 10.1
Visual graphic depicting the chemical process of (a) a water-splitting electrolyzer, and (b) a galvanic fuel cell. (Adapted from reference [6]. Copyright 2019 by the authors. Licensee MDPI, Basel, Switzerland. This article is an open access article distributed under the terms and conditions of the Creative Commons Attribution (CC BY) license.)

Metallic electrodes, for example, are an exceedingly popular choice for electrocatalysts. After all, the noble metal platinum sits as the best-known electrocatalyst for hydrogen evolution, [7]. Ruthenium oxide, meanwhile, operates as an exceedingly efficient electrocatalyst for oxygen evolution [8]. Other noble metals such as palladium and gold also demonstrate the capacity for electrochemistry, yet all are hampered by cost due to their scarcity. Many transition metals have shown promise for electrocatalytic activity yet need further treatment to show promising results [9]. Carbon-based electrocatalysts have long represented a popular framework for the study. Carbon nanomaterials often demonstrate a remarkably high surface area, with many carbon allotropes existing at the forefront of research. Organic polymers such as polyaniline are also attractive candidates due to their electroconductive nature. However, carbon materials serve mostly as a platform for electrocatalysts and must be doped with other compounds to be of much use. Metal-organic frameworks (MOFs) serve as the middle ground of these two extremes, incorporating metallic compounds into a carbon frame to improve their electrochemical properties. MOFs suffer greatly from aggregation, however, and require a substrate to see effective use. With all the above information taken into account, hydrogels and their derivatives are attractive candidates for electrocatalyst production. Hydrogels serve as an effective substrate for electrocatalytic materials, and in many cases, improve the overall efficiency of the reaction [10]. Hydrogels also imbue special physical properties that could not exist in standard electrode materials. For example, many solid-state electrodes, such as those in lithium-ion batteries, are inflexible. These storage devices are apt to break if too much force is applied, or worse, explode [11]. Hydrogel electrodes, meanwhile, are not only flexible but are known to have self-healing properties as well [12].

10.2 Synthesis of Hydrogels

Hydrogels derive their name from a combination of their chemical synthesis and structural constitution. The IUPAC defines gel as a "Nonfluid colloidal network or polymer network that is expanded throughout its whole volume by a fluid" [10]. The term gel was coined from the word gelatin, which in turn relates to the Latin root word *gelare*, meaning "to freeze". It should be noted, however, that this suffix is considered a misnomer, as gelation chemistry relates to the formation of crosslinked bonds rather than a temperature-fueled reduction in molecular motion. Meanwhile, the prefix hydro is about the fact that the gel's fluid medium is primarily composed of water. This holds for other types of gel as well, organogels have an organic compound as their dispersion medium, while aerogels maintain a gaseous medium after synthesis. Furthermore, gels that are dried at room temperature often collapse, forming a substance known as xerogel. The physical structure of gels is often derived from either organic precursors, organosilicon precursors, or entirely silicon-based precursors [13]. The gel precursor selected is often responsible for most of the gel's structural properties after solidification. Knowledge of a gel's composition is useless without further knowledge of its synthesis techniques. Gelation chemistry, also known as sol-gel chemistry, begins by dispersing multifunctional colloid particles within a given medium. These particles can vary from a mix of monomers to short-length polymers, and interact to form a semirigid structure known as a sol. As these particles interact within the sol, they combine to form ever-expanding crosslinked chains. This interaction is commonly known as percolation, and as percolation approaches critical mass, the material reaches its gel point. A material's gel point is often described as the point at which the polymer network expands throughout the whole of the system. While this is true, it should be noted that intramolecular reactions can still occur in a process known as aging. Aging can have many effects on the gel in question, causing shrinkage, increasing strength, and in extreme circumstances, causing cracking within the final product [13]. This process is a natural consequence of the crosslinking reactions used to form the gel. While there are a few ways to subvert this process, overall, aging is inevitable [14].

Crosslinking in gelation chemistry falls under two broad domains, chemical and physical. Chemical crosslinking involves the direct formation of covalent bonds between the different molecules in the gel. How these bonds are formed is defined by the polymerization process deployed. Chain-growth polymerization involves repeating units of the same monomer, often involving an initiator to kickstart the reaction. Alternatively, step-growth reactions make use of alternating reactions to fuel their growth, employing monomers with different functional groups. Some methods make use of condensation reactions, creating water as a byproduct of the overall reaction. Click chemistry represents a more modern approach to crosslinking, employing near-instantaneous reactions to create strong bonds in the sample [15]. Physical crosslinking, meanwhile, relies on the substantial interactions between one or more polymer chains. This process is generalized into two forms of entanglement: topological and cohesional. Topological entanglement is rather straightforward, molecules have mass and take up space, and as such, polymer chains can physically interact and intertwine. Visually, this process is likened to the act of wired earbuds tangling in your pocket. Cohesional entanglement is more complicated, having more to do with the electrostatic interactions of the material in question. Ionic bonding, hydrogen bonding, and π-π stacking are but a few examples of the electrostatic interactions responsible for cohesional crosslinking. This style of crosslinking is comparable to the formation of lamella and spherulites in linear polymer chains. It should also be noted that these physical interactions often relate to the functional properties of the gel itself, further influencing the functionality of the gel in question.

There are many reasons to synthesize hydrogel composites, alongside a variety of ways to synthesize these composites. Standard hydrogels such as gelatin are often contiguous, with little variance in the structure aside from its amorphous nature. However, inclusions from metallic ion clusters or metal-organic frameworks diversify the structure, enhancing certain properties. Other composites can be described as hydrogel sandwiches, layering different gels that impart different functions such as adhesion or flexibility. Some composites have a separate compound distributed across their surface, coating the gel with the material of choice. One interesting form of gel even consists of a hydrogel crosslinked throughout the pores of another – these are known as interpenetrating networks [15]. The sky is the limit when it comes to synthesizing hydrogel composites, with many methods at a chemist's disposal. However, when synthesizing hydrogels, the most important factor to consider is the function of the gel, and what properties exist to affect that function.

10.3 Factors Affecting the Properties of Hydrogel

The physiochemical structure of a hydrogel is one of the most important aspects to consider when exploring its applications. The physical structure of the gel influences flexibility, toughness, and strength. Concurrently, the chemical structure can dictate reactivity and electrochemistry, and a myriad of other factors. Nevertheless, the word "physiochemical" was purposely selected because these factors are intrinsically linked. The chemical composition of the gel precursors influences the formation of the gel, which in turn influences the function of the gel. For example, one team synthesized a strong, flexible, regenerative hydrogel for use as a biocompatible pressure sensor [16]. They achieved this by integrating reversible ionic reactions into the formulation of the gel, thus allowing fractured pieces to meld together. Meanwhile, a separate group synthesized a MOF-based hydrogel using similar methods [17]. This MOF-hydrogel exhibited anti-biofouling properties and takes advantage of the binding ability of MOFs to extract uranium from ocean water. In this section, the physiochemical factors affecting hydrogels will be further explored, demonstrating the adaptability common in hydrogels.

As mentioned previously, ionic interactions can play a large part in the properties and formation of a hydrogel [16]. It is well known that ionic bond strength is much weaker than covalent bond strength. However, in the field of materials, certain weaknesses can prove to be a strength all on their own. Overall, there are three major benefits to be explored when discussing ionic bonding in hydrogels: flexibility, self-healing capability, and conductivity. Interestingly, the first two properties to be discussed are products of the same intramolecular interactions. Within the overall polymer network, ionic sites link together to form ionic bonds. When stress is applied to the material, these ionic sites can slide by each other, breaking and reforming these loose bonds in rapid succession. When this stress is relieved, the bonds slide back near their original positions, allowing for high deformation recovery. Were this network entirely covalent, many bonds would irreversibly break, lowering the mechanical properties of the gel over time. This process is once again mirrored in regenerative materials. When an ionic hydrogel is cleaved in two and then pressed together, the ionic sites can reform with ease. While the strength recovered after is often weaker than the virgin gel, such properties are desirable in certain use cases. Ionic conductivity represents a different approach to the use of ionic compounds in a hydrogel. In such a gel, ions are used as charge carriers and are either incorporated into the structure itself or allowed to flow freely throughout the network. In the latter case, the gel acts as a substrate for the ions and is designed to ferry ions from one end of the electrolyte to the other. In this one study, a flexible, ionic hydrogel demonstrated its capability as a power source for wearable electronic devices [18].

Covalent bonding is essential to overall gel formation, forming the majority of a given gel's structure. Covalent bonds are present in every stage of synthesis, from monomer to hydrogel. As such, these bonds are responsible for the elasticity or rigidity of a gel depending on the context. Elasticity is a greatly desired property when considering hydrogels. Solid and liquid-state materials often lack elasticity and flexibility, while hydrogels exhibit these properties in spades. In one study, an ultra-tough and flexible hydrogel film was developed to mimic the properties of skin [19]. This gel could resist 1000% strain, exhibiting great toughness, and flexibility. As seen in Figure 10.2, this material closely mimicked human skin, while demonstrating its capacity as an electrode substrate. A separate team took a different approach, synthesizing a rigid, bio-based hydrogel that could potentially serve as a bio-scaffold for tissue regeneration [20]. As shown in Figure 10.3, the polarity of the gel material greatly affected the resultant structure of the material. In addition to flexibility, adhesion is a desirable property to consider in hydrogel synthesis. In one publication, a bio-based hydrogel was developed that could adhere to a wide variety of surfaces, including some surfaces underwater [21]. This property was attributed to the great abundance of hydroxyl and aromatic groups contained within the gel's structure. One of the most important factors when considering hydrogel factors is porosity. The porosity of the gel can be influenced by monomer size and crosslink density. The primary application of pore size in gels is the increase in surface area that it coincides with. An increase in surface area allows more room for chemical reactions, which can be advantageous for certain reactions. Meanwhile, pore size allows for size exclusion in other cases, allowing the selective flow of ions and other small molecules and gases.

While the physical properties of a hydrogel are very important, this point is moot if they fail to demonstrate abilities conducive to their application. In terms of this chapter, hydrogels are desired for their use in electrocatalytic devices. In such a case, the structures present in a fuel cell represent a

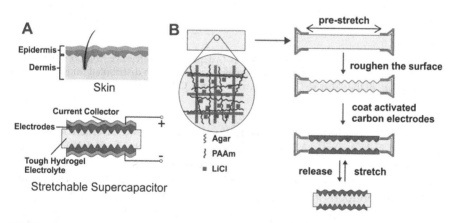

FIGURE 10.2
(a) Schematic illustration of the crosssectional view of the skin structure and the skin-inspired design of a stretchable supercapacitor based on the surface-microstructured tough hydrogel electrolyte. (b) Schematic illustration of the agar/polyacrylamide/LiCl (AG/PAAm/LiCl) tough hydrogel electrolyte for the fabrication of stretchable supercapacitor utilizing the pre-stretching and release approach. (Adapted with permission [19]. Copyright 2019, American Chemical Society.)

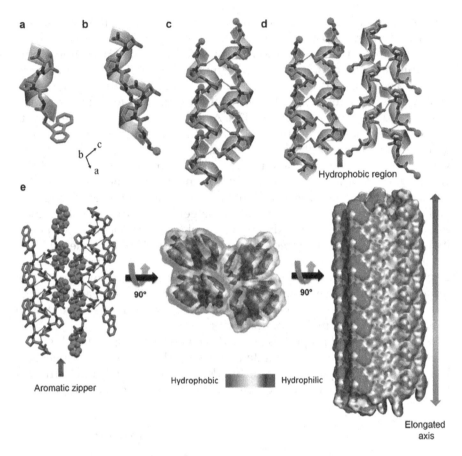

FIGURE 10.3
Single-crystal structure of Fmoc-Gly-Pro-Hyp. (Adapted from reference [20]. Copyright 2020, American Chemical Society. This article is an open access article distributed under the terms and conditions of the Creative Commons Attribution (CC BY) license.)

baseline for consideration. A basic fuel cell is composed of electrodes separated by an electrolyte. As such, when considering a hydrogel for electrode purposes, the material must contain redox-active sites where electrochemical reactions can occur. As far as both the electrodes and the electrolyte are concerned, conductivity is important. The electrodes of the material require excellent electron conductivity, while the electrolyte must be designed to facilitate ion flow. The different parts of the fuel cell can be further broken down depending on need. For example, hydrogels are often employed as binders due to their adhesion and flexibility. Meanwhile, the porosity of a hydrogel could be tuned to be selective on gas sizes, allowing it to act as a gas diffusion layer within the cell. Overall, there are many important properties to consider when selecting hydrogels for a specific application.

10.4 Hydrogel-based Electrocatalysts

Electrocatalysts are vital for the implementation of redox-based electrochemical storage and generation. Without electrocatalysts, advanced technologies such as smartphones and laptops would cease to exist. Fuel cells make use of advanced electrocatalysts to generate energy and water from hydrogen. Meanwhile, electrolysis generators use electricity to cleave water into hydrogen and oxygen. The reversibility of these reactions allows for adaptive, and more importantly, clean energy storage. Electrocatalysts are also important for the process of CO_2 reduction, wherein CO_2 waste is converted into synthetically useful chemicals.

10.4.1 Hydrogels for Oxygen Reduction Reactions

The oxygen reduction reaction (ORR) persists as an important half-reaction for the application of fuel cell technology. While electrolysis focuses on the splitting of water into its diatomic substituents, fuel cells are designed to transform the chemical energy stored in a dihydrogen bond into chemical energy. Oxygen reduction exists as the terminus of the overall reaction, combining hydrogen and oxygen to produce water as the final product of the reaction. The primary benefit of fuel cell technology is its potential for cyclic emission reduction. Purely green hydrogen is produced with H_2 and O_2 as its only gas emissions, fuel cells use these emissions to create water, which can then be used again for further hydrogen production. As such, the primary challenges facing fuel cell deployment are hydrogen production, material costs, and overall durability/reliability. Hydrogel applications for the first issue are discussed later in this chapter, however, the latter two remain the primary focus of this section. The high cost of fuel cells circles back to the use of platinum as a common electrocatalyst. As this problem was discussed in detail previously, it does not need to be elaborated here. Nevertheless, hydrogels and their composites demonstrate great potential to lower deployment costs by substituting platinum with a transition metal or carbon-based materials instead, as shown in Figure 10.4. Durability, meanwhile, is a concern affecting all fuel cells, not just those containing platinum. Redox reactions often occur in corrosive environments and at elevated temperatures. As such, materials must be designed with these factors in mind. One final consideration is that unlike many of the other half reactions discussed in this chapter, oxygen reduction is often measured using half-wave potential. This is because the rate and order of the half-reaction are more important when considering ORR efficiency. As a frame of reference, Pt/C generally has a half-wave potential of 0.84 V ± 0.03 V [22].

The literature on oxygen-reductive hydrogels is vast, with many excellent examples to choose from. One group synthesized a MOF containing polypyrrole hydrogel that demonstrated a unique crosslinking structure [23].

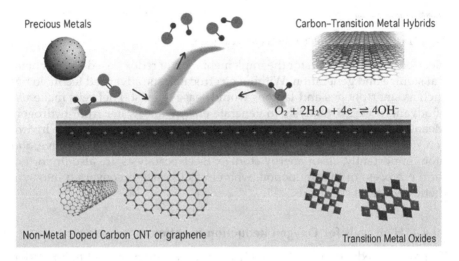

FIGURE 10.4
Demonstration of the ORR mechanism. (Adapted with permission [20]. Copyright 2015, American Chemical Society.)

Polypyrrole is a linear conductive polymer not known for its crosslinking potential. However, the inclusion of sulfate-active MOFs allowed for the ionic crosslinking of the sample to form a conductive hydrogel. Due to the crosslinking method used in this experiment, the resulting hydrogel demonstrated a unique architecture. While most gels form a sponge-like network, this material demonstrated a mycelial-looking network, as seen in Figure 10.5 [23]. The MOF used in this project is known as NiPcTs, a heavily aromatic compound with liquid crystal properties. While this study offered no $E_{1/2}$ data, the material was noted to have significantly improved stability and lower cost when compared with Pt/C. A similar study used cobalt phthalocyanine (CoPc) as a ligand binding site and crosslinker for the creation of electrocatalytic gel-like composites [24].

The creation of these electrocatalysts consisted of three primary steps. In the first step, CoPc was crosslinked with chitosan to produce an extremely porous hydrogel. The gel was further mixed with graphene oxide to further improve the material's electrocatalytic properties. Finally, the gel precursor was annealed at a high temperature to form the resultant electrocatalyst. The limiting current half-wave potential observed in this experiment was −223 mV, very similar to the observed $E_{1/2}$ of the platinum reference (−193 mV). The stability of the material was measured to stabilize at around 80% after continuous use cycles. Another study synthesized polyacrylamide (PAM) hydrogels to serve as a framework and dopant for ORR active electrode materials [25]. The hydrogel itself was suffused with iron and cobalt ions to further increase the compound's electrocatalytic activity. The

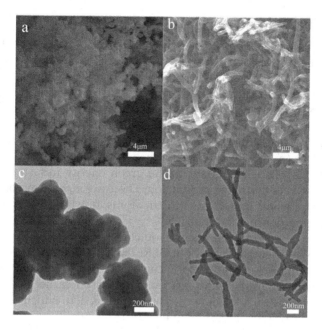

FIGURE 10.5
SEM images of (a) pristine PPy and (b) NiPcTs/Co/PPy hydrogel; (c) TEM images of pristine PPy and (d) NiPcTs/PPy hydrogel. (Adapted with permission [23]. Copyright 2020, Elsevier.)

metal-suffused gel was also mixed with melamine and phytic acid to further dope the compound. The resultant gel was dried and calcined to activate the material. All the materials selected above were chosen to maximize elemental doping in the finished product. PAM and melamine contributed nitrogen to the electrocatalyst, while phytic acid imbued the material with phosphorous. As a result, the electrocatalytic properties of the hydrogel-derived materials rivaled those of Pt/C, demonstrating an $E_{1/2}$ of 0.85 V and a Tafel slope of 108.69 mV/dec. As for stability, the composite retained 91.54% stability after 15 hours of continuous use, which was much superior to the 71.43% retention observed with Pt/C. In one final project, nitrogen-doped graphene oxide hydrogels were synthesized using the hydrothermal method [26]. While many electrocatalysts seek to cut costs by reducing metal inoculation, this research went one step further by eliminating metal from the material entirely. The initial solution was composed of the prior mentioned graphene oxide mixed with melamine. The melamine in this experiment served to crosslink the graphene oxide sheets and serve as a nitrogen source after annealing. Where the electrocatalytic potential is concerned, the material demonstrated a half-wave potential of 0.6 V. While stability tests were seemingly not performed on this material, it represents a potential future for metal-free electrocatalysts.

10.4.2 Hydrogels for Oxygen Evolution Reactions

Oxygen evolution reactions (OER) are vital for life as we know it. Plants and algae employ this reaction during photosynthesis, generating diatomic oxygen as a byproduct. It is this reaction that caused the Great Oxygen Event billions of years ago, and it is this reaction that will bring humanity into a new age of energy. Insofar as energy is concerned, the OER half-reaction is but one part of the water splitting whole, with hydrogen evolution reactions (HER) representing the other half. Overall, this reaction is viewed as $2H_2O(l) \rightarrow 2\ H_2(g) + O_2(g)$, with diatomic hydrogen being the desired product. Water catalysis is important to the green energy industry as it is the cleanest method of hydrogen production. There are many ways one could go about splitting water into its constituent atoms, however, electrolysis exists as the most popular method thus far.

With hydrogen being the desired outcome of this reaction, one could rightly ask why scientists care about OER. This line of thinking leads back to the duality of half-reactions and the concept of overpotentials. Water splitting is an example of the redox reactions discussed earlier in this chapter. The scheme above fails to show the complete reaction, which involves both ions in the electrolyte alongside an electron transfer. These reactions must resolve concurrently, and as such are only as fast as the slowest reaction. Oxygen evolution is the slower of the two reactions due to its sluggish reaction kinetics, so improving the efficiency of this half of the reaction improves the process. One way to improve the efficiency of an electrochemical reaction is to lower the activation overpotential of the reaction. All electrochemical reactions have a minimum required energy input to proceed – this concept is known as activation energy. However, these reactions often require more energy than the minimum input, the difference between the actual and theoretical input is known as the overpotential of the reaction.

Hydrogels have been found to serve as both an effective electrolyte and electrode for oxygen evolution reactions. One study of note designed a conductive hydrogel using polyaniline crosslinked with metal-coordinated phytic acid active sites [27]. Polyaniline is an attractive polymer for electrocatalysts in general due to the conductivity afforded by its impressive capacity for resonance. Phytic acid meanwhile is a naturally derived compound often found in the hulls and shells of grain and seed. Phytic acid's unique, phosphate-rich, cyclic structure allows for the compound to coordinate with metallic ions, with this overall process shown in Figure 10.6. The metallic species explored in this study were iron and cobalt, selected due to their economic viability and improved OER potential as compared with other transition metals. The hydrogel developed in this experiment demonstrated excellent charge transfer and stability, offering low resistance values between 23 and 36 Ω at 350 mV with little change in efficacy after 24-hour cycling.

A similar study used naturally sourced carrageenan as a precursor hydrogel for an electrocatalytic aerogel [28]. Carrageenan is a natural polymer

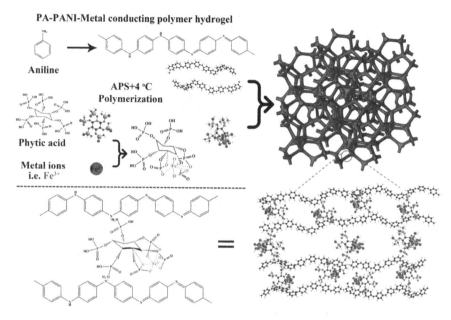

FIGURE 10.6
Schematic illustration of the synthesis of PANI–PA–metal (Hgel-M$_x$) hydrogels for electrocatalysis of OER. (Adapted with permission [27]. Copyright 2022, American Chemical Society.)

derived from some species of seaweed and was selected for two important properties. The first property of note was that I-carrageenan, a certain conformation of carrageenan, polymerizes in a double-helical shape. This double-helical shape allows for increased crosslinking within the gel. The second important property of carrageenan is the presence of sulfate groups along the backbone of the polymer. Light sulfur doping has previously been shown to increase OER selectivity, while the negative charge demonstrated by the sulfate group is an attractive site for cations [29]. To improve the electrocatalytic properties of the gel, the transition metal cations Co^{2+} and Ni^{2+} were dispersed in the gel, forming a CoNi MOF with the prior discussed sulfate groups. The resulting hydrogel was finally pyrolyzed to form the desired electrocatalyst. As far as electrocatalytic activity is concerned, the gel electrocatalyst demonstrated a low overpotential of 337 mV at 10 mA/cm² and a small Tafel slope of 77 mV/dec. These results are similar to that of RuO_2, a catalytic benchmark with an overpotential measured around 292 mV at 10 mA/cm² and a Tafel slope of 73 mV/dec in this study. Additionally, these electrodes were used to create a zinc-air battery that was powerful enough to charge a fan, as shown in Figure 10.7. Overall, this experiment suggests that biologically sourced materials could function as practical alternatives to noble metal-sourced electrocatalysts. In a similar vein, one group synthesized an extremely efficient OER catalyst using the bacteria *Geobacter*

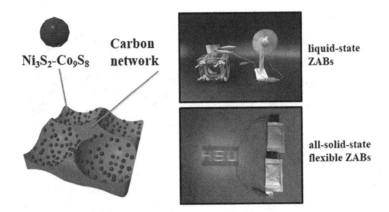

FIGURE 10.7
Schematics of the synthesized (Ni_3S_2–Co_9S_8/NCAs) material and the applications of zinc-air batteries. (Adapted with permission [28]. Copyright 2022, American Chemical Society.)

sulfurreducens [30]. *G. sulfurreducens* are a unique species of bacteria that can reduce compounds like graphene oxide as a part of their respiration process. Additionally, this bacterium and many others produce polymer films that allow them to stick to each other [31]. As such, cultures of these bacteria were incubated alongside lab-produced graphene oxide to produce reduced graphene oxide hydrogels. These hydrogels were subsequently dried and made into an electrode for further testing. Overall, it was discovered that the electrodes demonstrated an overpotential of 270 mV at 10 mA/cm^2, with a Tafel slope of 43 mV/dec.

10.4.3 Hydrogels for Hydrogen Evolution Reactions

As mentioned earlier, hydrogen is the desired product of the overall water-splitting reaction. This hydrogen can be used for energy storage [2], chemical synthesis [32], and, of course, energy generation [33]. In this aspect, hydrogels excel as a photocatalyst for hydrogen evolution reactions (HER). Indeed, much work has been done in this field, with many projects demonstrating desirable results. However, as far as hydrogel electrocatalysts for HER are concerned, the work done is scarce. While there are many examples to be found that incorporate hydrogels into electrocatalytic hydrogen evolution, they face stark issues. For example, many articles use hydrogels and their derivatives as a substrate for noble metal catalysts [34–36]. While such studies have merit, they do little to address the issues of cost and sustainability with which hydrogels are prized for. Meanwhile, other studies on the matter are outdated and are likely either surpassed by modern science or found lacking. Nevertheless, there are some articles of interest on this topic tangentially related to hydrogels as an HER electrocatalyst.

One study used aerophobic hydrogels to improve hydrogen evolution by using the gel as a coating to prevent gas accumulation on the electrode [37]. Such an application is important as the gas buildup is responsible for electrocatalytic inefficiency, especially at higher overpotentials. Other studies have developed multifunctional electrocatalysts that could facilitate overall water splitting [38]. While some are apt to assume that a lack of data means hydrogels are not useful for HER, these individuals could not be further from the truth. Hydrogels represent a new horizon in the field of HER electrocatalysts. Due to the similarities and reversibility of electrochemical redox reactions, many of the methods listed elsewhere could be tweaked to fit these applications. After all, there is a reason that most HER-compatible hydrogels are found in bifunctional electrocatalysts.

10.4.4 Hydrogels for CO_2 Reductions Reactions

Carbon dioxide production has posed a persistent problem for society since the Industrial Age. Carbon dioxide acts as a pervasive greenhouse gas and is the primary cause of global warming. Many countries currently aim to reduce or eliminate the atmospheric proliferation of CO_2 through carbon sequestration. In this process, CO_2 gas is collected and separated from the overall emissions of the plant or factory. The collected CO_2 gas is then sequestered, either biologically or chemically, preventing further CO_2 propagation in the atmosphere. The electrochemical reduction of CO_2 represents an interesting path for carbon sequestration, as these emissions can be chemically altered to a more useful state. Depending on the reduction path selected, CO_2 can be reduced into carbon monoxide (CO), formic acid (HCOOH), methane (CH_4), a variety of alcohols (ROH), and many other chemical subspecies. All the chemicals mentioned above are of interest both industrially and commercially for a variety of applications. Subsequently, the prospect of enhancing the efficiency of electrochemical CO_2 reduction has attracted much attention. Improving the efficiency of this reaction is not without challenge, as slow kinetics, poor stability, and high overpotentials inhibit the commercial utilization of this procedure. One important metric to keep in mind is the concept of faradaic efficiency. This concept relates to the loss of energy in a reaction through side reactions and heat. While heat is an unavoidable byproduct in redox reactions, side reactions can be reduced or prevented through chemical alteration of the electrocatalyst, which is detrimental to the electrocatalyst's function. Due to these variables, processes with high faradaic efficiency are preferred, especially when the complexity of the reactions is considered.

Due to the organic nature of CO_2, many biological systems are uniquely adapted to interact with the molecule. Many enzymes found in nature are designed to directly interact with CO_2 to produce other compounds, such as the carbon fixation seen during photosynthesis. One study proposed the use of an enzyme derived from the bacterium *Clostridium ljungdahlii* for CO_2

electrocatalytic reduction [39]. This enzyme, known as formate dehydrogenase (FDH), is used in the anaerobic respiration of some bacteria cells. While an intensive view of the biological mechanism of CO_2 reduction is beyond the scope of this paper, it can be summarized as transforming CO_2 into formate by the stabilization of unfavorable intermediates, shown in Figure 10.8. This protein is known to require the use of an electron carrier to facilitate reduction, as well as a membrane to anchor and localize the reaction. Polyaniline hydrogels suit this function in both categories, acting as an electron-conducting substrate to help immobilize and activate the protein. Synthesis of the protein gel composite was rather straightforward, the enzyme itself was extracted from the bacteria designed to overproduce it. The enzyme solution was then drop-casted onto the PANI electrode, with glutaraldehyde used as a crosslinking agent. As far as the CO_2 reduction itself, optimal formate production occurred at −0.06 V. Formate evolved at 1.42 μmol/h.cm² at a faradaic efficiency of 92.7%. The team attributed this high turnover rate to the efficient delivery of electrons throughout the hydrogel network. Another study took a similar path, using two separate enzymes to incorporate CO_2 into larger macromolecules as a proof of concept [40]. To synthesize this hydrogel, a standard polyvinyl alcohol (PVA) hydrogel was modified to incorporate 2,2-viologen; 2,2-viologen serves a similar function to polyaniline in the prior example, acting as an electrochemically active brace that can immobilize the enzymes to the substrate. The two enzymes bound to the gel were ferredoxin $NADP^+$ reductase (FNR) and NADPH-dependent crotonyl-CoA. These enzymes together bio-electrocatalytically transform $NADP^+$ to NADPH when given a charge. This electrically produced NADPH eventually drives the reductive carboxylation of crotonyl-CoA, sequestering CO_2 within the

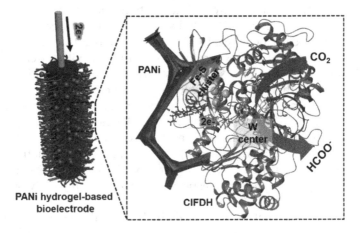

FIGURE 10.8
Schematic illustration of ClFDH-PANI electrode and direct electron transfer from conductive PANI hydrogel to ClFDH for electroenzymatic CO_2 conversion to formate. (Adapted with permission [39]. Copyright 2019, American Chemical Society.)

molecule. After much experimentation, the team was able to achieve an analytical yield of 57% with 91% faradaic efficiency. While smaller yields such as this may seem unimpressive at a glance, it is important to know that this reaction is renewable, as well as one of the first of its kind in the field.

While biologically sourced electroenzymatic CO_2 reduction is impressive, other teams have taken more traditional approaches to the matter. Polymer-metal complexes have long found use in CO_2 reduction in the form of both electrocatalysts and photocatalysts. One study dispersed said complexes within a polymer ion gel (PIG) to reduce CO_2 to CO with high faradaic efficiency [41]. PIGs are a variant of hydrogel that contains an ionic solution as their solvent. In the literature, rhenium chloride acted as the electrolyte suspended within a complex gel network. This complex gel was primarily composed of silicon, interspersed with organometallic complexes. The importance of the gel network was to bind the resulting complexes in such a way as to prevent unwanted HER side reactions. This goal proved successful, as the rhenium control sample that lacked the PIG support split its faradaic efficiency between H_2 production (44.2%) and CO production (51.9%). The [Re]-PIG complex meanwhile, was able to reach up to 98.7% efficiency with only trace hydrogen production.

10.4.5 Hydrogels as Multifunctional Electrocatalysts

Multifunctional electrocatalysts seek to unify the benefits of different electrocatalysts into a single material. As discussed previously, many hydrogels do well in one application, only to suffer in another application. A hydrogel for HER electrocatalysts may suffer when used as an OER electrocatalyst, or vice versa. A multifunctional electrocatalyst, meanwhile, would excel at both applications and potentially more. Bifunctional electrocatalysts are currently the most popular area of exploration, however, trifunctional electrocatalysts exist as well [42]. Such electrocatalysts are important because they help to mitigate overall production costs. One study designed a bifunctional, hydrogel-derived, electrocatalyst for HER-OER reactions with all the above in mind [43]. The hydrogel precursor synthesized was primarily composed of reduced graphene oxide crosslinked with alginate. Alginate is a naturally derived compound, often sourced from seaweed and other brown algae. Suspended within this gel were Ni-Fe nanoparticles to serve as the active sites for electrolysis. The bimetallic nature of the nanoparticles was crucial to the overall bifunctionality, as their differing coordination sites allowed for improved multi-directional activity. The final product was nitrogenated to further increase electrocatalytic activity within the active sites, with the entire process shown in Figure 10.9. OER and HER overpotentials were measured to be 298 mV at 50 mA/cm^2 and 94 mV at 10 mA/cm^2, respectively. Meanwhile, the Tafel slopes were 54 mV/dec and 90 mV/dec.

A similar green energy project was initiated using chitosan as the base polymer [38]. Chitosan is a natural polysaccharide derived from chitin, a

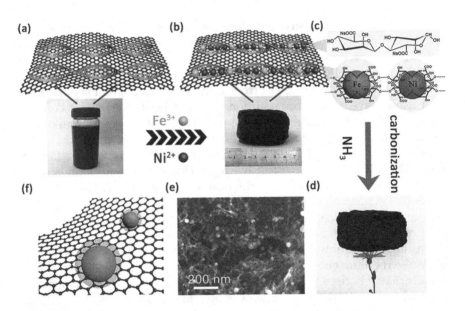

FIGURE 10.9
Schematic illustration for scalable synthesis of Ni_3FeN/r-GO catalysts derived from sustainable SA/r-GO hydrogels. (Adapted with permission [43]. Copyright 2018, American Chemical Society.)

material found in the shells of crustaceans and other arthropods. The chitosan hydrogel itself was designed as a precursor for carbon foam water-splitting electrodes. This hydrogel was synthesized using a dual crosslinking method to maximize certain mechanical properties. In the first step, a chitosan-NaOH hydrogel was developed to act as a substrate for further crosslinking. Next, the chitosan-NaOH hydrogel was crosslinked further with glutaraldehyde. Glutaraldehyde was selected to enhance and maintain the porosity of the gel, as well as increase strength and flexibility. After thermal processing, the carbon gel could either be doped with molybdenum for HER or a cobalt-iron-nickel cocktail for OER. The addition of molybdenum cations was achieved by solvent deposition, while the Co-Fe-Ni ions were added via the hydrothermal method. A simplified schema for the synthesis of this gel is shown in Figure 10.10. The final results produced a gel with an OER overpotential of 290 mV at 50 mA/cm² and an HER overpotential of 88 mV at 10 mA/cm². The Tafel slopes derived from this were 115 mV/dec and 56 mV/dec, respectively.

Not every bifunctional electrocatalyst needs to be for water splitting. In the case of rechargeable zinc-air batteries, the cathode acts as a bifunctional catalyst for both OER and ORR-style reactions [44]. Zinc-air batteries exhibit properties in common with both traditional batteries and fuel cells. As such, the redox-active electrodes must exhibit efficient reaction kinetics. One study devised particularly thin electrodes that exhibited an inverse opal structure

Hydrogels for Electrocatalytic Activities

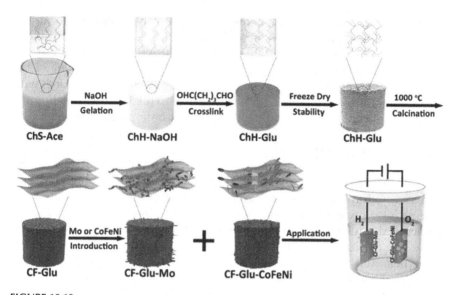

FIGURE 10.10
The schematic illustration of the synthesis and application of the chitosan hydrogel-derived CF-Glu-Mo and CF-Glu-CoFeNi composites. (Adapted with permission [38]. Copyright 2021, Elsevier.)

[45]. While the electrodes were undoubtedly important in this study, their enhanced electrocatalytic activity was attributed in part to the hydrogel electrolyte used. Traditional zinc-air battery electrolytes suffer either from low ionic conductivity or a complete lack of flexibility. The hydrogel electrolyte used in this publication, however, suffered neither of these drawbacks. This electrolyte consisted of a polyacrylic acid-polyacrylamide copolymer suffused throughout with an alkaline solution. The hydrogel electrolyte exhibited a low OER overpotential of 410 mV at 10 mA/cm² with a Tafel slope of 101 mV/dec. As mentioned previously. The ORR potential was measured from the half-wave potential. In doing so, it was found that the sample demonstrated a current density of 5.35 mA/cm² at 0.2V – a current density approaching that of Pt/C (5.27 mA/cm² at 0.83V).

10.5 Conclusion and Perspectives

As humanity grapples with rising energy demands and climate change, diverse and adaptable energy sources are required. Renewable energy sources are of particular importance as sustainability rises as a public concern. While there are currently many alternative energy sources available on the market, these sources are often limited by cost, location, or low energy production per area of occupied space. Alternatively, electrochemical fuel cells naturally sit at the forefront of renewable energy research. Fuel cells can be implemented anywhere they are needed, all the while outputting impressive amounts of power due to the energy density of their fuel. Due to the modularity of fuel cells, it is practical to install as many as are needed, with output remaining adjustable with energy needs. This is not to mention the fact that fuel cells are practically emission-free, only producing water as a byproduct. This wastewater byproduct can later be electrolyzed to reproduce hydrogen and oxygen for reuse in further fuel cells. Hydrogels, meanwhile, serve as a potent material for electrode synthesis. Hydrogels exhibit many properties useful in their role as an electrocatalyst. The high porosity observed within the gel network encourages electro-ionic flow. Meanwhile, the gel's interconnected network allows for high degrees of flexibility in electrochemical implementation. Other important properties to consider when discussing hydrogel electrocatalysts are their self-healing abilities, adhesive properties, and biocompatibility. Furthermore, this chapter has demonstrated that hydrogels excel in direct electrocatalytic processes as well. Many of these hydrogels and their derivatives not only rival traditional materials but surpass them as well. Even more impressive, most of the electrocatalysts discussed thus far are made from biologically sourced and renewable materials. Chitosan, alginate, and cellulose are all materials that can be collected as refuse. Meanwhile, transition metal oxides are common and inexpensive compared with rare metal resources. Hydrogels allow for a unification of these materials in ways not possible through alternative synthesis routes. Due to the biocompatibility innate in these materials, alternative electrocatalysts involving enzymes are possible. Such materials are of great importance in global carbon management generated by current energy sources. Even further, these biologically active electrocatalysts allow for the synthesis of chemical feedstock from CO_2 waste. To conclude, hydrogels represent a prominent field in the further creation of electrocatalysts for electrocatalytic activities. Consequently, further developments in the field are expected to lower the cost of fuel cell technologies and pave the way for a greener energy solution for humankind.

References

1. N. Metzger, X. Li, Technical and economic analysis of fuel cells for forklift applications, *ACS Omega*. 7 (2022) 18267–18275.
2. E.M. Gray, C.J. Webb, J. Andrews, B. Shabani, P.J. Tsai, S.L.I. Chan, Hydrogen storage for off-grid power supply, *Int. J. Hydrogen Energy*. 36 (2011) 654–663.
3. S. Fleischmann, J.B. Mitchell, R. Wang, C. Zhan, D. Jiang, V. Presser, V. Augustyn, Pseudocapacitance: From fundamental understanding to high power energy storage materials, *Chem. Rev.* 120 (2020) 6738–6782.

4. Rusdianasari, Y. Bow, T. Dewi, HHO Gas Generation in hydrogen generator using electrolysis, *IOP Conf. Ser. Earth Environ. Sci.* 258 (2019) 12007.
5. O.Z. Sharaf, M.F. Orhan, An overview of fuel cell technology: Fundamentals and applications, *Renew. Sustain. Energy Rev.* 32 (2014) 810–853.
6. Arredondo, Aguilar-Lira, Perez-Silva, J. Rodriguez, Islas, Hernandez, Characterization and application of agave salmiana cuticle as bio-membrane in low-temperature electrolyzer and fuel cells, *Appl. Sci.* 9 (2019) 4461.
7. A.M. Feltham, M. Spiro, Platinized platinum electrodes, *Chem. Rev.* 71 (1971) 177–193.
8. H. Ma, C. Liu, J. Liao, Y. Su, X. Xue, W. Xing, Study of ruthenium oxide catalyst for electrocatalytic performance in oxygen evolution, *J. Mol. Catal. A Chem.* 247 (2006) 7–13.
9. Z. Chen, X. Duan, W. Wei, S. Wang, B.-J. Ni, Recent advances in transition metal-based electrocatalysts for alkaline hydrogen evolution, *J. Mater. Chem. A.* 7 (2019) 14971–15005.
10. Y. Guo, J. Bae, Z. Fang, P. Li, F. Zhao, G. Yu, Hydrogels and hydrogel-derived materials for energy and water sustainability, *Chem. Rev.* 120 (2020) 7642–7707.
11. R. Zalosh, P. Gandhi, A. Barowy, Lithium-ion energy storage battery explosion incidents, *J. Loss Prev. Process Ind.* 72 (2021) 104560.
12. K. Peng, J. Zhang, J. Yang, L. Lin, Q. Gan, Z. Yang, Y. Chen, C. Feng, Green conductive hydrogel electrolyte with self-healing ability and temperature adaptability for flexible supercapacitors, *ACS Appl. Mater. Interfaces.* 14 (2022) 39404–39419.
13. J. Livage, Sol-gel processes, *Curr. Opin. Solid State Mater. Sci.* 2 (1997) 132–138.
14. J.P. Fernández-Hernán, B. Torres, A.J. López, J. Rams, The role of the sol-gel synthesis process in the biomedical field and its use to enhance the performance of bioabsorbable magnesium implants, *Gels.* 8 (2022) 426.
15. N.K. Devaraj, M.G. Finn, Introduction: Click Chemistry, *Chem. Rev.* 121 (2021) 6697–6698.
16. S.D. Sahoo, T.K. Vasudha, V. Muthuvijayan, E. Prasad, Chitosan-based self-healable and adhesive hydrogels for flexible strain sensor application, *ACS Appl. Polym. Mater.* 4 (2022) 9176–9185.
17. Z. Bai, Q. Liu, H. Zhang, J. Yu, R. Chen, J. Liu, D. Song, R. Li, J. Wang, Antibiofouling and water—stable balanced charged metal organic framework-based polyelectrolyte hydrogels for extracting uranium from seawater, *ACS Appl. Mater. Interfaces.* 12 (2020) 18012–18022.
18. P. Lv, L. Shi, C. Fan, Y. Gao, A. Yang, X. Wang, S. Ding, M. Rong, Hydrophobic ionic liquid gel-based triboelectric nanogenerator: Next generation of ultrastable, flexible, and transparent power sources for sustainable electronics, *ACS Appl. Mater. Interfaces.* 12 (2020) 15012–15022.
19. L. Fang, Z. Cai, Z. Ding, T. Chen, J. Zhang, F. Chen, J. Shen, F. Chen, R. Li, X. Zhou, Z. Xie, Skin-inspired surface-microstructured tough hydrogel electrolytes for stretchable supercapacitors, *ACS Appl. Mater. Interfaces.* 11 (2019) 21895–21903.
20. M. Ghosh, S. Bera, S. Schiffmann, L.J.W. Shimon, L. Adler-Abramovich, Collagen-inspired helical peptide coassembly forms a rigid hydrogel with twisted polyproline II architecture, *ACS Nano.* 14 (2020) 9990–10000.
21. S. He, B. Guo, X. Sun, M. Shi, H. Zhang, F. Yao, H. Sun, J. Li, Bio-inspired instant underwater adhesive hydrogel sensors, *ACS Appl. Mater. Interfaces.* 14 (2022) 45869–45879.

22. M. Ruan, J. Liu, P. Song, W. Xu, Meta-analysis of commercial Pt/C measurements for oxygen reduction reactions via data mining, *Chinese J. Catal.* 43 (2022) 116–121.
23. H. Li, J. Yin, Y. Meng, S. Liu, T. Jiao, Nickel/cobalt-containing polypyrrole hydrogel-derived approach for efficient ORR electrocatalyst, *Colloids Surfaces A Physicochem. Eng. Asp.* 586 (2020) 124221.
24. Y. Fu, D. Xu, Y. Wang, X. Li, Z. Chen, K. Li, Z. Li, L. Zheng, X. Zuo, Single atoms anchored on cobalt-based catalysts derived from hydrogels containing phthalocyanine toward the oxygen reduction reaction, *ACS Sustain. Chem. Eng.* 8 (2020) 8338–8347.
25. H.-J. Niu, S.-Y. Lin, Y.-P. Chen, J.-J. Feng, Q.-L. Zhang, A.-J. Wang, Hydrogel derived FeCo/FeCoP embedded in N, P-codoped 3D porous carbon framework as a highly efficient electrocatalyst for oxygen reduction reaction, *Appl. Surf. Sci.* 536 (2021) 147950.
26. Q. Xiang, Y. Liu, X. Zou, B. Hu, Y. Qiang, D. Yu, W. Yin, C. Chen, Hydrothermal synthesis of a new kind of N-doped graphene gel-like hybrid as an enhanced ORR electrocatalyst, *ACS Appl. Mater. Interfaces.* 10 (2018) 10842–10850.
27. C. Tang, B. Thomas, M. Ramírez-Hernández, E.M. Mikmeková, T. Asefa, Metal-functionalized hydrogels as efficient oxygen evolution electrocatalysts, *ACS Appl. Mater. Interfaces.* 14 (2022) 20919–20929.
28. X. Wang, Y. Yang, R. Wang, L. Li, X. Zhao, W. Zhang, Porous Ni3S2–Co9S8 carbon aerogels derived from carrageenan/NiCo-MOF hydrogels as an efficient electrocatalyst for oxygen evolution in rechargeable Zn–air batteries, *Langmuir.* 38 (2022) 7280–7289.
29. X. Cui, Z. Chen, Z. Wang, M. Chen, X. Guo, Z. Zhao, Tuning sulfur doping for bifunctional electrocatalyst with selectivity between oxygen and hydrogen evolution, *ACS Appl. Energy Mater.* 1 (2018) 5822–5829.
30. S. Kalathil, K.P. Katuri, A.S. Alazmi, S. Pedireddy, N. Kornienko, P.M.F.J. Costa, P.E. Saikaly, Bioinspired synthesis of reduced graphene oxide-wrapped geobacter sulfurreducens as a hybrid electrocatalyst for efficient oxygen evolution reaction, *Chem. Mater.* 31 (2019) 3686–3693.
31. R.M. Donlan, Biofilm formation: A clinically relevant microbiological process, *Clin. Infect. Dis.* 33 (2001) 1387–1392.
32. E. Cartier, J.H. Stathis, Atomic hydrogen-induced degradation of the SiSiO2 structure, *Microelectron. Eng.* 28 (1995) 3–10.
33. S.V.M. Guaitolini, I. Yahyaoui, J.F. Fardin, L.F. Encarnacao, F. Tadeo, A review of fuel cell and energy cogeneration technologies, in: 9th Int. Renew. Energy Congr. IREC 2018, 2018: pp. 1–6.
34. R. Du, X. Fan, X. Jin, R. Hübner, Y. Hu, A. Eychmüller, Emerging noble metal aerogels: state of the art and a look forward, *Matter.* 1 (2019) 39–56.
35. R. Du, J.-O. Joswig, X. Fan, R. Hübner, D. Spittel, Y. Hu, A. Eychmüller, Disturbance-promoted unconventional and rapid fabrication of self-healable noble metal gels for (photo-)electrocatalysis, *Matter.* 2 (2020) 908–920.
36. Q. Shi, C. Zhu, D. Du, C. Bi, H. Xia, S. Feng, M.H. Engelhard, Y. Lin, Kinetically controlled synthesis of AuPt bi-metallic aerogels and their enhanced electrocatalytic performances, *J. Mater. Chem. A.* 5 (2017) 19626–19631.
37. M. Bae, Y. Kang, D.W. Lee, D. Jeon, J. Ryu, Superaerophobic polyethyleneimine hydrogels for improving electrochemical hydrogen production by promoting bubble detachment, *Adv. Energy Mater.* 12 (2022) 2201452.

38. J. Ding, L. Zhong, Q. Huang, Y. Guo, T. Miao, Y. Hu, J. Qian, S. Huang, Chitosan hydrogel derived carbon foam with typical transition-metal catalysts for efficient water splitting, *Carbon N. Y.* 177 (2021) 160–170.
39. S.K. Kuk, K. Gopinath, R.K. Singh, T.-D. Kim, Y. Lee, W.S. Choi, J.-K. Lee, C.B. Park, NADH-free electroenzymatic reduction of CO2 by conductive hydrogel-conjugated formate dehydrogenase, *ACS Catal.* 9 (2019) 5584–5589.
40. L. Castañeda-Losada, D. Adam, N. Paczia, D. Buesen, F. Steffler, V. Sieber, T.J. Erb, M. Richter, N. Plumeré, Bioelectrocatalytic cofactor regeneration coupled to CO2 fixation in a redox-active hydrogel for stereoselective C–C bond formation, *Angew. Chemie Int. Ed.* 60 (2021) 21056–21061.
41. S. Sato, B.J. McNicholas, R.H. Grubbs, Aqueous electrocatalytic CO2 reduction using metal complexes dispersed in polymer ion gels, *Chem. Commun.* 56 (2020) 4440–4443.
42. J. Yang, X. Wang, B. Li, L. Ma, L. Shi, Y. Xiong, H. Xu, Novel iron/cobalt-containing polypyrrole hydrogel-derived trifunctional electrocatalyst for self-powered overall water splitting, *Adv. Funct. Mater.* 27 (2017) 1606497.
43. Y. Gu, S. Chen, J. Ren, Y.A. Jia, C. Chen, S. Komarneni, D. Yang, X. Yao, Electronic structure tuning in Ni3FeN/r-GO aerogel toward bifunctional electrocatalyst for overall water splitting, *ACS Nano.* 12 (2018) 245–253.
44. D. Liu, Y. Tong, X. Yan, J. Liang, S.X. Dou, Recent advances in carbon-based bifunctional oxygen catalysts for zinc-air batteries, *Batter. Supercaps.* 2 (2019) 743–765.
45. K. Tang, C. Yuan, Y. Xiong, H. Hu, M. Wu, Inverse-opal-structured hybrids of N, S-codoped-carbon-confined Co9S8 nanoparticles as bifunctional oxygen electrocatalyst for on-chip all-solid-state rechargeable Zn-air batteries, *Appl. Catal. B Environ.* 260 (2020) 118209.

11
Hydrogels for Metal-Ion Batteries

Swathi Pandurangan, Dhavalkumar N Joshi,
Arun Prasath Ramaswamy, and Vinod Kumar

CONTENTS

11.1 Introduction 213
11.2 Utilization of Hydrogels for the Emerging Metal-Ion Battery 214
    11.2.1 Zinc-Ion Batteries 215
    11.2.2 Lithium-Ion Batteries 217
    11.2.3 Sodium-Ion Batteries 218
    11.2.4 Potassium-Ion Batteries 220
    11.2.5 Calcium-Ion Batteries 221
    11.2.6 Aluminum-Ion Batteries 222
    11.2.7 Magnesium-Ion Batteries 224
11.3 Functional Hydrogel Preparation Techniques and Chemistry 225
11.4 Conclusion and Future Perspectives 227
References 228

11.1 Introduction

Hydrogels are emerging materials for next-generation electrochemical energy storage devices that have attracted significant interest among the research community [1]. Hydrogels are highly crosslinked three-dimensional (3D) networks that assimilate water and swell without dissolving. Due to their versatility, hydrogels are prominent materials in multiple applications such as biomedical engineering, water purification, drug delivery, and tissue engineering [2]. Hydrogels possess tunable mechanical stability, high thermodynamic equilibrium, and stable ionic transfer property [3]. They are widely applied in various electronic applications owing to their high strength and rigidity. In energy storage devices, hydrogel gelation chemistry contributes to multifunctional, rationally designed flexible electrodes and electrolytes. The numerous synthesis approaches of hydrogels enable tunable multifunctional architectures by tailoring them with different polymers and copolymers. The use of aqueous electrolytes in metal-ion batteries causes harmful changes such as (1) the debilitation of water and capacity fading,

(2) dissolving of cathodes in aqueous solution resulting in diminished battery performance, (3) uneven temperature variation, and (4) severe dendrite growth upon creating a short circuit and flammability. Hydrogel holds the unique characteristics of both soft and wet by combining the advantage of both aqueous and no aqueous electrolytes. This fascinating property provides highly stable leakage-proof hydrogel, which subdues dendrite growth, avoids short circuits, and is exceptionally resistant to temperature, enhancing good interfacial compatibility with the electrode and electrolyte. However, the application of hydrogel electrolytes provides only the physical frameworks that promote rapid ion mobility but does not participate in improving electrochemical performance. Hence, the commercial applications of hydrogels are still in development due to the limited approach.

Employing molecularly tunable functional groups on polymer chains with varied architecture via polymer chemistry engineering aspects can significantly improve multiple electrochemical properties in metal-ion batteries. The tuning of varied functional groups specifically increases particular properties such as the following: (1) highly thin and flexible; (2) availability in various shapes; (3) low temperature tolerant; (4) super soft and rigid; and (v) thermoresponsive. The organic and inorganic functional additives or binders can empower and increase the mechanical stability of the hydrogel-based electrolytes. The multifunctional varied architecture hydrogel can provide high ionic conductivity, suitable electrochemical properties, and stable ion transference quality with superior structural flexibility. The addition of inorganic functional groups in the polymer hydrogel (i.e., 2D materials, graphene oxide, metal-ions, nanoparticles, Mxene, and biomolecules) effectively increases the interaction of metal ions with different functional groups (hydroxyl, carboxyl, and amino) in metal-ion batteries. In this chapter, the utilization of hydrogel for electrodes, electrolytes, and binders in various metal-ion batteries is discussed in detail, as well as their future perspective in technology.

11.2 Utilization of Hydrogels for the Emerging Metal-Ion Battery

The high ionic conductivity and dimensional stability of the hydrogel electrolyte are suitable for various energy storage devices, including supercapacitors and batteries. Hydrogels are used as electrodes, electrolytes, and binders in numerous types of metal-ion batteries, such as zinc (Zn^+), lithium (Li^+), potassium (K^+), aluminum (Al^+), calcium (Ca^+), and magnesium (Mg^+) batteries. The multifunctional structural frameworks of diverse hydrogels, in conjunction with metal-ion batteries, enormously improve the transport of electrons per metal cation, enhancing energy density. Due to their unique

Hydrogels for Metal-Ion Batteries 215

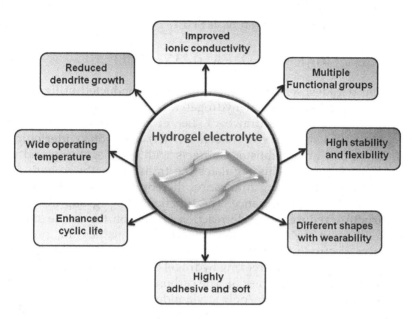

FIGURE 11.1
The multiple benefits of utilizing hydrogel as electrolytes in metal-ion batteries.

properties such as improved ionic conductivity, reduced dendrite growth, multiple functional groups, a wide operating temperature, high stability and flexibility, enhanced cycle life, different shapes with wearability, and highly adhesive, hydrogels act as a low-cost, versatile material with huge dominance (Figure 11.1). Detailed discussions of hydrogels in different metal-ion batteries are given in the following sections.

11.2.1 Zinc-Ion Batteries

As an alternative to lithium-ion batteries, low-cost zinc-ion batteries hold a greater theoretical capacity of 820 mAh/g with a standard redox potential of (−0.763 V vs SHE). Zinc-ion batteries are compatible with the environment and highly safe in aqueous electrolytes. However, due to the high freezing point of water in zinc metal batteries, the diffusion kinetics between the electrode and electrolyte are hampered in low-temperature operation circumstances. Apart from that, zinc undergoes dendritic formation due to the non-uniform deposition of zinc at the positive electrode (Equation 11.1). Also, the utilization of water-based aqueous electrolytes creates unwanted side reactions such as the hydrogen evolution reaction (HER) and the oxygen evolution reaction (OER), as shown in Equations (11.2) and (11.3).

$$4Zn^{2+} + 6OH^- + SO_4^{2-} + 5H_2O \rightarrow Zn_4(OH)_6SO_4 \cdot 5H_2O \qquad (11.1)$$

$$2H^+ + 2e^- \rightarrow H_2 \text{ (OV vs SHE)} \tag{11.2}$$

$$2H_2O \rightarrow 4H^+ + O_2 + 4e^- \text{ (1.23 V vs SHE)} \tag{11.3}$$

In recent decades, serious research has been conducted utilizing functionally designed crosslinked hydrogel with tightly packed 3D porous structures to address this drawback. Lea et al. [4] designed a hydrogel electrolyte based on polyvinyl alcohol (PVA) and polyzwitterionic salt using cation and anion transfer channels with two negatively and positively charged groups. This regulates the transport paths of both anion and cation without causing hindrance, thereby improving the high rate capacity and cyclic performance. The functional SO_3^- group in the polymeric structure reduces the side reaction of Zn^{2+} on the anode side. The uniform diffusion of Zn^{2+} promotes smooth stripping/platting, protecting the positive electrode (anode) from dendrite growth formation. Recently, Yan et al. prepared an anti-freezing hydrogel electrolyte with a CO_2^- source applicable in zinc-ion batteries [5]. The carbon dioxide derived crosslinked with potassium polyacrylate/polyacrylamide utilized sustainable hydrogel effectively, enabling wettability, high flexibility, and a stable discharge capacity of 80.40% retention at −20°C (Figure 11.2a). Chen and co-workers modified sorbitol hydrogel electrolytes from wheat straw as a sustainable and environmentally friendly adaptable material for high-performance zinc-ion batteries [6]. The presence of an abundant hydroxyl group and longer molecular chain, which is higher than the common cryoprotectants

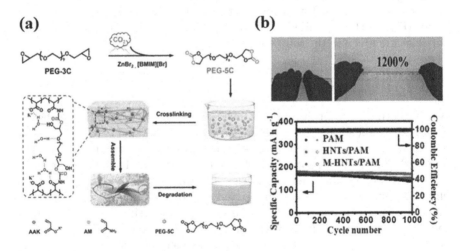

FIGURE 11.2
(a) Synthesis procedure of CO_2-derived polyethylene glycol dicyclic carbonate (PEG-5C) and crosslinking of PEG-5C with potassium acrylate (AAK) and acrylamide (AM) to form sustainable hydrogen. (b) Image of high mechanical strength during stretching to 1200% and coulombic efficiency behavior of ZIB at 10 C. (Adapted with permission [7]. Copyright [2021], Elsevier.)

(which modify the cellulose), provides superior mechanical stability, strong adhesion, high transparency, good flexibility, rich porosity, and ultra-low freezing point. This hydrogel system represents a "water in salt" system with a high ionic conductivity of 19.7 mS/cm at −40°C. Similarly, Cheng and his co-workers made [7] polyacrylamide hydrogel electrolyte crosslinked by ultra-stable halloysite nanotubes in zinc-ion batteries. The halloysites are crosslinked by 3-methacryloxypropyltrimethoxysilane (MPS) with polyacrylamide (PAM) (M-HNTs/PAM), which enables high strength and ultra-high stability as shown in Figure 11.2(b) with a larger strain of 1200% at the strength of 49 kPa. M-HNTs/PAM hydrogel holds stable cyclic stability and high coulombic efficiency of 92.7% for over 1000 cycles at 10 C (Figure 11.2b). A zinc-ion battery with conventional hydrogels unavoidably tends to freeze at sub-zero temperatures owing to the high relative freezing point of the water. This deteriorates the ionic conductivity and mechanical stability, producing poor cyclic stability and safety issues for the polymer electrolyte. Superior flexibility toward wearable battery application should be considered a design aspect in hydrogel electrolytes [8]. At low temperatures, hydrogel freezing occurs due to the formation of a strong hydrogen bond (HB) between the polymer chains and water molecules [9]. Organic additives and concentration salts are exhibited to prevent the breaking of HBs of water to attain a stable performance of zinc-ion batteries at sub-zero conditions. Wang et al. made a hydrogel polymer electrolyte by PAN gel polymer electrolyte in NiCo||Zn batteries with a wide electrochemical voltage window at the sub-zero temperature of −20 to 50°C. The prepared PAN hydrogel possesses high stability at anti-freezing temperatures with improved heat resistance and elasticity [10]. The improvement in mechanical stability is attributed to the reduction in water content, which helps in the freezing point of the hydrogel by concentrated ions.

11.2.2 Lithium-Ion Batteries

Among all the metal-ion batteries, the lithium-ion battery dominates battery technology due to its high energy density (100–260 Wh/kg), lightweight characteristics, power density, and flexible applications in various electronics industries. However, using inorganic liquid electrolytes in Li-ion batteries creates safety issues such as volatility, flammability, volume expansion, short circuits, and explosions. So, researchers have focused much more on solid-state electrolytes to overcome the above limitations. However, applying ceramic or inorganic solid electrolytes forms undesirable dendrite growth on the surface. The use of solid pellet electrolytes, often non-flexible and thick, increases the electrode-to-electrolyte interphase that deforms the smooth Li-ion migration of the batteries. The best possible way to address the above issues is the usage of highly flexible polymers that satisfy the need for flexibility, stability, and good electrochemical performance. This polymer electrolyte extends the application to all flexible electronic devices in various fields. Hydrogels act as an emerging electrode and electrolyte material among these polymer electrolytes due to their stretchability, self-healing nature, and mechanical stability. Hydrogels offer superior ionic conductivity similar to aqueous electrolytes due to the strong intermolecular interactions of the functional polymers. It is also highly applicable as the anode/cathode electrolytes owing to its good porous structure that completely protects from the loss of active materials by volume expansion. Peng et al. recently prepared a hydrogel composite anode utilizing cobalt molybdate ($CoMoO_4$) nanotubes by incorporating them with polyaniline (PANI) polymer hydrogel [11]. The $CoMoO_4$ @ PANI anode was prepared via in-situ polymerization for lithium-ion batteries. The rod-like hierarchy protects and alleviates the volume expansion of $CoMoO_4$ by stripping and platting lithium. This hydrogel anode provides a large capacity of 805 mAh g^{-1} with a coulombic efficiency of 99%. Jiang and co-workers prepared a fully stretchable solid-state lithium-ion battery using a polyacrylamide (PAM) hydrogel electrolyte [12]. This hydrogel is designed as a "water in salt" based aqueous electrolyte. The interpenetrated Ag-nanowire network enhances the energy transportation capacity from the electrode to the electrolyte. The prepared electrolyte holds the unique stretchable lithium-ion battery with a specific capacity of 119 mAh/g and the unstretched position delivers a higher discharge capacity of 240 mAh/g. This Ag-NW network interconnected PAM hydrogel demonstrated a new way for integrating highly stretchable solid-state hydrogel electrolyte-based lithium-ion batteries. Hydrogel electrolytes provide a wide range of opportunities to tune novel varied functional materials in chemistry. However, desired and stable mechanical performance is inauspicious for the flexible hydrogel in a lithium-ion battery application. To address the above limitations, Zhu et al. designed a highly mechanically stable flexible hydrogel electrolyte, as shown in Figure 11.3, containing n-butyl acrylate-silicone copolymer (PBA-Si) and carboxymethyl cellulose (CMC) [13]. This hydrogel electrolyte film exhibits high ionic conductivity of 1.15 mS/cm. At the same, it shows a high mechanical stretchable ratio of >300%. The high stretching ability elastomeric polymer was prepared by simple casting technique as shown in Figure 11.3(a); in order to improve the ionic conductivity, they designed a self-moisture absorbent "solution crystal" of LiTFSI with the interaction of H_2O molecules, as shown in Figure 11.3(b), which rapidly increases the lithium-ion transfer path in the hydrogel electrolyte. Figure 11.3(c) and Figure 11.3(d) show the digital photographs of the dried and wet film.

11.2.3 Sodium-Ion Batteries

In recent years, sodium-ion batteries have emerged as an alternative to the dominating lithium-ion battery technology. Sodium attracted significant interest among the research community due to its low cost, natural

Hydrogels for Metal-Ion Batteries 219

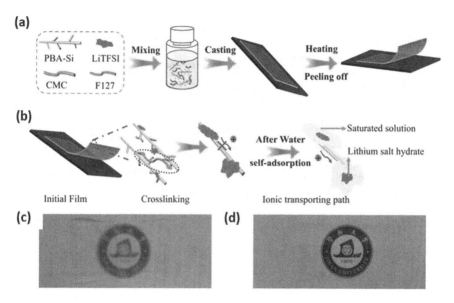

FIGURE 11.3
Preparation schematic of the (a) PBA-Si/CMC/F127/LiTFSI film; (b) formation of a "sea island"-type lithium-ion transport pathway in the flexible electrolyte film; (c) and (d) digital photographs of dry and wet HEF. (Adapted with permission [13]. Copyright [2022], Elsevier.)

raw material availability (2.6 wt.% of Na in the earth crust is sodium and 0.005 wt.% is only lithium), and corresponding electrochemical performance similar to lithium. However, despite its low cost, the sodium-ion faces hindrance due to its high ionic radius Na$^+$ (1.02 Å), which affects the intercalation chemistry of the Na$^+$ ion insertion and reinsertion of the host electrode. Designing high interplanar spacing with alloys, carbons, and metal oxides enables superior Na$^+$ intercalation/deintercalation without disarranging the host framework. Singh et al. have reported composite graphitic carbon nitride with reduced graphene oxide (g-C$_3$N$_4$@RGO) via a hydrothermal process with a high redox potential of 0.6–2.2 V, shown in Figure 11.4(a) [14]. The hydrogel composite (g-C$_3$N$_4$@RGO) annealed at high temperature restricts the agglomeration and enables a higher discharge capacity of 170 mAh/g at 0.1 C with a stable coulombic efficiency of 96% for almost 1000 cycles (Figure 11.4b). Liu et al. made a low-temperature tolerance aqueous sodium-ion battery (ASIB). This work depicts the first ever ASIB operating at the low operating temperature of −30°C [15]. Figure 11.4(c) demonstrates the employed fumed SiO$_2$ gel framework, Na$_2$SO$_4$, and methyl alcohol as an anti-freezing agent by preventing the freeing of water molecules in the hydrogel. The Na$_2$SO$_4$- SiO$_2$ novel hydrogel additives increase the lithium-ion conductivity with a discharge capacity of 78.9 mAh/g at −20°C (Figure 11.4d).

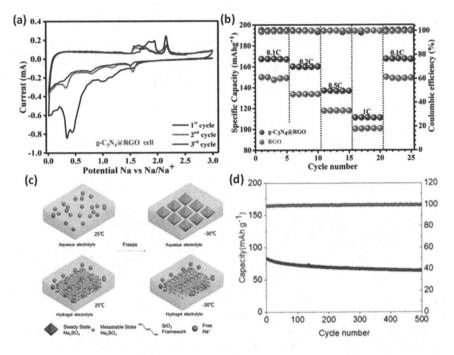

FIGURE 11.4
(a) Redox potential of g-C_3N_4@RGO cell for the first three cycles. (b) Specific capacity comparison for g-C_3N_4@RGO and RGO. (c) The freezing process comparison for aqueous electrolyte Na_2SO_4-SiO_2 novel hydrogel from 25°C to −30°C (d) The long term cyclic capacity of (1 C = 0.13 A/g) rate operating at −20°C. (Adapted with permission [14] and [15]. Copyright [2022], Elsevier.)

11.2.4 Potassium-Ion Batteries

Potassium-ion batteries provide an alternate way to coercive battery technology. However, the exploration of potassium-ion batteries as a practical usage is still under development. The diffusion and intercalation of potassium-ions are hindered due to the larger ionic radius of (K^+:1.38 Å) compared with that of sodium (Na^+:1.02 Å) and lithium (Li^+:0.68 Å) [16]. Potassium-ion batteries hold similar electrochemical properties as lithium; in particular, the K^+/K redox couple exhibit lower potential than the lithium-ion battery in non-aqueous electrolytes [17]. However, the advancement of potassium-ion batteries is lacking in materials that do not facilitate the insertion chemistry of larger K^+ ions. Anode materials that promote K^+ ion insertion are phosphorous, graphite, oxides (K_2Ti_8O, $K_2Ti_4O_9$, Co_3O_4- Fe_2O_3 composites), and sulfides (MoS_2 and SnS_2) [18–22]. Identifying the excellent insertion of the K^+ cathode material is a fruitful challenge that can collapse the host framework. Cathode material preparation is a tedious process due to hygroscopicity, thus hindering the choice of cathode material. However, researchers aimed to produce some feasible cathode materials such as $K_3V_2(PO_4)_3$ [23], $KVPO_4F$,

KVOPO$_4$ [24], and KVP$_2$O$_7$ [25]; among the broad classification of materials polyanionic compounds possess high voltage and good mechanical stability.

Goodenough et al. have prepared a p-type organic material with polypyrrole to avoid the problem due to the intercalation of cations [26]. Polypyrrole is a green sustainable material with less cost and is an excellent alternative to the inorganic material. Compared with other organic materials, conducting polymers operate at a high voltage range with better suitability for the K$^+$ cathode. Chen et al. have made polypyrrole hydrogel-enabled free-standing cathode film [27]. It enables stable conductivity with good flexibility. In addition, hydrogel as a cathode eradicates the usage of aluminum foil and carbon black for active material loading. A hydrogel cathode provides many electron-conducting channels with high mechanical stability. A polypyrrole hydrogel cathode, through a convenient one-step route, maintains the potential window of 2–4.2 with an 86% retention capacity of 100 cycles at 0°C.

11.2.5 Calcium-Ion Batteries

Calcium-ion batteries, Ca^{2+}, recently became the most promising electrochemical energy storage system, also called post-lithium-ion batteries. Among all the metal-ion batteries such as (Na, K, Mg, and Al), Ca-ion batteries are the most abundant in the earth's crust metal, at nearly 4.1% of the earth's crust. Calcium anode batteries provide similar volumetric capacities (2073 mAh/cm^3) to that of lithium anode (2062 mAh/cm^3), but the gravimetric capacity is much lower, 1337 mAh/g. However, the research on calcium-ion batteries is still in the early development stage. The research work related to Ca metal batteries with various configurations has been analyzed to identify the operating conditions of the Ca^{2+} ion, such as voltage (1–4 V), capacities (60–250 mAh/g), and current densities (2–500 mA/g), which are still lower than lithium metal batteries. The development of calcium metal cathodes provides superior discharge capacity, but the calcium metal energy densities are currently less than the state-of-the-art lithium. Few hydrogel-based electrolyte works are explored in calcium metal batteries. Among that, Tang et al. prepared a hydrogel electrolyte by employing PVA by dissolving 10 wt.% in the calcium salt (Ca (NO$_3$)$_2$ [28]. The full cell performance was also analyzed with a sulfur/carbon (S/C) anode and a Ca$_{0.4}$MnO$_2$ cathode. The molecular dynamic simulation (MD) and experimental study were analyzed to illustrate the evolution of calcium salt (Ca(NO$_3$)$_2$ in aqueous and salt electrolytes; around 10.9% of the water molecules interact with Ca^{2+} while others rapidly couple through hydrogen bonding, which is shown in Figure 11.5(a). Figure 11.5(b) describes the interaction of the highly concentrated PVA hydrogel electrolyte with a capacity of 560 mAh/g at a current rate of 0.1 C. The prepared stable electrolyte with high salt concentration decreased the reaction of water activity in the hydrogel, thereby increasing the electrochemical stability window (Figure 11.5c). Compared with the conventional calcium aqueous electrolyte S/CjCa(NO$_3$)$_2$ in the Ca$_{0.4}$MnO$_2$ battery, prepared

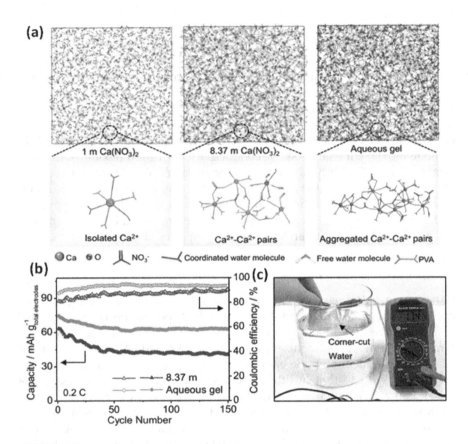

FIGURE 11.5
(a) Local structure evolution by molecular dynamics simulation (MD) at 10 ns for a 1 m Ca(NO$_3$)$_2$ electrolyte, a saturated Ca(NO$_3$)$_2$ electrolyte, and an aqueous gel electrolyte. (b) The coulombic efficiency of the full cell S/C||Ca$_{0.4}$MnO$_2$ full cells at 0.2 C. (c) Fabricated pouch cell corner cut soaking in water still delivers voltage of 1.74 V. (Adapted with permission [28]. Copyright The Authors, some rights reserved; exclusive licensee Nature Portfolio. They are distributed under a Creative Commons Attribution License 4.0 (CC BY).)

hydrogel electrolyte S/C|PVA gel in the Ca$_{0.4}$MnO$_2$ battery exhibits better and stable cyclic stability at a current density of 0.2 C over 150 cycles. The average coulombic efficiency was 83%, with a 93% capacity retention rate. However, by maintaining the same atmospheric conditions, the conventional aqueous electrolyte S/C||Ca(NO$_3$)$_2$ in the Ca$_{0.4}$MnO$_2$ battery exhibited only 65% of retention capacity with coulombic efficiency of 86%.

11.2.6 Aluminum-Ion Batteries

Compared with all multivalent metal battery systems such as Zn, Mg, Ca, and Al, aluminum-ion batteries (AIBs) have gained considerable attention due to their meager material costs, long cycle life, and stable performance. Solid-state AIBs can avoid issues such as side reactions, leakage, and so on when compared with aqueous electrolytes and can improve the electrochemical stability and safety performance of AIBs. Wang et al. designed a flexible hybrid polymer hydrogel electrolyte for AIBs – the synthetic polymer PAM was grated into gelatin chains via free radical polymerization. The fabricated full cell $MoO_3||VOPO_4$ exhibits a stable discharge capacity of 55 mAh/g with excellent flexibility and extreme mechanical stability (squeezing, twisting, folding, and cutting) under different positions. [29], similarly, Sun et al. [30] prepared the first-ever work on designing a sodium acrylate (ANa) PANa-CMC electrolyte in AIB. First, ANa was synthesized by adjusting the AA monomer with NaOH solution. Subsequently, the sodium carboxymethyl CMC, the crosslinker MBA, and initiator APS, as prepared precursor solution, were continuously stirred. The obtained gel solution was in-situ polymerized at 50°C for 14 hours to form a PANa-CMC hydrogel electrolyte. The introduction of CMC in the hydrogel protects the AL anode from self-corrosion with the improved mechanical stability of the GPE hydrogel. The prevention of self-corrosion in the AL anode by CMC significantly improves the higher current density at 10 mA/cm². The in-situ polymerized PANa-CMC electrolyte in an Al-air battery delivers a capacity of 1.72 V in comparison with a pure PANa electrolyte, which delivers only 1.60 V. Huang et al. designed [31] a novel poly(N-isopropylacrylamide) (PNIPAM) quasi solid-state hydrogel-based Al^{3+}/H^+ conductive electrolyte. This hydrogel electrolyte posse's distinctive quality of both hydrophilic amide group (–CONH) and hydrophobic isopropyl group (–$CH(CH_3)_2$). Hence, the rich microspore's structure, with the unique properties of the hydrogel, absorbs electrolyte swelling quickly, giving a high conductivity of 1.8 S/cm. At high temperatures, the hydrogen bonds in hydrogel weaken and shrink the three-dimensional electrolytes. Simultaneously, the interaction between the hydrophobic isopropyl groups strengthens the electrolyte. The fabricated full-cell battery 2D-layered $Al_xVOPO_4·2H_2O$ cathode and MoO_3 anode exhibit a high specific capacity of 125 mAh/g at 0.1 A/g over 10,000 cycles with no capacity fading [31]. Recently, aluminum-ion batteries contributed to the development of wearable electronic applications in accordance with Xiong et al., who fabricated a fiber-shaped aqueous Al-ion battery [32]. They utilized a manganese hexacyanoferrate (MnHCF) cathode with a graphene oxide decorated MoO_3 cathode and a $PVA/Al(CF_3SO_3)_3$ hydrogel electrolyte, shown in Figure 11.6(a). The MnHCF cathode by simple co-precipitation reaction is based on a rocking chair energy storage mechanism. The contributed functional groups in the GO and the MoO_3 accommodate the change due to the volume expansion. The fabricated fiber-shaped entire cell enables profitable increases in the Al^{3+} insertion into GO/MoO_3 cathode, thereby increasing the discharge capacity of 42 mAh/cm³ and high energy density of 30.6 mWh/cm³ at 0.5 A/cm³ by maintaining good stability with 91.6% over 100 cycles at 1 A/cm³; additionally, the wearable fiber

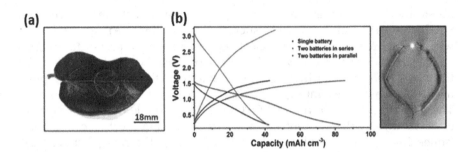

FIGURE 11.6
(a) The optical image of PVA/Al(CF$_3$SO$_3$)$_3$ hydrogel electrolyte. (b) A fiber-shaped full cell delivers a discharge capacity of 42 mAh/cm^3 for a single battery, which increases to 80 mAh/cm^3 for two batteries in parallel. (Adapted with permission [33]. Copyright The Authors, some rights reserved; exclusive licensee Wiley. Distributed under a Creative Commons Attribution License 4.0 (CC BY).)

shaped entire cell delivers excellent flexibility and durability, as presented in Figure 11.6(b).

11.2.7 Magnesium-Ion Batteries

Similar to other metal-ion batteries, magnesium-ion batteries have also gained attention in the research community due to their lower cost, high theoretical volumetric energy density (3833 mA h cm^{-3}), low electrode potential (-2.37 V vs NHE), and the abundant resource of magnesium to replace conventional lithium-ion batteries in the future. Based on the commercial semicrystalline polymer polyethylene oxide (PEO), Shao et al. [33] and his co-workers have synthesized a new polymer electrolyte membrane in magnesium-ion batteries. The incorporated Mg nanoparticles and (Mg(BH$_4$)$_2$) dispersed well in the PEO matrix in the mass ratio of 1:1:8, thereby enhancing the electrochemical performance with the stable discharge capacity of 90 mAh/g at 90°C. The coulombic efficiency was maintained at 100% over 150 charge/discharge cycles. Compared with conventional polymers (PEO, PVDF, PVP, etc.), synthesized ionic liquids provide superior lithium-ion conductivity at room temperature, high chemical stability, and a wide electrochemical potential window. Considering the merits of the ionic liquid, Wang et al. [34] prepared a gel polymer electrolyte by utilizing 1-ethyl-3-methylimidazolium trifluoromethanesulfonate (EMITf) ionic liquid with PVA and Mg(Tf)$_2$ in Mg-ion batteries. The introduction of EMITf ionic liquid drastically improves the ionic conductivity from 2.10×10^{-4} S/cm with a wide electrochemical window close to 5.0 V, showcasing the suitable usage of environmentally friendly PVA in magnesium-ion batteries.

11.3 Functional Hydrogel Preparation Techniques and Chemistry

Hydrogel materials are made up of highly crosslinked hydrated polymers that are more hydrophilic functional groups that absorb and store water. They possess the unique characteristic of being super soft and wet; the absorbed water inside the hydrogel dissolves the ions of aqueous electrolytes. Depending on the desired application area, the functionality framework, mechanical stability, and performance vary. Nowadays, hydrogels are mainly applied in electrochemical devices such as batteries and supercapacitors because of their high ionic conductivity and stretchability. Hydrogels are usually made up of many functional polymers that are highly reactive to water, such as carboxyl, hydroxyl, carbonyl, amino, and other biopolymers. Hydrophilic polymers such as polyacrylamide (PAAm), PVA, polyacrylic acid (PAA), PANI, cellulose, chitosan, and so on [35] are mostly used. Typical hydrogel electrolytes are prepared with acid, alkaline, or neutral aqueous solution by adding different salts such as –COOH, –COO, –OH, or –CONH$_2$ [36]. Hydrogel possesses three-dimensional hierarchical structures with many desired properties such as ionic conductivity, electronic conductivity, ions/electrons permeability, stretchability, and flexibility. Regarding energy storage devices, hydrogels are prepared with interpenetrating, crosslinking, and semi-interpenetrating networks, which help improve the porosity and large surface area. Crosslinking is a widely used technique via free radical polymerization, condensation polymerization, and click chemistry. Chemical crosslinking (i.e., covalent bonding) leads to the formation of highly crosslinked structures between polymers and monomers [37]. These techniques allow metal ions to be crosslinked by utilizing functional polymers to form flexible free-standing hydrogels.

In the application of energy storage materials, hydrogels act as three-dimensional templates that form unique interconnected nanostructures that enable rapid mass transport and electron transport. As per the above hydrogel preparation techniques, chemical activation further upgrades them to attain the desired electrochemical properties (porosity, high surface area, ionic/electron conductivity). Compared with conventional preparation methods (solvothermal/hydrothermal), the templating technique can cover all extended functional precursors and materials. The interconnected micro/nano 3D porous structured hydrogels possess highly controllable doping chemical homogeneity even in extremely low-temperature conditions. This 3D carbon metal oxide and ceramic structure holding was prepared via simple heat treatment of hydrogel under oxidizing atmosphere. This oxide framework can be utilized in preparing electrode materials for batteries under oxidizing (LiCoO$_2$) and reducing (LiFePO$_4$) atmosphere [38]; this can also be made with conventional solid-state electrolytes. Table 11.1 shows different

TABLE 11.1
Comparative Gelation Chemistry Employed in Preparing Hydrogel Electrolytes with Different Metal-Ion Batteries

S.No	Polymer and preparation method	Gelation chemistry employed	Application in metal-ion batteries
1.	PMC 3D crosslinked hydrogel electrolyte.	C=C double bond initiated the free radical crosslinking polymerization in the PMC hydrogel electrolyte.	Ultra-long cycle life with a coulombic efficiency of 99.5% over 5000 cycles in zinc-ion batteries [39].
2.	PAM, agarose, and carboxymethyl cellulose (CMC) used to form the crosslinked self-healable hydrogel electrolyte (SHE).	Acrylamide monomers initialize the free radical polymerization along with the agarose molecular and CMC chain polymer networks.	More dynamic hydrogen promotes a high discharge capacity in zinc-manganese dioxide batteries exhibiting a discharge capacity of 304 mAh/g with a retention capacity of 83.1% [3].
3.	Polyacrylamide (PAM) gelatin acrylamide and N,N-methylenebisacrylamide were utilized to form the PAM hydrogel electrolyte.	In-situ free radical polymerization initiated by the crosslinkers at 60°C.	A flexible calcium-ion battery is fabricated with ITO||PAM||ZnHCF, which delivers a discharge capacity of 71.2 mAh/g [40].
4.	Fumed silica, Na_2SO_4 solution, and methyl alcohol to produce the Na_2SO_4-SiO_2 hydrogel electrolyte.	The simple stirring procedure was employed to produce Na_2SO_4-SiO_2 hydrogel electrolyte ionic bonds formed between SiO_2 and SO^{2-}_4; three-dimensional hydrogel electrolyte immobilizes the metastable Na_2SO_4 state by 3D SiO_2 structures.	The additive methyl alcohol in the hydrogel acts as an anti-freezing function that prevents water freezing. Fabricated sodium-ion full cell performance with a stable discharge capacity of 78.9 mAh/g at −20°C [15].
5.	Acrylamide, KCF_3SO_3, and KFSI dissolved in deionized water to obtain water in salt hydrogel electrolyte.	Water in the salt hydrogel electrolyte formed through simple polymerization. The polymer network with a hierarchical framework undergoes fast K-ion transport.	The hydrogen bond between the polymerized network and the water phase helps in superior ionic conduction in potassium-ion batteries with a high discharge capacity of 135 mAh/g over 3000 cycles. [41].

(*Continued*)

TABLE 11.1 (CONTINUED)

Comparative Gelation Chemistry Employed in Preparing Hydrogel Electrolytes with Different Metal-Ion Batteries

S.No	Polymer and preparation method	Gelation chemistry employed	Application in metal-ion batteries
6.	Self-healing polymer hydrogel film PBA-Si, CMC, F127, and LiTFSI	Synthesized PBA-Si provides the molecular level stretching ability of the hydrogel. CMC acts as a crosslinker. PBA-Si hydrogel was obtained via esterification.	The self-healing hydrogel exhibits superior ionic conductivity of 1.15 ms cm^{-1}. The fabricated full $LiMn_2O_4$ cell maintains the discharge capacity retention of 73.2% over 1000 charge/discharge cycles [42].
7.	Water-based aqueous solid-state polymer electrolyte prepared by eco-friendly polymer PVA and xanthan gum	Two different kinds of water-based gel polymers were prepared on the anolyte and catholyte sides by a simple solution casting technique.	The fabricated aluminum air cells were discharged at 200 $\mu A/cm^2$ higher than 1 V for 10 hours. The gel electrolyte is made without using any organic solvents, achieved between the anode and cathode site, and is separated by the electrolyte with lower water content at the anode side [43].

gelation chemistries for preparing various hierarchical hydrogel materials in metal-ion batteries. However, the overall hydrogel preparation technique is not limited to this. Various hydrogel preparation techniques, such as incorporating interpenetrating polymers, block copolymers, and changing the density of crosslinkers and doping levels, provide a remarkable opportunity to produce rationally designed hydrogel materials to meet the challenging aspects of futuristic energy storage applications.

11.4 Conclusion and Future Perspectives

Hydrogel electrolytes play a crucial role in developing all solid-state flexible metal-ion batteries. They enhance the ion diffusion kinetic mechanism with improved electrochemical stability. The ultra-high flexible hydrogel electrolyte rapidly increases the electrolyte interface, which has a significant impact on providing high energy density metal-ion batteries. Hydrogels possess distinctive advantages of unique functionally aligned polymer networks that offer superior mechanical stability to conventional nanomaterials. Moreover, the working ion transport mechanism and the polymer dynamics of hydrogels contribute to the future development of all portable flexible electronic applications in various fields. The interconnected hierarchical polymer network emphasizes better interaction with the active electrode material, thereby increasing the charge/discharge performance and high coulombic efficacy in aqueous and non-aqueous metal-ion batteries. Nonetheless, the applicability of the hydrogel electrolyte is yet to be standardized due to its discrepant ionic conductivity and electrochemical performance. Functionalized aligned polymer networks well accommodate inorganic metal oxide frameworks for their versatile purpose in metal-ion batteries. This chapter shows insights into works on hydrogels with favorable merits in metal-ion batteries. However, there are a few constraints that remain challenging, such as electrochemical limitations in aqueous electrolytes and a wide range of operating temperatures in water-based solvents. Understanding hydrogel kinetics in ion transport mechanisms, mechanical stability, improving electrochemical performance, and electrode-to-electrolyte interface chemistries is a key challenge. Multifunctional hierarchal polymers with novel gelation chemistry and high ion transport pathways with inimitable properties indicate cost-effective, eco-friendly, multiple-shaped wearable portable electronic applications. The all-solid-state flexible hydrogel electrode and electrolyte materials will play a significant role in fabricating flexible, high-energy-density batteries in portable electronic industries in the near future.

References

1. P. Yang, J.-L. Yang, K. Liu, H.J. Fan, Hydrogels enable future smart batteries, *ACS Nano.* 16 (2022) 15528–15536.
2. Y. Guo, Z. Fang, G. Yu, Multifunctional hydrogels for sustainable energy and environment, *Polym Int.* 70 (2021) 1425–1432.
3. W. Ling, F. Mo, J. Wang, Q. Liu, Y. Liu, Q. Yang, Y. Qiu, Y. Huang, Self-healable hydrogel electrolyte for dendrite-free and self-healable zinc-based aqueous batteries, *Mater Today Phys.* 20 (2021) 100458.
4. L. Li, L. Zhang, W. Guo, C. Chang, J. Wang, Z. Cong, X. Pu, High-performance dual-ion Zn batteries enabled by a polyzwitterionic hydrogel electrolyte with regulated anion/cation transport and suppressed Zn dendrite growth, *J Mater Chem A Mater.* 9 (2021) 24325–24335.
5. X. Ou, Q. Liu, J. Pan, L. Li, Y. Hu, Y. Zhou, F. Yan, CO_2-sourced anti-freezing hydrogel electrolyte for sustainable Zn-ion batteries, *Chem Eng J.* 435 (2022) 135051.
6. Y. Quan, W. Zhou, T. Wu, M. Chen, X. Han, Q. Tian, J. Xu, J. Chen, Sorbitol-modified cellulose hydrogel electrolyte derived from wheat straws towards

high-performance environmentally adaptive flexible zinc-ion batteries, *Chem Eng J.* (2022) 137056.
7. P. Xu, C. Wang, B. Zhao, Y. Zhou, H. Cheng, A high-strength and ultra-stable halloysite nanotubes-crosslinked polyacrylamide hydrogel electrolyte for flexible zinc-ion batteries, *J Power Sources.* 506 (2021) 230196.
8. S. Huang, L. Hou, T. Li, Y. Jiao, P. Wu, Antifreezing hydrogel electrolyte with ternary hydrogen bonding for high-performance zinc-ion batteries, *Adv Mater.* 34 (2022) 2110140.
9. T. Sun, H. Du, S. Zheng, J. Shi, Z. Tao, High power and energy density aqueous proton battery operated at –90°C, *Adv Funct Mater.* 31 (2021) 2010127.
10. H. Wang, J. Liu, J. Wang, M. Hu, Y. Feng, P. Wang, Y. Wang, N. Nie, J. Zhang, H. Chen, Concentrated hydrogel electrolyte-enabled aqueous rechargeable NiCo//Zn battery working from –20 to 50°C, *ACS Appl Mater Interfaces.* 11 (2018) 49–55.
11. Z. Peng, H. Zhang, I. Ali, J. Li, Y. Ding, L. Deng, T. Han, H. Zhu, X. Zeng, D. Cheng, A rod-on-tube CoMoO4@ hydrogel composite as lithium-ion battery anode with high capacity and stable rate-performance, *J Alloys Compd.* 858 (2021) 157648.
12. X. Cao, D. Tan, Q. Guo, T. Zhang, F. Hu, N. Sun, J. Huang, C. Fang, R. Ji, S. Bi, High-performance fully-stretchable solid-state lithium-ion battery with a nanowire-network configuration and crosslinked hydrogel, *J Mater Chem A Mater.*10 (2022) 11562–11573.
13. X. Lia, S. Zenga, W. Lia, H. Lina, H. Zhonga, H. Zhua, Y. Maia, Stretchable hydrogel electrolyte films based on moisture self-absorption for flexible quasi-solid-state batteries, *Chem Eng J.* 439 (2022) 135741.
14. A. Patel, H. Gupta, S.K. Singh, N. Srivastava, R. Mishra, D. Meghnani, R.K. Tiwari, A. Tiwari, V.K. Tiwari, R.K. Singh, Superior cycling stability of saturated graphitic carbon nitride in hydrogel reduced graphene oxide anode for Sodium-ion battery, *FlatChem.* 33 (2022) 100351.
15. Y. Cheng, X. Chi, J. Yang, Y. Liu, Cost attractive hydrogel electrolyte for low temperature aqueous sodium-ion batteries, *J Energy Storage.* 40 (2021) 102701.
16. R. Rajagopalan, Y. Tang, X. Ji, C. Jia, H. Wang, Advancements and challenges in potassium-ion batteries: A comprehensive review, *Adv Funct Mater.* 30 (2020) 1909486.
17. T. Masese, K. Yoshii, Y. Yamaguchi, T. Okumura, Z.-D. Huang, M. Kato, K. Kubota, J. Furutani, Y. Orikasa, H. Senoh, Rechargeable potassium-ion batteries with honeycomb-layered tellurates as high voltage cathodes and fast potassium-ion conductors, *Nat Commun.* 9 (2018) 1–12.
18. J. Han, M. Xu, Y. Niu, G.-N. Li, M. Wang, Y. Zhang, M. Jia, C. ming Li, Exploration of K 2 Ti 8 O 17 as an anode material for potassium-ion batteries, *Chem Commun.* 52 (2016) 11274–11276.
19. Y. Wu, Y. Xu, Y. Li, P. Lyu, J. Wen, C. Zhang, M. Zhou, Y. Fang, H. Zhao, U. Kaiser, Unexpected intercalation-dominated potassium storage in WS2 as a potassium-ion battery anode, *Nano Res.* 12 (2019) 2997–3002.
20. I. Sultana, M.M. Rahman, S. Mateti, V.G. Ahmadabadi, A.M. Glushenkov, Y. Chen, K-ion and Na-ion storage performances of Co 3 O 4–Fe 2 O 3 nanoparticle-decorated super P carbon black prepared by a ball milling process, *Nanoscale.* 9 (2017) 3646–3654.

21. X. Ren, Q. Zhao, W.D. McCulloch, Y. Wu, MoS2 as a long-life host material for potassium-ion intercalation, *Nano Res.* 10 (2017) 1313–1321.
22. V. Lakshmi, Y. Chen, A.A. Mikhaylov, A.G. Medvedev, I. Sultana, M.M. Rahman, O. Lev, P. v Prikhodchenko, A.M. Glushenkov, Nanocrystalline SnS 2 coated onto reduced graphene oxide: Demonstrating the feasibility of a non-graphitic anode with sulfide chemistry for potassium-ion batteries, *Chem Commun.* 53 (2017) 8272–8275.
23. J. Han, G.-N. Li, F. Liu, M. Wang, Y. Zhang, L. Hu, C. Dai, M. Xu, Investigation of K 3 V 2 (PO 4) 3/C nanocomposites as high-potential cathode materials for potassium-ion batteries, *Chem Commun.* 53 (2017) 1805–1808.
24. K. Chihara, A. Katogi, K. Kubota, S. Komaba, KVPO 4 F and KVOPO 4 toward 4 volt-class potassium-ion batteries, *Chemical Communications.* 53 (2017) 5208–5211.
25. W.B. Park, S.C. Han, C. Park, S.U. Hong, U. Han, S.P. Singh, Y.H. Jung, D. Ahn, K. Sohn, M. Pyo, KVP2O7 as a robust high-energy cathode for potassium-ion batteries: Pinpointed by a full screening of the inorganic registry under specific search conditions, *Adv Energy Mater.* 8 (2018) 1703099.
26. H. Gao, L. Xue, S. Xin, J.B. Goodenough, A high-energy-density potassium battery with a polymer-gel electrolyte and a polyaniline cathode, *Angewandte Chemie.* 130 (2018) 5547–5551.
27. F. Chen, J. Wang, J. Liao, S. Wang, Z. Wen, C. Chen, A hydrogel-enabled free-standing polypyrrole cathode film for potassium-ion batteries with high mass loading and low-temperature stability, *J Mater Chem A Mater.* 9 (2021) 15045–15050.
28. X. Tang, D. Zhou, B. Zhang, S. Wang, P. Li, H. Liu, X. Guo, P. Jaumaux, X. Gao, Y. Fu, A universal strategy towards high–energy aqueous multivalent–ion batteries, *Nat Commun.* 12 (2021) 1–11.
29. P. Wang, Z. Chen, H. Wang, Z. Ji, Y. Feng, J. Wang, J. Liu, M. Hu, J. Fei, W. Gan, A high-performance flexible aqueous Al ion rechargeable battery with long cycle life, *Energy Storage Mater.* 25 (2020) 426–435.
30. Z. Zhang, C. Zuo, Z. Liu, Y. Yu, Y. Zuo, Y. Song, All-solid-state Al–air batteries with polymer alkaline gel electrolyte, *J Power Sources.* 251 (2014) 470–475.
31. H. Wang, P. Wang, Z. Ji, Z. Chen, J. Wang, W. Ling, J. Liu, M. Hu, C. Zhi, Y. Huang, Rechargeable quasi-solid-state aqueous hybrid Al3+/H+ battery with 10,000 ultralong cycle stability and smart switching capability, *Nano Res.* 14 (2021) 4154–4162.
32. T. Xiong, B. He, T. Zhou, Z. Wang, Z. Wang, J. Xin, H. Zhang, X. Zhou, Y. Liu, L. Wei, Stretchable fiber-shaped aqueous aluminum-ion batteries, *EcoMat.* (2022) e12218.
33. Y. Shao, N.N. Rajput, J. Hu, M. Hu, T. Liu, Z. Wei, M. Gu, X. Deng, S. Xu, K.S. Han, Nanocomposite polymer electrolyte for rechargeable magnesium batteries, *Nano Energy.* 12 (2015) 750–759.
34. J. Wang, S. Song, R. Muchakayala, X. Hu, R. Liu, Structural, electrical, and electrochemical properties of PVA-based biodegradable gel polymer electrolyte membranes for Mg-ion battery applications, *Ionics.* 23 (2017) 1759–1769.
35. G. Li, X. Zhang, M. Sang, X. Wang, D. Zuo, J. Xu, H. Zhang, A supramolecular hydrogel electrolyte for high-performance supercapacitors, *J Energy Storage.* 33 (2021) 101931.

36. C. Zhong, Y. Deng, W. Hu, J. Qiao, L. Zhang, J. Zhang, A review of electrolyte materials and compositions for electrochemical supercapacitors, *Chem Soc Rev.* 44 (2015) 7484–7539.
37. N. Kassenova, S. Kalybekkyzy, M.V. Kahraman, A. Mentbayeva, Z. Bakenov, Photo and thermal crosslinked poly (vinyl alcohol)-based nanofiber membrane for flexible gel polymer electrolyte, *J Power Sources.* 520 (2022) 230896.
38. H. Shi, Z. Fang, X. Zhang, F. Li, Y. Tang, Y. Zhou, P. Wu, G. Yu, Double-network nanostructured hydrogel-derived ultrafine Sn–Fe alloy in three-dimensional carbon framework for enhanced lithium storage, *Nano Lett.* 18 (2018) 3193–3198.
39. P. Lin, J. Cong, J. Li, M. Zhang, P. Lai, J. Zeng, Y. Yang, J. Zhao, Achieving ultra-long lifespan Zn metal anodes by manipulating desolvation effect and Zn deposition orientation in a multiple cross-linked hydrogel electrolyte, *Energy Storage Mater.* 49 (2022) 172–180.
40. P. Wang, H. Wang, Z. Chen, J. Wu, J. Luo, Y. Huang, Flexible aqueous Ca-ion full battery with super-flat discharge voltage plateau, *Nano Res.* 15 (2022) 701–708.
41. Y. Li, Z. Zhou, W. Deng, C. Li, X. Yuan, J. Hu, M. Zhang, H. Chen, R. Li, A superconcentrated water-in-salt hydrogel electrolyte for high-voltage aqueous potassium-ion batteries, *ChemElectroChem.* 8 (2021) 1451–1454.
42. X. Lia, S. Zenga, W. Lia, H. Lina, H. Zhonga, H. Zhua, Y. Maia, Stretchable hydrogel electrolyte films based on moisture self-absorption for flexible quasi-solid-state batteries, *Chem. Eng. J.* 439 (2022) 135741.
43. M.F. Gaele, T.M. di Palma, Rechargeable aluminum-air batteries based on aqueous solid-state electrolytes, *Energy Technol.* 10 (2022) 2101046.

Part III

Emerging Applications, Challenges, and Future Perspectives

12
Hydrogels for Metal-Air Batteries

Debasrita Bharatiya, Biswajit Parhi, Sachit Kumar Das, and Sarat K Swain

CONTENTS

12.1 Introduction 235
12.2 Characterizations 239
12.3 Types and Uses of Hydrogels in Metal-Air Batteries 240
    12.3.1 Natural/Biological Polymer Hydrogels 240
    12.3.2 Synthetic/Manmade Polymer Hydrogels 242
    12.3.3 Double Networks by Natural and Synthetic Polymer Hydrogels 244
    12.3.4 Metal-based Polymer Hydrogels 245
    12.3.5 Graphene-based Hydrogels 245
    12.3.6 Mixed/Hybrid Hydrogels 246
12.4 Future Perspectives of Hydrogel Uses in Metal-Air Batteries 250
12.5 Concluding Remarks 251
Acknowledgments 252
References 252

12.1 Introduction

Current green-sustainable technology and its research have influenced the development of many smart materials that show suitable uses in pharmaceutical [1], energy, electronic [2], and electrochemical water splitting [3] sectors with their renewability and economic aspects over conventional technologies. The introduction of metal-ion batteries has been satisfying energy needs in recent times. The influx of structure regulation, the choice of electrolyte and catalyst, experimental environment adjustment, and the insertion of selective defects/dopants have boosted the performance of designed metal-air-based batteries and have overcome the power deficiency [4]. Even the rational design and selection of metal anodes have become an important reason for this success [5]. The selection of electrolytes, catalysts, metals, high design flexibility, rapid discharging/charging capability, proper cell configuration, and clear flow concepts are the major components required to know the energy efficiency, current density, and powder density of metal-air batteries [6]. Energy needs and consumption have increased in the last few years, and to satisfy these needs, intercalated Li-ion batteries have played their role with high energy efficiency, but their limitation lies with the high cost. To overcome these issues, metal-air batteries have come into the picture.

Metal-air batteries are composed of metal as an anode, a membrane separator, and air as an open-cell structured porous cathode, which generates electricity through the redox reaction. The anodic air can be oxygen. The use of metal-air batteries has been dominated by portable and wearable electronic devices, electric vehicles, and other electric devices. It is also reported that metal-air batteries are providing better energy density than the currently popular Li-ion batteries [1]. The oxygen from the environment is availed with better energy density compared with stirred oxygen. In metal-air batteries, the anode end is attached to metal whereas the cathode contains air in a proper electrolyte. The metal anode can have alkaline earth metals, alkali metals, or first-row transition metals, whereas the electrolyte can be aqueous or non-aqueous, which depends on the nature of the anode. The cathode end allows the continuous supply of oxygen from the atmosphere as it contains an open porous architecture. This battery is a combination of traditional batteries and fuel cells. However, limitations with metal-air batteries are oxidation and the decomposition of most electrolytes, which can be eradicated with the choice of suitable electrolytes and solvents. Li et al. discussed different types of solvents as electrolytes such as tetramethylene glycol dimethyl ether (TEGDME), dimethoxyethane (DME), dimethyl sulfoxide (DMSO), and acetonitrile with their poor electrophilic characteristics and stability [7]. These types of batteries are among the most promising batteries due to features like high energy density, safety, low cost, and so on. The theoretical specific energy density of zinc-air batteries is around 1218 Wh/kg, while for lithium-air batteries, it is around 5928 Wh/kg [8].

There are many metal-air batteries developed with different anodic conditions such as Li-air (oxygen), Zn–air, Al-air, and V-air batteries [9, 10]. The use of Li-air batteries is a risk due to the high cost of Li, its hazardous nature, the limited supply of this metal, and the high theoretical specific energy density of 5200 Wh/kg. To avoid this issue, metals like Zn and Al came into the picture due to their low cost, low equilibrium potential, durability, non-flammability, abundance, flat discharge potential, and inertness with a high value of energy density of 1080 Wh/kg. Metal-air batteries have a higher energy density of about 3~30 times higher as compared with lithium-ion batteries because the oxygen stored in the reaction can be collected directly from the environment as the oxygen used in the reaction is already stored in the battery in advance [6]. The electrochemical reversibility and stability of zinc are much better compared with magnesium, lithium, and aluminum [11, 12]. The conversion of chemical energy into electrical energy and vice versa during charge/discharge has been observed in metal-air batteries. Metal-air batteries are predominantly chosen by scientists to design several electronic devices due to their high energy density, low toxicity, rich availability, environmentally friendly nature, and good life cycle time for energy storage technology, transportation, and power source, including electric grids, wind turbines, and photovoltaic panels. Figure 12.1 depicts the above-mentioned advantageous features of metal-air batteries in diverse applications.

The capacitance, powder density, energy efficiency, and activity of any type of battery can be studied by using their electrochemical analysis. Among many such characteristic studies, the oxygen reduction reaction (ORR) and the oxygen evolution reaction (OER) are important phenomena for the energy conversion and storage capacity of the designed batteries or fuel cells [3]. The choice of catalyst addition for the electrochemical study is important and done by metal-organic frameworks (MOF) and carbon-based electro-catalysts over Pt, Ru, and Ir oxides due to their low-cost, bifunctional, and durable properties. To overcome challenges like instability, high cost, non-abundance, and toxicity, concepts such as high conductivity, the requirement of solid electrolyte, availability of abundance metal, preference over low toxic and inert metals, selection of carbon-based catalyst to avoid high costs, favorable electrode interface, and preparation of a reasonable flow concept are

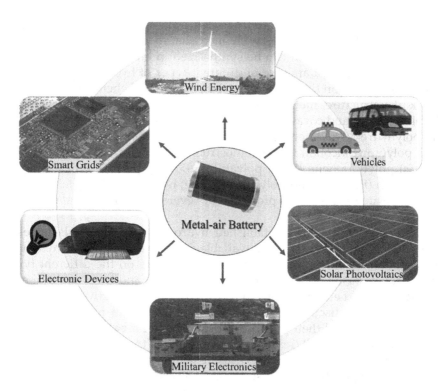

FIGURE 12.1
Illustration of different applications of metal-air batteries.

required [9]. This discussion leads toward the selection of a smart electrolyte that will enhance the flow rate, stable, chemically active, anti-corrosive, and possess low toxicity.

The choice of electrolyte or electrode binders is very important for metal-air batteries [13]. They should be active similar to metals and possess favorable conductivity, as well as excellent capability, and originate from basic salts such as KOH, NaOH, NASICON type, perovskite-type, polymer-based, hydrogel-based, or mixed types. Hydrogels have been found to be quite useful for electrochemical experiments with their stability, solid form and anticorrosive and active chemical interactive power. It has been seen that hydrogels are extensively used in tissue engineering, soft electronics, actuators, supercapacitors, and energy applications due to their entanglement, electrostatic interactions, water-rich cavity, and covalent chemical crosslinking [14]. The other factors responsible for being a good electrolyte are high ionic conductivity, dimensional stability, operability, and mechanical electrodes. With high aspect ratios, large surface area, high chemical stability, tailorable architectures, conductive nanostructures, 3D frameworks, multi-functionalities, stimuli-responsive, self-healing, flexibility, charge transport, and good interface contact with electrodes, hydrogel electrolytes are widely used in many metal-air batteries [15, 16]. Hydrogels are a class of materials that are fabricated from monomers or polymers to form 3D crosslinked networks with interfacial spaces by chemical bonds or physical interactions and can be synthesized in one or multiple steps [17, 18]. The development and use of different types of hydrogels in metal-air batteries depend on their interaction by physical and chemical modes, network-like structure, metal coordination, and other chemical crosslinking by electrostatic and covalent bonding [19]. Neutral synthetic polymers such as polyhydroxyethylmethacrylate (PHEMA), poly(ethylene oxide) (PEO), and poly(vinyl alcohol) (PVA), nanostructured conductive polymeric gels (CPGs) polythiophene (PTh), polyaniline (PANI), poly(3,4-ethylenedioxythiophene) (PEDOT), and polypyrrole (PPy) hydrogels are widely used in energy-related applications promoted by several electrochemical processes [20, 21].

Numerous papers have established hydrogel-based materials for electronic uses, but articles are not available on the participation of hydrogels in metal-air batteries. In this discussion, we focus on the different types of hydrogels based on their chemical composition and their role in metal-air batteries and energy storage purposes. In this perspective, we critically investigate and discuss types of hydrogels, the factors responsible, their general design, and their purpose of use in metal-air batteries with advantages, limitations, and remedies. Furthermore, the discussion includes future perspectives and hydrogel uses in metal-air batteries in electrochemical-related applications. The use of graphene and polymer-based hydrogels has invaded current battery technology and reduced the risks in the use of conventional batteries.

12.2 Characterizations

Studying hydrogels and their characteristic behavior for metal-air batteries is important before their practical applications. The primary investigations performed are X-ray diffraction (XRD), scanning electron microscopy (SEM), atomic force microscopy (AFM), high-resolution transmission electron microscopy (HRTEM), and Fourier transform infrared (FTIR) to critically analyze crystallinity, topography, height and thickness, morphology/external texture, and chemical/functional groups in hydrogels, respectively. These techniques are helpful to gather information before reaching the electrochemical responses. Electrochemical analysis is an important technique to identify current density, charge density, and capacitance of the designed hydrogel, which indicates the suitability of designing metal-air batteries and their long-term use in the development of electronic devices. This experiment is normally performed with cyclic voltammetry (CV) and the charge-discharge cycle test technique to maintain its standard parameters.

Shi et al. discussed polymeric hydrogels and studied their morphology and correlation with energy storage and conversion factors. The team focused on the development of PANI-based CPGs by using phytic acid as the crosslinker and dopant. The SEM and HRTEM micrographs of these designed hydrogels suggested a coral-like dendritic hollow-porous nanofiber with an average diameter of 80 nm caused by slow chemical oxidation and self-assembly polymerization. The enhanced high gravimetric and volumetric pseudocapacitance in the case of polymer hydrogel improved electrochemical performances. The nanostructures and porosity are also helpful for the betterment of electrochemical tests. In this work, PANI hydrogel in 1M H_2SO_4 electrolyte exhibited specific capacitance (sc) of ~480 F/g at a current density of 0.2 A/g due to the ease of electronic and ionic transport brought by the hierarchically porous structure. The team also discussed the PPy hydrogel as electrodes sandwiched within a PVA–H_2SO_4 gel-like electrolyte. The CV curve suggested a good electrochemical performance of the designed hydrogel with a specific capacitance of 380 F/g, excellent rate capability, and areal capacitance as high as 6.4 F/cm^2 at a mass loading of 20 mg/cm^2 due to the high surface area and porous nature confirmed from TEM micrographs. Even the interpenetrating or tangling-like structure of these hydrogels also enhanced the mechanical strength, making these materials more suitable for practical applications [22].

Iwakura et al. designed nickel/zinc (Ni/Zn) batteries with the incorporation of crosslinked potassium polyacrylate polymer hydrogel and KOH aqueous solution as electrolytes and studied the charge-discharge behavior. It is observed that the designed battery cell with polymer hydrogel as the electrolyte showed a remarkable charge-discharge curve and cycle performance compared with the KOH solution with the inhibition of the Zn electrode. The electrolyte was 1 g of hydrogel and 0.01 dm^3 of a 7.3M KOH (aq),

Ni as cathode, and Zn as anode with the electrodeposition of zinc on porous carbon paper (TORAY, TGPH-060) in a 7.3M KOH/0.7M ZnO aqueous solution (25°C) at 60 mA/cm^2 for 1 h used as the experimental condition. The remarkable charge-discharge result lies with the involvement of high ionic conductivity of ca. 6×10^{-1} S/cm at 25°C caused by polyacrylate hydrogel due to a high water-absorbing capacity [18].

The mechanical study and its uses are very important for scaffolding and other template-like uses. Mechanical strength and consistency play an important role in designing many wearable electronics and electrical appliance accessories. Metal coordination and multifunctional and self-healing hydrogels performed excellently for these uses. It is reported that Zn-based PPy hydrogel had very good strength and elasticity. The elasticity enhances the structural changes in the hydrogel whereas the strength can assure its proper use for specific designs of electronics. Other factors such as compression strength, tensile strength, electrical conductivity, stimuli-responsive attitude, and storage modulus give these hydrogels a better performance [23].

12.3 Types and Uses of Hydrogels in Metal-Air Batteries

The need and use of metal-air batteries have gained popularity in recent years. The preparation of such batteries is dependent on the energy conversion phenomenon. To give a better charge storage attitude, hydrogels came into the picture to challenge low electrocatalytic activities influenced by smart preparation methods [20, 24]. High energy density, surface area, environmental protection, low toxicity, ease of handling, and low cost ensured the use of these hydrogels for metal-air batteries [25, 26]. Exploring different compositions of materials, this discussion is focused on hydrogel-based metal-air batteries depending on composition, catalytic efficiency, and electrochemical activity. Depending on the composition and derivation, hydrogels can be classified into natural/biological polymers, synthetic/manmade polymers, double network by natural and synthetic polymers, graphene-based, and mixed/hybrid. Figure 12.2 illustrates the different types of hydrogels based on composition and use in metal-air batteries.

12.3.1 Natural/Biological Polymer Hydrogels

Natural hydrogels are naturally derived hydrogels that are obtained from mammalian extracellular matrices like xanthenes, fibrin, hyaluronate, and collagen, and insects and crustaceans like chitin and chitosan. The renewable, biocompatible, low toxicity, environmentally friendly, abundantly available, and easy extraction of natural polymers are used electrochemically. Ionic conduction is in a polymer framework, which makes these polymers

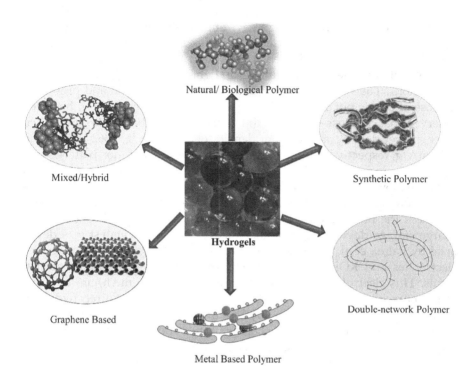

FIGURE 12.2
Schematics of different hydrogels for metal-air batteries.

show good activity as membranes and electrolytes in metal-air batteries. Liu et al. designed chitosan/metal oxides of SiO_2, SnO_2, or ZnO hydrogel membranes (HGMs) prepared by fabrication methods, which act as electrolyte membranes in Al-air batteries, and studied their electrochemical responses. The addition of metal oxides improves the hardness of the hydrogel, making it more useful as an electrolytic membrane. The high ionic conductivity ($\sigma = L/R_b A$) and better discharge performance of CS/metal oxide hydrogels are primarily influenced by the higher water uptake capacity of chitosan and the anti-corrosive metal oxide ion formation, respectively [27]. Zhang et al. reported an interesting bacterial cellulose/potassium hydroxide/potassium iodide (BC/KOH/KI) hydrogel electrolyte and its uses in zinc-air batteries. The 3D networks, flexibility, hydrophilicity, and stable chemical properties of the hydrogel have a mechanical strength of 2.1 MPa, lean narrow charge-discharge voltage gap of 0.46 V at 5 mA/cm², energy efficiency of 73%, and ionic conductivity 54 mS/cm [28]. The high energy density enhanced specific capacitance, and better charge storage capacity of metal-air batteries are incited using low-cost natural polymers as hydrogel electrolytes [29].

Cellulose seems to be used as a potent material for the design of metal-air batteries such as Zn-air, Al-air, and so on, which exhibits porosity, bending flexibility, rich availability, benign environmentally, and ionic conductivity. Fu et al. reported a cellulose-based alkaline-exchange electrolyte membrane from natural cellulose and used it as a solid electrolyte for the development of rechargeable zinc-air batteries in lightweight and flexible electronic applications. The –OH⁻ ions improved the ionic conduction and prevented the loss of ohmic polarization with the crosslinking increment. It has been observed that cellulose-based Zn-air batteries possess better specific capacitance, power density, and current density of 492 mAh/g, 2362 mW/g, and 4650 mA/g, respectively, compared with bare Zn-air and Li-ion batteries [30]. Wang et al. designed excellent batteries derived from Zn-air electrodes and carbon nanowires derived from cellulose as electrocatalysts. The participation of homogeneous and covalently bonded cellulose-based electrocatalysts enhances the number of electrocatalytic sites making Zn-air batteries for charge storage devices [31]. Another interesting electrocatalyst was developed by Palma et al. taking xanthene natural materials as high conductive electrolytes and KOH-HCl, which performs better cell discharge mechanisms. This electrolyte provides a cell capacitance >70 mAh/cm² at a discharge current density of 10 mA/cm² due to the formation of direct hydrogen bonding in the polymeric chain [32]. In particular, naturally abundant polymeric hydrogels are rich in ionic conductivities and discharge the ions at a lower current density, making them good metal-air batteries.

12.3.2 Synthetic/Manmade Polymer Hydrogels

Synthetic and stimuli-responsive polymeric hydrogels are prepared from different sources such as PVA, PEO, PMMA, and PANI, and seem to provide better performance in electrochemical catalytic behaviors. They act as good electrolytes and as cell membranes in metal-air batteries instigated by covalent and hydrogen bond interaction, crosslinking properties, ionic conductivity, and synthetic availability. PVA has been used as an electrode coating, which improved the discharge capability in metal-air batteries of Zn-air and Al-air. Chen et al. discussed PVA/poly(diallyl dimethylammonium chloride) (PDDA) film coating on Zn electrodes that improved its discharging capacity of 317 mAh/g after 60 cycles. These coatings also gathered the mass of Zn during testing and allowed for further use after the preparation of batteries. PVA-based hydrogel provides an energy density of 2905 Wh/L, which is better compared with uncoated Zn electrodes [33]. Wang et al. designed aqueous-based rechargeable metal-air batteries along with superabsorbent sodium polyacrylate (PAN) hydrogel as an electrolyte. This hydrogel electrolyte has a stretching capacity of over ten times at −50°C without any freezing depending on temperature stimuli condition. The design has been done in such a way that the usefulness of this wearable electronic shows its suitability in humid summers and cold winters, limiting the ionic conductivity in any harsh conditions. A highly concentrated electrolyte is frost-resistant because the high concentration of the ions reduces the proportion that facilitates mechanical strength and prevents the formation of H-bonds, making it more useful for the design of a rechargeable battery [34].

Leng et al. designed an interesting stimuli-based polyzwitterionic hydrogel (PZHE) from a [2-(methacryloyloxy) ethyl] dimethyl-(3-sulfopropyl) (SBMA) monomer as the electrolyte in Zn-air batteries with ionic conductivity of 32.0 mS/cm. This hydrogel was used by adapting the high content in hydrophilic groups. The observed Zn^{2+} transference number of 0.65 indicates Zn deposition longevity. Abundant hydrophilic distribution, ionic groups, immobilization, and homogeneous distribution of the zwitterionic polymer provide a long cycle life of over 3500 h and enabled good processability, good adhesion between electrode and electrolyte, and self-healing properties, making it a promising candidate for metal-air battery preparation. The current density observed for Zn-air/PZHE is 3 mA cm^{-2} with hysteresis stability of 220 mV for more than a 50-h cycle, proving reduction in the corrosive mechanism of the battery cell, making it a next-generation wearable electronic [35]. Polymer-derived additives such as PDDA/PVA/ion-containing polymer and α-hydroxy naphthoic acid are used as promising candidates in metal-air batteries – improving the deposition and dissolution of metals. The fast ionic conduction of these additives also hinders the movement of air-electrode in an aqueous medium and diminishes electrode corrosion, giving a suitable platform as electrolytes in batteries [36]. The low-cost, oxygen-rich, segment movement of PEO exhibits ionic conductivity of 10^{-3} S/cm, but PVA PEO substituted PEO provides the synergistic effect, causing better performance, more mechanical stability, and less ionic conductivity than bare PEO – making it a better electrolyte for batteries. Wei et al. elaborately discussed many artificial organic polymers of fibrous and microporous nylon, polyethylene (PE), polypropylene (PP), and polyvinyl chloride (PVC) and their high storing capacity of more electrolytes depending on the capillary forces. This phenomenon leads to the maintenance of a high charge-discharge capacity, good insulating capacity, better mechanical strength, and anti-corrosion resistance [37].

Huang et al. designed a smart electrolyte of proton-incorporated polyacrylamide (PAM) crosslinked by vinyl hybrid silica nanoparticles (VSNPs). The hydrogel is quite effective for battery production governed by crosslinking, toughness, tunable ionic conductivity, and good super-stretchability. The hydrogen bonds present in the hydrogel do not break, rather, they recombine to promote energy distribution in the metal-air batteries. The electrolyte bears a mechanical strain of up to 1500% with the addition of silica nanoparticles [38]. Quasi-solid polymer hydrogel electrolytes such as PVA. PEO, PAM, and PANs improve the stability of metal anodes and the mechanical strength of the batteries with the capability of retaining liquid solutions by avoiding leakage for efficient ionic conductivity [39]. This solution acts as a medium for ion transport and a separator in metal-air batteries while developing an anti-corrosive property with better performance in energy storage applications.

12.3.3 Double Networks by Natural and Synthetic Polymer Hydrogels

Incorporating natural hydrogels with synthetic hydrogels is called double networks or hybrid hydrogels. They give better electronic and mechanical properties. The interpenetration method and synergistic effect help the hydrogel to inherit the benefits of each component and offer superior properties compared with bare component-based material. These hydrogels are mainly developed by two-step and one-step polymerization methods, respectively. The preparation of poly(N-isopropyl acrylamide) (PNIPAM)/PANI and PNIPAM/PPy hydrogels follows the two-step polymerization method where PNIPAM is first constructed then it acts as a supporting matrix for in-situ polymerization of the double network type polymer [20]. The merging of solid and liquid electrolytes improves the mechanical strength and tunes the ionic conductivity in the batteries with improvements in electrode and electrolyte interfacial interaction. Chan et al. synthesized a double-network hydrogel prepared from iota carrageenan (IC) and polyacrylamide (AAm) electrolyte used in Zn batteries. The IC helps the immobilization of ions contributed by environmental attracting and conducting behavior lacking mechanical strength. However, the attachment of PAAm polymer in the double network type structure improves mechanical strength and conductivity, respectively [40]. The ionic conductivity of 2.15×10^{-3} S/cm at room temperature is observed for this hydrogel electrolyte and it seems to be a promising candidate as an electrolyte [40]. Zhang et al. synthesized an interesting double network hydrogel electrolyte PAMs-CNF/KOH/KI prepared from synthetic PAMs and natural cellulose nanofiber (CNF) via in-situ polymerization. The strong mechanical strength and excellent water retention properties of PAMs and the porous structure and nanosize of CNF facilitate anti-dehydration, anti-freezing, ionic conductivity, physical entanglement, non-covalent bonding with excellent mechanical properties, and discharge-charge performance of Zn-air batteries. The addition of KI promotes energy efficiency and long life cycles of batteries, causing a better performance as a hydrogel electrolyte [41]. Sun et al. synthesized an interesting double-network hydrogel-based poly(2-acrylamide-2-methyl propane sulfonic acid) (PAMPS)/methyl cellulose (MC). The excellent water absorption-retention property of PAMPS and good water solubility of MC improve the crosslinking ability of network formation by enhancing the mechanical strength, anti-freezing, toughness, stretchability, elasticity, recoverability, and anti-fatigue nature of prepared hydrogels. The ionic conductivity, specific capacitance, and energy capacitance of these hydrogel-based metal-air batteries are 10^5 mS/cm at RT, 764.7 mAh/g, and 850.2 mWh/g, respectively. PAMPS-K/MC hydrogel provides a good storage property of up to 64% after 96 h, proving itself a good candidate as an electrolyte for the development of metal-air batteries [42]. The double network-based electrolyte exhibits good mechanical strength, anti-fatigue, anti-freezing, good toughness, and stretchability, making this a promising

Hydrogels for Metal-Air Batteries

candidate for metal-air batteries depending on its porosity, size, crosslinking ability, and covalent-non-covalent interactions.

12.3.4 Metal-based Polymer Hydrogels

Metal-polymer-based hydrogels refer to a special class of hydrogel derived by ligand-substitution and polymerization reactions between metals and polymers. The availability of coordinate sites, stability, and long-lasting, anticorrosive properties make the metals useful in supercapacitor applications. The hierarchical polymeric porous structure of polymers invites the metal particles for further interaction via synthesis methods. The uniform distribution, fast kinetics, long cycle life, high structural integrity, and unique compositional features provide a strong platform for metal-polymer-based hydrogels, boosting energy storage and electrocatalysis [25]. Zhao et al. discussed the development of Pt/PANI hetero-structured hydrogel and its electrocatalytic properties. The polymeric pores and catalytically active Pt enhance the electrocatalytic behavior of the hydrogel, making it useful for battery preparation [20]. The promotion of a reaction between oxygen and OH– ions in ORR/OER electrocatalysts and the participation of metal oxides such as IrO_2 and RuO_2 have been extensively used in metal-air batteries due to their low cost, favorable polarization, enhanced electrochemical responses, ionic conductivity caused by the presence of metals, and high OER catalytic activity. The availability of metal oxides as electrode additives in Zn-air batteries improves electrochemical performance with anti-corrosion, stability, mitigated dendritic growth at the Zn electrode, self-discharging capacity, and good ionic conductivity [36].

12.3.5 Graphene-based Hydrogels

Non-metal and carbon-based hydrogels generally refer to graphene and graphene-based nanocomposites that are widely used in energy storage applications. Recently, graphene oxide and its derivatives have prompted research interests in fuel cells, supercapacitors, and rechargeable metal-air battery applications owing to properties such as quantum confinement, a high surface area, long life, high durability, multifunctional, hydrophilicity-hydrophobicity, a flat surface, a uniform hexagonal structure, and hierarchical porosity. Up to now, the most active electrocatalysts are air cathodes. These metal-free carbon materials show significant electrocatalytic behavior responding to different metal electrodes and increasing the number of active sites in themselves. Wang et al. designed 2–20 nm sized graphene quantum dot (GQD) hydrogel as an electrocatalyst and studied its use in metal-air batteries. The low cost increased catalytic active sites, mitigated agglomerations, incited electronic transfer by the unsaturated bond, caused agglomeration of

GQDs, and facilitated electron transfer. The GQD-based hydrogel shows a better discharge performance with a current density of 20 mA/cm^2, making this electrocatalyst a more useful candidate for ORR in both fuel cells and metal/air batteries [43].

Tam et al. designed B-doped graphene quantum dots anchored on a graphene hydrogel (GH-BGQD) via a one-step hydrothermal route. The unique structure, porosity, large area, production of numerous reactive sites, and smart electrolyte mass deposition made this electrolyte more captive as an electrolyte toward OER and ORR. This electrolyte exhibits admirable trifunctional electrocatalytic activity and a hydrogen evolution reaction with excellent long-term stability compared with conventional Pt/C and Ir/C catalysts. The GH-BGQD-based metal-air electrode shows current density, specific capacity, and power density of 10 mA/cm^2, 687 mAh/g, and 112 mW/cm^2, respectively, with an open-circuit voltage of 1.40 V and a discharge voltage of 1.23 V for 100 h acting as a new generation electrocatalyst for battery applications [44]. The carbon-multifunctional materials at nano dimensions with a biocompatible and oxygen-rich group make graphene-based hydrogels more convenient for use in metal-air batteries as a result of their flat and large surface, porous cavities, and π-π interaction causing ionic conductivity.

12.3.6 Mixed/Hybrid Hydrogels

Mixed hydrogels are attributed to the combination of polymer-metal, polymer-polymer, and polymer-carbon-based hydrogels that contribute to supercapacitor application. Among all the types, mixed/hybrid hydrogels seem to show superiority in electrochemical responses. Hybrid hydrogels are tremendously useful for high energy density and durable batteries and flexible and wearable devices. The main challenge the metal-air battery deals with is resistance capability against harsh highly alkaline environments, low water retention uptake, and poor mechanical and low conductivity issues. However, hybrid hydrogels overcome these issues with the help of superior mechanical strength, multifunctional groups of synthetic polymers, the porosity and small size of natural polymers, and oxygen-rich non-metallic carbon alignments and crosslinking networks.

Zhang et al. developed a hybrid hydrogel prepared from synthetic sodium polyacrylate and natural starch. The idea of considering natural cellulose is mainly due to its abundance, low toxicity, improved strength, high molecular entanglement, and promotion of hydrogen bonding. The designed battery shows the physical deformation of 100% stretching during the electrochemical test with excellent expandability-practicability. This hydrogel-based sandwiched zinc-air battery gives high energy density and energy efficiency of 67.5 mW/cm^2 and 68%, respectively, which is useful for energy-consuming gadgets [45]. Kim et al. designed a highly active catalyst derived from an agarose gel-mediated strategy for the selective electrodeposition of MnO_2 and Co_3O_4 through a facile electrodeposition method. This electrocatalyst improves OER and ORR electrocatalytic activities of the battery. The sectionalized bifunctional configuration of transition metal oxide and the natural polymer-based catalyst extend the performance of electrodes, making them useful for energy storage applications [46]. Features of a good catalyst are lower cost, non-noble material, high power density, inexpensive, controlled structure, economical, and highly efficient in behavior, which shows efficacy for charge storage batteries [47]. A similar type of electrocatalyst was developed by Yang et al. who designed a trifunctional electrocatalyst prepared from metallic iron/cobalt and a synthetic polymer of PPy hydrogel. This homogenous and crosslinked electrocatalyst improves the electrocatalytic properties by employing ORR, OER, and HER methods. In contrast to the conventional catalyst, these electrocatalysts encompass better electrocatalytic behavior with a current density of 5 mA/cm^2 and 62% energy efficiency for rechargeable Zn-air batteries [48]. Flexibility, high aspect ratio and surface area, nanosized structure, improved mechanical properties for strain accommodation, and shortest route for charge/mass/ion transport help in the decrement of the interfacial impedance between the electrode and the electrolyte, making the hydrogels suitable for electrochemical uses.

Green chemistry synthetic methods influence the designs of several other hydrogels derived from natural sources and they combine with other precursors for the production of electrochemical accessories. The design of a hydrogel derived from natural xanthan gum polysaccharide and 4–24 wt% HCl solutions improved the ionic conductivities of 10^{-2} S/cm with an anodic efficiency of 80% in Al-air batteries showing notable electrochemical responses. The natural environment, the existence of functional groups, and the adsorption effects of xanthane on metal electrodes help in the reduction of corrosion that occurs in the cell with high anodic efficiency and limited dendrite formation in HCl acidic environments [49]. Zhao et al. synthesized and demonstrated hydrogel electrolytes derived from bacterial cellulose microfibers (BC) and PVA via in-situ polymerization. This electrolyte helps in the increment of the tensile strength of 0.951 MPa nine times compared with bare PVA due to the crosslinking affinity of both the natural and synthetic polymers causing double network, high water retention capability, and more porosity in the hydrogel showing ionic conductivity of 80.8 mS/cm at RT. An interesting feature observed with the addition of the Co_3O_4@Ni cathode to the hydrogel further improves electrochemical and catalytic properties with long cycling performance. The details of different types of hydrogels, their uses, and their electrochemical parameters are listed in Table 12.1. Figure 12.3 shows several advantages of hydrogels and their uses in the development of new-generation batteries and energy storage devices.

TABLE 12.1
Details of Hydrogels, Their Uses, and Important Electrochemical Parameters

Types of hydrogels	Composition	Uses in metal-air batteries	Ionic Conductivity	Current Density	Power Density	Specific Capacitance	References
Polymer	PVA-g-PAA/KCl	Electrolyte for electrochemical capacitor	41.6 mS/cm	1.0 A/g		80 F/g	[14]
Polymer	PAA	Electrolyte for Ni/Zn battery	6×10^{-1} S/cm				[18]
Double network	Pt/PANI	Electrocatalyst in metal-ion batteries			High		[20]
Polymer	PANI	Electrochemical capacitors		0.2 A/g		480 F/g	[22]
Polymer-metal oxide	SiO_2-CS-HGM	The electrolyte in Al-air coin cell	9×10^{-5} S/cm	1.0 mA/cm^2		289 mAh/g	[27]
Polymer-metal	BC-KOH-KI	Electrolyte for zinc-air batteries	54 mS/cm	10 mA/cm^2	129 mW/cm^2	794 mAh/g	[28]
Polymer-metal	[P[(CZn-AAm)]	Electrolytes for zinc-ion batteries	2.15×10^{-3} S/cm	0.25 mA/cm^2		272 mAh/g	[40]
Double network	PAM-CNF/KOH/KI	Electrolyte for zinc-air batteries	223 mS/cm	2 mA/cm^2	43 mW/cm^2	743 mAh/g	[41]
Double network	PAMPS-K	Electrolyte in Zn-air batteries	105 mS/cm	1 mA/cm^2	73.9 mW/cm^2	765 mAh/g	[42]
Graphene	GQD/graphene	Electrocatalyst in metal-air batteries		20 mA/cm^2			[43]
Graphene	GH-BGQD	Electrocatalyst in Zn-air batteries		10 mA/cm^2	112 mW/cm^2	687 mAh/g	[44]

(Continued)

TABLE 12.1 (CONTINUED)
Details of Hydrogels, Their Uses, and Important Electrochemical Parameters

Types of hydrogels	Composition	Uses in metal-air batteries	Ionic Conductivity	Current Density	Power Density	Specific Capacitance	References
Polymer-metal	PA-Na-starch/KOH	Sandwiched Zn-air batteries	82 mS/cm	1 mA/cm^2		673 mAh/g	[45]
Metal-polymer	Agarose/MnO$_2$-Co$_3$O$_4$	Electrocatalyst in metal-air battery		1 mA/cm^2			[46]
Mixed	Xanthan-HCl	Electrochemical accessories for Al-air cells	1.9×10^{-2} S/cm	3 mA/cm^2	3 W/kg		[49]
Mixed	BC/PVA	Electrolyte for Zn-air batteries	80.8 mS/cm	0.5 mA/cm^2			[50]

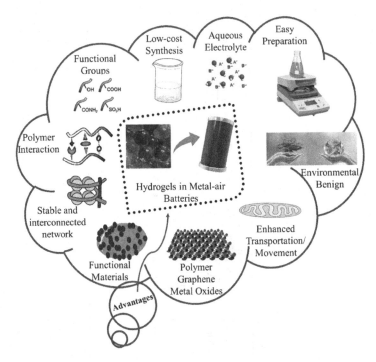

FIGURE 12.3
Advantages and factors of hydrogels for metal-air batteries.

12.4 Future Perspectives of Hydrogel Uses in Metal-Air Batteries

Energy shortages can be resolved by the development of metal-air batteries instead of conventional Li-air batteries. However, a few limitations are found in using traditional metal-air batteries such as electrolyte leakage and water evaporation [26]. These problems are discarded by using current-time smart hydrogels as electrolytes. Figure 12.4 suggests the use and risk of conventional batteries, parameters to control the risk, and types of hydrogel-based materials. The performance of metal-air batteries is influenced by the ionic conduction of hydrogels as the purpose of hydrogels is to store a large number of conductive ions and increase ion transport by adding additives. This phenomenon also promotes the crosslinking attitude of polymers. This also enhances the mechanical strength of polymer-based hydrogels through the availability of large deformation, stretching, winding, and so on. The hydrogels also develop the flexibility of the material as an electrolyte by maintaining mechanical strength due to their crosslinking attitude. To avoid the environmental attacks caused by temperature, composition, and humidity,

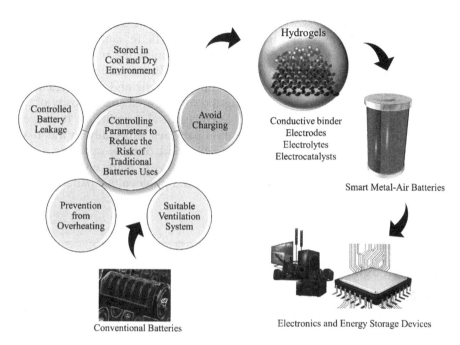

FIGURE 12.4
Risks in conventional battery use and the development of hydrogels for metal-air batteries.

hydrogel electrolytes can be developed by maintaining a freezing point and also due to their self-adjustment ability, interface compatibility, and adsorption efficiency.

12.5 Concluding Remarks

The introduction of green and sustainable energy acts as a promising candidate for the energy storage of batteries, whereas in conventional technologies of batteries, problems of low efficiency, high cost, and safety remain unsolved. In conclusion, hydrogels have been predominately and effectively used in metal-air batteries from various perspectives as membranes, electrolytes, and electrocatalysts. Synthesis methods such as polymerization, adsorption, sol-gel, and Hummer's method have elevated the electrocatalytic and electrochemical behavior of hydrogels. The upgraded electrochemical and catalytic behavior of hydrogels depends on the composition, choice of precursors, cost-effectiveness, environmentally benign nature, abundance, and ease of handling. These hydrogels incited energy density, charge density, specific capacitance, and power density with the development of better performance accessories for metal-air batteries. These hydrogels challenge many harsh acidic and basic conditions for better catalytic activities. Enhanced mechanical strength, good ionic conductivity, water absorbing capacity, adsorption capacity, interfacial adaptability, crosslinking, and the network-like structure of the hydrogel cause catalytic and electrochemical activity of metal-air batteries, affecting the surface area and porosity of the hydrogels. Possessing many good electrochemical properties, hydrogels have a great potential for next-generation batteries, especially in metal-air batteries as flexible, rechargeable, wearable, and durable appliances and energy storage devices to manage the flow of energy in many energy generators, including electric grids, wind turbines, and photovoltaic panels.

Acknowledgments

The authors acknowledge the research support grants awarded by the University Grant Commission (UGC) India and the DS Kothari Post-Doctoral fellowship to Dr. Biswajit Parhi. Dr. Debasrita Bharatiya acknowledges Veer Surendra Sai University of Technology (VSSUT), Burla, India, for the support in research facilities as a research associate.

References

1. K. Y. Lee and D. J. Mooney, Hydrogels for tissue engineering *Chem. Rev.* 101, 1869 (2001).
2. B. Parhi, D. Bharatiya, and S. K. Swain, Surfactant free green synthesis of GOSiMa hybrid nanocomposite for charge storage application *Ceram. Int.* 46, 27184 (2020).
3. M. Zhang, Q. Dai, H. Zheng, M. Chen, and L. Dai, Novel MOF-Derived Co@N-C bifunctional catalysts for highly efficient Zn–air batteries and water splitting *Adv. Mater.* 30, 1705431 (2018).
4. Y. Huang, Y. Wang, C. Tang, J. Wang, Q. Zhang, Y. Wang, and J. Zhang, Revealing the role of crystal orientation of protective layers for stable zinc anode *Adv. Mater.* 31, 1 (2019).
5. H. F. Wang, C. Tang, and Q. Zhang, A review of precious-metal-free bifunctional oxygen electrocatalysts: Rational design and applications in Zn–air batteries *Adv. Funct. Mater.* 28, 1 (2018).
6. X. Han, X. Li, J. White, C. Zhong, Y. Deng, W. Hu, and T. Ma, Metal-air batteries: From static to flow system *Adv. Energy Mater.* 8, 1801396 (2018).
7. Y. Li and J. Lu, Metal–air batteries: Will they be the future electrochemical energy storage device of choice? *ACS Energy Lett.* 2, 1370 (2017).

8. J. Fu, Z. P. Cano, M. G. Park, A. Yu, M. Fowler, and Z. Chen, Electrically rechargeable zinc-air batteries: Progress, challenges, and perspectives *Adv. Mater.* 29, 1604685 (2017).
9. Y. Liu, P. He, and H. Zhou, Rechargeable solid-state Li–air and Li–S batteries: Materials, construction, and challenges *Adv. Energy Mater.* 8, 1701602 (2018).
10. Y. Shao, F. Ding, J. Xiao, J. Zhang, W. Xu, S. Park, J. G. Zhang, Y. Wang, and J. Liu, Making Li-air batteries rechargeable: Material challenges *Adv. Funct. Mater.* 23, 987 (2013).
11. D. Bharatiya, B. Parhi, and S. K. Swain, Preparation, characterization and dielectric properties of GO based ZnO embedded mixed metal oxides ternary nanostructured composites *J. Alloys Compd.* 869, 159274 (2021).
12. D. Capsoni, M. Bini, S. Ferrari, E. Quartarone, and P. Mustarelli, Recent advances in the development of Li–air batteries *J. Power Sources* 220, 253 (2012).
13. P. Pei, K. Wang, and Z. Ma, Technologies for extending zinc–air battery's cyclelife: A review *Appl. Energy* 128, 315 (2014).
14. Z. Wang, F. Tao, and Q. Pan, A self-healable polyvinyl alcohol-based hydrogel electrolyte for smart electrochemical capacitors *J. Mater. Chem. A* 4, 17732 (2016).
15. Y. S. Zhang and A. Khademhosseini, Advances in engineering hydrogels *Science (80)* 356, 1 (2017).
16. A. Koschella, M. Hartlieb, and T. Heinze, A "click-chemistry" approach to cellulose-based hydrogels *Carbohydr. Polym.* 86, 154 (2011).
17. Q. Peng, J. Chen, T. Wang, X. Peng, J. Liu, X. Wang, J. Wang, and H. Zeng, Recent advances in designing conductive hydrogels for flexible electronics *InfoMat* 2, 843 (2020).
18. C. Iwakura, H. Murakami, S. Nohara, N. Furukawa, and H. Inoue, Charge–discharge characteristics of nickel/zinc battery with polymer hydrogel electrolyte *J. Power Sources* 152, 291 (2005).
19. Y. Guo, J. Bae, F. Zhao, and G. Yu, Functional hydrogels for next-generation batteries and supercapacitors *Trends Chem.* 1, 335 (2019).
20. F. Zhao, Y. Shi, L. Pan, and G. Yu, Multifunctional nanostructured conductive polymer gels: Synthesis, properties, and applications *Acc. Chem. Res.* 50, 1734 (2017).
21. Y. Shi, L. Peng, Y. Ding, Y. Zhao, and G. Yu, Nanostructured conductive polymers for advanced energy storage *Chem. Soc. Rev.* 44, 6684 (2015).
22. Y. Shi and G. Yu, Designing hierarchically nanostructured conductive polymer gels for electrochemical energy storage and conversion *Chem. Mater.* 28, 2466 (2016).
23. Y. Guo, J. Bae, Z. Fang, P. Li, F. Zhao, and G. Yu, Hydrogels and hydrogel-derived materials for energy and water sustainability *Chem. Rev.* 120, 7642 (2020).
24. M. J. Tan, B. Li, P. Chee, X. Ge, Z. Liu, Y. Zong, and X. J. Loh, Acrylamide-derived freestanding polymer gel electrolyte for flexible metal-air batteries *J. Power Sources* 400, 566 (2018).
25. Z. Fang, A. Zhang, P. Wu, and G. Yu, Inorganic cyanogels and their derivatives for electrochemical energy storage and conversion *ACS Mater. Lett.* 1, 158 (2019).
26. P. Zhang, K. Wang, P. Pei, Y. Zuo, M. Wei, X. Liu, Y. Xiao, and J. Xiong, Selection of hydrogel electrolytes for flexible zinc–air batteries *Mater. Today Chem.* 21, 100538 (2021).

27. Y. Liu, Q. Sun, X. Yang, J. Liang, B. Wang, A. Koo, R. Li, J. Li, and X. Sun, High-performance and recyclable Al-air coin cells based on eco-friendly chitosan hydrogel membranes *ACS Appl. Mater. Interfaces* 10, 19730 (2018).
28. Y. Zhang, Y. Chen, X. Li, M. Alfred, D. Li, F. Huang, and Q. Wei, Bacterial scellulose hydrogel: A promising electrolyte for flexible zinc-air batteries *J. Power Sources* 482, 228963(2021).
29. N. Chawla, Recent advances in air-battery chemistries *Mater. Today Chem.* 12, 324 (2019).
30. J. Fu, J. Zhang, X. Song, H. Zarrin, X. Tian, J. Qiao, L. Rasen, K. Li, and Z. Chen, A flexible solid-state electrolyte for wide-scale integration of rechargeable zinc–air batteries *Energy Environ. Sci.* 9, 663 (2016).
31. R. Wang, J. Cao, S. Cai, X. Yan, J. Li, W. M. Yourey, W. Tong, and H. Tang, MOF@cellulose derived Co–N–C nanowire network as an advanced reversible oxygen electrocatalyst for rechargeable zinc–air batteries *ACS Appl. Energy Mater.* 1, 1060 (2018).
32. T. M. Di Palma, F. Migliardini, M. F. Gaele, and P. Corbo, Physically cross-linked xanthan hydrogels as solid electrolytes for Al/air batteries *Ionics* 25, 4209 (2019).
33. X. Chen, Z. Zhou, H. E. Karahan, Q. Shao, L. Wei, and Y. Chen, Recent advances in materials and design of electrochemically rechargeable zinc–air batteries *Small* 14, 1801929 (2018).
34. H. Wang, J. Liu, J. Wang, M. Hu, Y. Feng, P. Wang, Y. Wang, N. Nie, J. Zhang, H. Chen, Q. Yuan, J. Wu, and Y. Huang, Concentrated hydrogel electrolyte-enabled aqueous rechargeable NiCo//Zn battery working from −20 to 50°C *ACS Appl. Mater. Interfaces* 11, 49 (2019).
35. K. Leng, G. Li, J. Guo, X. Zhang, A. Wang, X. Liu, and J. Luo, A safe polyzwitterionic hydrogel electrolyte for long-life quasi-solid state zinc metal batteries *Adv. Funct. Mater.* 30, 2001317 (2020).
36. J. Yi, P. Liang, X. Liu, K. Wu, Y. Liu, Y. Wang, Y. Xia, and J. Zhang, Challenges, mitigation strategies and perspectives in development of zinc-electrode materials and fabrication for rechargeable zinc–air batteries *Energy Environ. Sci.* 11, 3075 (2018).
37. Y. Wei, Y. Shi, Y. Chen, C. Xiao, and S. Ding, Development of solid electrolytes in Zn–air and Al–air batteries: from material selection to performance improvement strategies *J. Mater. Chem. A* 9, 4415 (2021).
38. Y. Huang, M. Zhong, F. Shi, X. Liu, Z. Tang, Y. Wang, Y. Huang, H. Hou, X. Xie, and C. Zhi, An intrinsically stretchable and compressible supercapacitor containing a polyacrylamide hydrogel electrolyte *Angew. Chemie - Int. Ed.* 56, 9141 (2017).
39. C. Wang, J. Li, Z. Zhou, Y. Pan, Z. Yu, Z. Pei, S. Zhao, L. Wei, and Y. Chen, Rechargeable zinc-air batteries with neutral electrolytes: Recent advances, challenges, and prospects *EnergyChem* 3, 100055 (2021).
40. C. Y. Chan, Z. Wang, Y. Li, H. Yu, B. Fei, and J. H. Xin, Single-ion conducting double-network hydrogel electrolytes for long cycling zinc-ion batteries *ACS Appl. Mater. Interfaces* 13, 30594 (2021).
41. Y. Zhang, H. Qin, M. Alfred, H. Ke, Y. Cai, Q. Wang, F. Huang, B. Liu, P. Lv, and Q. Wei, Reaction modifier system enable double-network hydrogel electrolyte for flexible zinc air batteries with tolerance to extreme cold conditions *Energy Storage Mater.* 42, 88 (2021).

42. N. Sun, F. Lu, Y. Yu, L. Su, X. Gao, and L. Zheng, Alkaline double-network hydrogels with high conductivities, superior mechanical performances, and antifreezing properties for solid-state zinc–air batteries *ACS Appl. Mater. Interfaces* 12, 11778 (2020).
43. M. Wang, Z. Fang, K. Zhang, J. Fang, F. Qin, Z. Zhang, J. Li, Y. Liu, and Y. Lai, Synergistically enhanced activity of graphene quantum dots/graphene hydrogel composites: A novel all-carbon hybrid electrocatalyst for metal/air batteries *Nanoscale* 8, 11398 (2016).
44. T. Van Tam, S. G. Kang, M. H. Kim, S. G. Lee, S. H. Hur, J. S. Chung, and W. M. Choi, Smart windows: Electro-, thermo-, mechano-, photochromics, and beyond *Adv. Energy Mater.* 9, 1902066 (2019).
45. Y. Zhang, Y. Chen, M. Alfred, F. Huang, S. Liao, D. Chen, D. Li, and Q. Wei, Alkaline sodium polyacrylate-starch hydrogels with tolerance to cold conditions for stretchable zinc-air batteries *Compos. Part B Eng.* 224, 109228 (2021).
46. G. P. Kim, H. H. Sun, and A. Manthiram, Design of a sectionalized MnO_2 Co_3O_4 electrode via selective electrodeposition of metal ions in hydrogel for enhanced electrocatalytic activity in metal-air batteries *Nano Energy* 30, 130 (2016).
47. D. Bharatiya, S. Kumar, S. Raghunandan, and P. Paik, Dielectrics of graphene oxide decorated with nanocomposite silica-coated calcium copper titanate (CCTO) nanoparticles *J. Mater. Sci.* 54, 6272 (2019).
48. J. Yang, X. Wang, B. Li, L. Ma, L. Shi, Y. Xiong, and H. Xu, Novel iron/cobalt-containing polypyrrole hydrogel-derived trifunctional electrocatalyst for self-powered overall water splitting *Adv. Funct. Mater.* 27, 1606497 (2017).
49. F. Migliardini, T. M. Di Palma, M. F. Gaele, and P. Corbo, Solid and acid electrolytes for Al-air batteries based on xanthan-HCl hydrogels *J. Solid State Electrochem.* 22, 2901 (2018).
50. N. Zhao, F. Wu, Y. Xing, W. Qu, N. Chen, Y. Shang, M. Yan, Y. Li, L. Li, and R. Chen, Flexible hydrogel electrolyte with superior mechanical properties based on poly(vinyl alcohol) and bacterial cellulose for the solid-state zinc–air batteries *ACS Appl. Mater. Interfaces* 11, 15537 (2019).

13

Hydrogels for Flexible/Wearable Batteries

Runwei Mo

CONTENTS

13.1 Introduction 257
13.2 Types of Hydrogels 259
13.3 Properties of Hydrogels 260
13.4 Material Design and Synthetic Approach 261
    13.4.1 Self-Assembly of Graphene Hydrogel via Supramolecular Interactions 261
    13.4.2 Hydrogel Formation from a Conductive Filler Suspension 262
    13.4.3 Crosslinking Conductive Polymers by Dopant Molecules 263
13.5 Recent Advancements in Flexible/Wearable Batteries 263
    13.5.1 Hydrogel Electrolytes for Flexible Li-Ion/Na-Ion Batteries 263
    13.5.2 Hydrogel Electrolytes for Flexible Multivalent Metal-Ion Batteries 265
13.6 Challenges 266
13.7 Conclusion and Future Perspectives 268
References 269

13.1 Introduction

With the rapid development of emerging wearable electronic devices, higher requirements are placed on energy storage devices. An energy storage device not only needs to have excellent electrochemical performance but also needs to have good mechanical stability. Batteries and supercapacitors are currently the most successful commercial energy storage devices. Compared with supercapacitors, batteries have a higher energy density, which is considered to be the most potential energy storage device in wearable electronics. Therefore, to meet the requirements of wearable electronic devices, it is very important to develop reliable and durable high-performance flexible batteries [1–3].

Different from traditional batteries, flexible batteries need to consider both mechanical stability and electrochemical performance. Therefore, it is necessary to develop the structural design of flexible batteries. In terms

DOI: 10.1201/9781003351566-16   *257* of electrode design, on the one hand, it is necessary to provide more active sites and good electrochemical stability. On the other hand, it also needs to have good matching with the current collector. Furthermore, in terms of the current collector design, it is necessary to provide high charge transport capacity and fast ion migration channels. It also needs to have excellent flexibility, including bending, stretching, and even twisting. In addition to the pursuit of electrochemical properties such as high energy density, power density, and long cycle life, flexible batteries for wearable electronic devices also require high safety, low cost, biocompatibility, bendability, a lightweight nature, and mechanical stability [4, 5]. In order to achieve the goal of high safety standards, the materials used for flexible batteries need to be nonflammable, have green environmental protection, and be non-toxic [5].

In recent years, hydrogels as functional materials have been considered to have an increasingly important role in the structure of flexible batteries. Hydrogels possess a variety of unique properties, mainly including tunable tensile strength, biocompatibility, elastic modulus, and multipurpose functionality, which can address key issues in flexible batteries. Therefore, hydrogels with micro-nano structures are considered to have broad application prospects in flexible batteries [6, 7]. It is worth noting that hydrogels also possess soft and wet properties, which are mainly attributed to the interconnected network structure between the water-filled polymer chains. This also enables hydrogels to be used as electrolytes in flexible batteries. More importantly, by designing the composition and structure of the hydrogel, its electromechanical and physicochemical properties can be optimized, mainly including specific surface area, ion/charge transfer rate, electrical conductivity, pore structure, and mechanical stability [6]. Based on these unique properties of hydrogels, this opens a new avenue for applications in high-performance flexible energy storage devices [7].

In recent years, researchers have carried out a lot of research work on the electrical, chemical, and mechanical properties of hydrogels, and have made good progress [8]. However, there are still some problems that need to be solved. For example, hydrogels are often prone to poor mechanical properties during the integration of flexible devices, which is mainly attributed to the low binding strength of individual polymer chains and weak noncovalent interactions [8]. However, to alleviate the above problems, several approaches have been attempted, including multiple interactions, the inclusion of nanoparticles in a hydrogel matrix, double networks, topologies, and so on [9–11]. As flexible devices often need to consider biocompatibility, there is still a problem that most synthetic polymer hydrogels cannot satisfy: both good biocompatibility and excellent mechanical properties. Therefore, it is important to develop flexible energy storage devices with good biocompatibility and excellent mechanical properties. In addition, it is worth noting that the rapid development of flexible energy storage devices also brings environmental problems. In recent years, research on the renewable utilization of hydrogels made from natural resources (such as alginate, cellulose, chitosan, etc.) has aroused widespread interest, which can effectively solve the increasingly serious global energy crisis [12].

In this chapter, the latest research results of hydrogels as key materials in the field of flexible batteries are summarized. Based on an in-depth understanding of the types and properties of hydrogels, the synthesis strategies of hydrogels are introduced, including hydrogel formation from a conductive filler suspension, self-assembly of graphene hydrogel via supramolecular interactions, and crosslinking conductive polymers by dopant molecules. In addition, the latest research results of hydrogels as electrolytes in the field of flexible batteries are summarized. Finally, the challenges and prospects in this field are analyzed.

13.2 Types of Hydrogels

Hydrogels are formed by the crosslinking force between hydrogels and polymer chains, therefore, they can be divided into three types: physical hydrogels, chemical hydrogels, and hybrid hydrogels. The formation of physical hydrogels is mainly based on non-covalent forces to achieve crosslinking between polymer chains, including ionic forces, hydrogen bonds, polyelectrolyte complexation, Van der Waals interactions, and hydrophobic forces. The weak interactions between polymer chains in physical hydrogels lead to their reversible responses to environmental changes. Generally speaking, when physical hydrogels are subjected to external stimuli, they tend to exhibit fragile, disordered, and mechanically weak properties. For example, physical hydrogels readily dissolve in water and organic solvents when they encounter heating, which is also mainly attributed to weaker interactions.

Unlike physical hydrogels, chemical hydrogels are formed by covalent bonding between polymer chains, so they are also called permanent hydrogels. It is worth noting that chemical hydrogels do not exhibit reversible reactions like physical hydrogels. This is because there are strong covalent bonds between the macromolecular chains in the chemical hydrogel, which makes the hydrogel not dissolve in the surrounding medium. The preparation of chemical hydrogels is mainly formed by the chemical crosslinking of small molecules (such as genipin, formaldehyde, diglycidyl ether, etc.) in various ways. The condensation reaction is achieved by the formation of covalent bonds between small molecules and polymers. In the specific condensation reaction process, the crosslinking agent can be selected in a variety of ways, in which a mono-functional group containing the crosslinking agent can usually be used, and it can also contain two functional groups [13, 14]. In addition, another reaction mechanism during the preparation of chemical hydrogels is the free radical mechanism. In this mechanism, there are certain requirements for the selection of crosslinking agents. Specifically, in order to be able to bond with the polymer chains on both sides, the crosslinking agent needs to contain at least two double bonds. Generally speaking, the hydrogels prepared by chemical crosslinking have outstanding mechanical, electrical, and chemical properties [15].

In the preparation process of hydrogels, there are both chemical crosslinking and physical crosslinking methods, which are called hybrid hydrogels. The properties of hybrid hydrogels can be regulated by different proportions of chemical crosslinking and physical crosslinking during hydrogel preparation. After the rational design, the properties of hybrid hydrogels can combine the advantages of chemical hydrogels and physical hydrogels, thereby avoiding the problems existing in chemical and physical hydrogels. Therefore, hybrid hydrogels have attracted extensive interest among researchers in recent years.

13.3 Properties of Hydrogels

A very important property of hydrogels is the swelling property, which is the basis for other properties. Based on the description of the Flory–Rehner theory, swelling can be expressed by a function, which is a function of the elastic properties of the polymer chains and their compatibility with water molecules [16]. Generally speaking, the swelling of hydrogels occurs in three stages: the first stage is the primary bound water, in which the water diffuses into the hydrogel network. The second stage is secondary bound water, in which the relaxation of the polymer chains occurs. The third stage is free water, in which the expansion of the hydrogel network occurs. It is worth noting that the hydrogels prepared by selecting polymers with hydrophilic pendant groups have a wide range of applications in the biomedical field. The reason for their analysis is that the presence of hydrophilic groups can not only facilitate the interaction with biological tissues but also fully absorb water. Usually, when the hydrogel is in a fully swollen state, it often exhibits various properties, such as softness, viscoelasticity, elasticity, and biocompatibility.

When stimulated by the external environment, hydrogels exhibit different responses. The different conditions of the external environment can be divided into three categories. The first is biological stimuli, such as enzymes. The second is chemical stimulation, such as ionic strength and pH. The third is physical stimuli, such as light and temperature. Under the stimulation of different external environmental conditions, the influence on the hydrogel is also different. Among them, physical stimuli tend to be external, while biological and chemical stimuli tend to be internal. In addition, there is a type of hydrogel that is a pH-responsive hydrogel, in which the hydrogel contains covalent crosslinks of ionic pendant groups and can respond to changes in pH. The analytical reason is that under the condition of high pH, the hydrogel with anionic groups swells, which leads to electrostatic repulsion. In contrast, under the condition of low pH, the hydrogel with cationic groups swells. Based on the change mechanism of water-polymer or polymer-polymer interactions, the hydrogel exhibits swelling/de-swelling behavior when the temperature changes, which is called temperature-sensitive hydrogel [17]. Temperature-sensitive hydrogels can be divided into positive temperature-sensitive and negative temperature-sensitive according to the different temperature response mechanisms [18, 19]. Exhibiting swelling behavior at temperatures above the upper critical solution temperature is called positive temperature sensitivity. Conversely, exhibiting swelling behavior at temperatures below the lower critical solution temperature is called negative temperature sensitivity.

13.4 Material Design and Synthetic Approach

13.4.1 Self-Assembly of Graphene Hydrogel via Supramolecular Interactions

Graphene hydrogels are prepared by self-assembly based on the reaction mechanism of supramolecular interactions. Usually, other materials can be introduced during the preparation of graphene hydrogels to form hybrid hydrogel composites. The more commonly used method is the hydrothermal synthesis method, which is to prepare graphene hydrogels or graphene-based composite hydrogels from high-temperature aqueous solutions under high vapor pressure [20]. In recent years, researchers have carried out a lot of research on graphene hydrogels. Shi et al. successfully prepared graphene hydrogels with a crosslinked three-dimensional network structure by hydrothermal synthesis [21]. Specifically, graphene oxide (GO) was first uniformly dispersed in an aqueous solution and poured into an autoclave. Finally, the autoclave was placed in a drying oven and heated at 180°C for 12 hours. During this reaction process, GO self-assembles to form graphene hydrogels under supramolecular interactions, including electrostatic interactions, hydrogen bonding, π-π stacking, and so on. It is worth noting that the graphene hydrogel can be regulated by changing the GO concentration. In order to further improve the conductivity of graphene hydrogels, hydroiodic acid or hydrazine can be selected for further treatment to achieve the purpose of removing oxidized groups [22].

To realize the versatility of graphene hydrogels, other functional materials can be introduced during the preparation of graphene hydrogels, which also enriches the structural design of graphene-based composite hydrogels. For example, Cong et al. successfully introduced Fe_3O_4 nanoparticles into graphene hydrogels via a hydrothermal reaction [23]. The uniform distribution of Fe_3O_4 nanoparticles in graphene hydrogels is realized based on the interaction principle of ferrous ion reduction of GO. It is worth noting that the as-prepared Fe_3O_4/graphene hydrogel exhibits magnetic properties. In addition, Liu et al. successfully prepared TiO_2/graphene hydrogels by the same synthetic method using the principle of self-assembly [24]. The prepared TiO_2/graphene hydrogel exhibits excellent mechanical strength. Different functions can be achieved by introducing different materials into graphene hydrogels, which lays the foundation for the practical application of graphene-based hydrogels.

13.4.2 Hydrogel Formation from a Conductive Filler Suspension

Electrically conductive hydrogels are widely used as an important type of hydrogel. Currently, the most common method is to introduce conductive materials, such as conductive polymers [25], metal nanoparticles [26], and carbon-based materials [27], into the preparation process of hydrogels to form electrically conductive hydrogels. Among them, carbon materials are widely used as conductive materials in the preparation of electrically conductive hydrogels due to their excellent properties. The results show that the introduction of carbon materials can not only improve electrical conductivity but also improve their mechanical properties [28]. As mentioned in the previous section, graphene, a typical 2D material among carbon materials, is often added to hydrogels as a conductive material. Recently, researchers successfully synthesized graphene oxide (GO)/polyacrylic acid (PAA) hydrogels by thermally induced self-assembly [29]. The as-prepared GO/PAA hydrogel exhibits excellent mechanical properties, which is mainly attributed to the fact that the functional groups on the GO surface can form hydrogen bonds with PAA on the one hand, and on the other hand, make it have good dispersibility in aqueous solution. The researchers used a similar approach to composite with other hydrogel matrices, such as polyvinyl alcohol, poly(acrylamide), and poly(N-isopropyl acrylamide) [30, 31].

In addition, carbon nanotubes, as a classical one-dimensional material among carbon materials, have received extensive attention as conductive additive materials. Recently, Zhang et al. successfully introduced one-armed carbon nanotubes into hydrogels by means of supramolecular interaction and crosslinking [32]. It is worth noting that in order to achieve the dispersibility of single-walled carbon nanotubes in PVA solution, the researchers innovatively used PDA to modify the surface of single-walled carbon nanotubes so that the prepared composite hydrogel showed excellent properties. In the preparation process of carbon-based composite hydrogels, carbon materials usually only act as conductive fillers, so two aspects play a decisive role in the performance of composite hydrogels. On the one hand, it is necessary to ensure the high dispersibility of the carbon material in the pre-gel solution, and on the other hand, it is necessary to construct a structurally stable polymer chain.

13.4.3 Crosslinking Conductive Polymers by Dopant Molecules

In addition to the above-mentioned introduction of additive materials in the hydrogel preparation process, researchers have also attempted to form 3D hydrogel networks by directly crosslinking conductive polymer chains. It is worth noting here that the insulating polymer matrix is not used throughout hydrogel preparation [33]. For the first time, Macdiarmid et al. successfully prepared a hydrogel containing only conducting polymer by constructing a sufficiently concentrated solution of polyaniline (PANI) [34]. The results show that the gelation process in high-concentration PANI solution conditions is caused by the one-dimensional nematic arrangement of the related amorphous PANI and the physical crosslinking of the crystalline domains.

In addition to gelation at high concentrations, researchers are also actively exploring other conditions for gelation. Bao et al. successfully achieved gelation by doping polyaniline with phytic acid and conducted an in-depth study of its gelation mechanism [33]. The results show that the hydrogel has a 3D hierarchical micro-nano structure, which is mainly attributed to the fact that each phytic acid molecule can react with multiple polyaniline polymer chains by protonation of the amino groups on polyaniline. Besides PANI mentioned above, polypyrrole (PPy) is another conducting polymer that can form electrically conductive hydrogels by self-crosslinking. Yu et al. innovatively proposed the successful synthesis of conductive PPy hydrogels with micro-nano structures by constructing a water/organic dual-phase interface [35]. It is worth noting that this method can control the morphology of PPy hydrogels, including hollow spherical structures and hierarchical network structures, by changing the molar ratio of the crosslinking agent phytic acid to the monomeric pyrrole. In addition, the structure of PPy hydrogel can also be regulated by changing the types of reaction raw materials, for example, by adding copper phthalocyanine-3,4',4'',4'''-tetrasodium tetrasulfonate to the reaction raw materials salt, and successfully prepared PPy hydrogels with a fibrous structure. This method can control the morphology and structure of the hydrogel by changing the types and molar ratios of the crosslinking agent and the conductive polymer monomer. Therefore, this method has attracted extensive research interest among researchers.

13.5 Recent Advancements in Flexible/Wearable Batteries

13.5.1 Hydrogel Electrolytes for Flexible Li-Ion/Na-Ion Batteries

Due to their high energy density and long cycle life, Li-ion and Na-ion batteries have become prime candidates for energy storage devices in electronic devices [36, 37]. The rapid development of wearable electronic devices puts forward higher requirements for battery technology. In addition to the pursuit of long cycle life and higher energy density, the development of flexible Li-ion or Na-ion batteries has attracted extensive interest among researchers [38, 39]. In addition, in order to solve the problems of flammability and high price in traditional organic electrolytes, hydrogel electrolytes have become an emerging research field. Zhong et al. successfully prepared PAAm hydrogel-based electrolytes and assembled them into aqueous rechargeable sodium-ion batteries [40]. The research results showed that the battery exhibited excellent electrochemical performance, especially since its capacity retention rate was as high as 90% after 100 charge-discharge cycle tests. Cao et al. also successfully prepared a chitosan–Li$^+$ supramolecular hydrogel electrolyte and assembled it into a lithium-ion capacitor [41]. It is worth noting that the Li-ion capacitor exhibits excellent mechanical stability and can achieve multiple twists and bends. The researchers also tested the electrochemical performance of the lithium-ion capacitor, and the results showed that the area capacity can reach 10 mF/cm under the condition of a current density of 1.8 mA/cm^2. More importantly, the lithium-ion capacitor exhibits excellent cycle stability, which can achieve 10000 charge-discharge cycles. [41]

In order to meet the requirements of wearable electronic devices for energy storage devices, flexible batteries need to exhibit stable electrochemical performance under mechanical stress conditions. Recently, Liu et al. successfully synthesized a PVA-LiNO$_3$ hydrogel electrolyte and assembled it into a quasi-solid aqueous rechargeable lithium-ion battery [42]. The research results show that the capacity retention rate of the battery is still as high as 79.8% after 500 charge-discharge cycle tests under a current density of 0.5 A/g. More importantly, the battery exhibited good mechanical flexibility. Even under mechanical stress conditions, such as extrusion, bending, folding, and twisting, the battery still exhibits stable electrochemical performance. It is worth noting that the battery can still work normally after many cuts without short-circuit problems. This indicates that the battery can meet the power supply requirements of wearable electronic devices as a flexible energy storage device [42]. Apart from the flexible Li-ion battery, the researchers also applied this research strategy to flexible Na-ion batteries [43]. For example, Guo et al. successfully prepared a highly safe and flexible aqueous Na-ion battery, in which the anode material and cathode material were nano-sized NaTi$_2$(PO$_4$)$_3$@C and Na$_{0.44}$MnO$_2$, respectively [43]. The battery not only exhibited excellent electrochemical performance, such as high power density, high volume energy, and long cycle life, but also showed good mechanical stability. In particular, it can still have stable electrochemical performance under the action of external force.

In addition to designing hydrogels with good mechanical properties, self-healing hydrogels have attracted extensive attention in recent years as an emerging research field due to their unique properties. Tang et al. successfully prepared a novel water-in-salt hydrogel electrolyte with self-healing ability, which mainly contained chitosan, PAAm, and tetraethoxysilane [44].

It is worth noting here that the as-prepared hydrogel electrolyte not only provides a wide electrochemical window of 2.6 V and a high ionic conductivity of 51.3 mS/cm but also exhibits excellent self-healing ability. The researchers also tested the electrochemical performance of the self-healed hydrogel electrolyte. Compared with the hydrogel electrolyte before self-healing, the ionic conductivity of the hydrogel electrolyte after self-healing is almost unchanged. It exhibits excellent cycling stability after being assembled into Li-ion batteries. In particular, its energy density is as high as 110.7 Wh/kg after 2000 charge-discharge cycle tests [44]. Therefore, the above research work provides a good example of the application of hydrogels in flexible batteries and presents a potential solution for designing high-performance hydrogel-based flexible Li-ion/Na-ion batteries in the future.

13.5.2 Hydrogel Electrolytes for Flexible Multivalent Metal-Ion Batteries

In addition to the above-mentioned flexible Li-ion batteries or Na-ion batteries, multivalent metal batteries based on multivalent reaction mechanisms have also attracted extensive attention from researchers [45, 46]. Researchers have successfully fabricated a novel PVA-based self-healing hydrogel and assembled it into a zinc-ion battery [47]. The hydrogel exhibits excellent self-healing ability, which can be achieved even when a parallel mismatch of the cut surfaces. The batteries assembled from the hydrogel exhibited excellent cycling stability and good rate capability. The capacity retention rate is still as high as 97.1% after 1000 charge-discharge cycles at a current density of 1 A/g. The power density of the battery is as high as 5089 W/kg. More importantly, stable electrochemical performance can still be obtained after repeated self-healing treatments. It is worth noting here that the PANa-based hydrogel exhibits excellent mechanical properties and good self-healing ability, where the strength and tensile strains of the hydrogel after healing are 205 kPa and 1000%, respectively. The main reason is that the ionic interaction between the PANa chain and iron ions significantly enhances the self-healing ability of the hydrogel. The researchers tested the performance of the PANa hydrogel in an alkaline environment, and the results showed that the hydrogel has excellent alkali resistance stability. In addition, Zhi et al. also successfully developed a PAAm-based hydrogel electrolyte and applied it in zinc-ion batteries [48]. The as-prepared hydrogels exhibit highly compressible properties, which are mainly attributed to the abundant H-bonds and strong covalent bonds in the polymer matrix. The assembled batteries exhibit excellent rate performance and cycling stability. The specific capacity is as high as 277.5 mAh/g under the condition of a charge-discharge rate of 1 C, and it can meet the long-cycle test requirements of 1000 times under the condition of a charge-discharge rate of 4 C.

The properties of hydrogel electrolytes play an important role in the performance of flexible batteries. Therefore, studies on the failure mechanisms of hydrogel electrolytes have attracted extensive interest. The failure mechanism of hydrogel electrolytes is mainly divided into two types, one is the tension failure mechanism that occurs in the hydrogel electrolyte, which belongs to the mechanism mentioned above. The other is the lamination failure mechanism of hydrogels. To solve the problem of lamination failure, researchers have carried out research work on the structural design of hydrogels [49]. Specifically, this hydrogel electrolyte is fabricated by laminating hydrogels with opposite charges. It is worth noting that the hydrogel electrolyte exhibits excellent adhesion properties, which are mainly attributed to the efficient energy dissipation in the hydrogel matrix and the interaction between opposite charges in the hydrogel interface. The researchers also made changes to the electric field, enabling effective control of the adhesion strength in the interface. The hydrogel electrolyte exhibited excellent mechanical properties and stable electrochemical performance after being assembled into a battery. Stable power output can be achieved even after 2000 bending and tensile strains. More importantly, this unique structure can effectively prevent the self-discharge problem that is often prone to occur during long-term storage, thus also illustrating the excellent electrochemical stability of the battery. In addition to being a flexible energy storage device for wearable electronic devices, flexible high-energy batteries can also be integrated with flexible piezoresistive sensors, thereby realizing the combination of flexible battery technology and flexible sensing technology [48]. This new technology has broad application prospects for intelligent medical electronic devices.

13.6 Challenges

To further optimize the electrochemical and mechanical properties of flexible energy storage devices, an in-depth analysis of the advantages and disadvantages of current flexible energy storage technologies is important. For example, compared with reported organic electrolytes, hydrogel electrolytes have a narrower chemically stable potential window, which also severely limits further improvements in energy density. In addition, it is worth noting that the wide chemically stable potential window can also facilitate the selection of electrode materials with more ion insertion/extraction potentials in the electrolyte. However, the choice of electrode materials in hydrogel electrolytes is often limited by the water-splitting reaction [36]. Therefore, developing a hydrogel-electrolyte-based flexible energy storage technology with a wide chemically stable potential window is an important research direction in the future. In addition to the chemically stable potential window mentioned above, another urgent problem to be solved is the optimization of polymer host materials, which mainly needs to avoid side effects in electrochemical reactions and ensure good compatibility of solvent water and electrolyte salts. In order to further improve the electrochemical performance of hydrogel electrolyte-based flexible energy storage devices, especially the charge-discharge performance under high-rate conditions, it is necessary for the hydrogel electrolyte to provide sufficient ionic conductivity to realize fast ion migration during the process of electrochemical reaction. Therefore, in the design of hydrogel electrolytes, there is an urgent need to improve the tolerance of high concentrations of electrolyte salts and good affinity for charged electrolyte ions.

Besides the electrochemical properties of hydrogel electrolytes, the mechanical properties of hydrogel electrolytes are also a very important concern for flexible energy storage devices. Compared with pristine rigid or weak hydrogels, double-crosslinked hydrogels with strong reversible and covalent bonds, such as dynamic boronate, ionic, and hydrogen bonds, exhibit superior elasticity and mechanical strength. In addition, to realize the multifunctional characteristics of flexible energy storage devices, it is necessary to develop the self-healing function of hydrogel electrolytes and the integration technology with other smart functions. In the process of designing self-healing hydrogel electrolytes, it is necessary to explore the interactions between polymer molecules and their correlation with external stimuli. It is worth noting here that in the process of assembling the hydrogel electrolyte into an energy storage device, the fabrication process has a great influence on the comprehensive performance of the energy storage device. In terms of the fabrication of hydrogel electrolyte-based energy storage devices, the fabrication process of organic electrolyte-based energy storage devices that have been commercialized can be used for reference. Compared with the fabrication process of organic electrolyte-based energy storage devices, the fabrication process of hydrogel electrolyte-based energy storage devices has obvious advantages in terms of fabrication cost and device safety. The high safety and good biocompatibility of hydrogel-electrolyte-based flexible energy storage devices make them have broad application prospects in the field of medical smart electronic devices.

The hydrogel electrolyte can provide a high concentration of electrolytic ions and high ionic conductivity during the electrochemical reaction, which makes it possible to achieve high-rate charge/discharge after being assembled into a battery. However, due to the limitation of electrode materials, the actual rate performance of hydrogel electrolyte-based flexible batteries is quite different from the theoretical value. Therefore, it is very important to develop new electrode materials that can fully exploit the advantages of hydrogel electrolytes for realizing high-performance hydrogel electrolyte-based flexible energy storage devices. At this stage, researchers have carried out substantial research on monovalent ion flexible energy storage devices (such as lithium-ion batteries, sodium-ion batteries, etc.), and have made good progress in improving electrochemical performance and mechanical stability. However, research on multivalent ion flexible energy storage devices (such as aluminum ion batteries, magnesium ion batteries, etc.) is still relatively lacking. It is worth the effort in multivalent ion flexible batteries, which provide more avenues for the development of high-performance flexible energy storage devices that can be used in wearable electronic devices.

13.7 Conclusion and Future Perspectives

In this chapter, recent developments in hydrogel electrolytes for flexible batteries in wearable electronics were discussed. Hydrogels not only possess excellent electrochemical properties but also have good mechanical properties, which make them promising as electrolytes in flexible energy storage devices. Due to the unique properties of hydrogels, the limitations of traditional liquid electrolytes in flexible energy storage devices are overcome. Hydrogel electrolytes with highly stretchable properties play an important role in the development of high-performance flexible energy storage devices. Furthermore, as the application fields of wearable electronic devices become more and more extensive, this puts forward higher requirements for the functionality of flexible energy storage devices. To adapt to multifunctional flexible energy storage devices, it is urgent to develop multifunctional hydrogel electrolytes. For example, extreme condition tolerance, self-healing ability, sol-gel transition protection, electrochromic behavior, and other additional features. Therefore, the research work on hydrogels should pay more attention to the following aspects in the future.

1. In order to adapt to long-term practical applications, the mechanical properties of hydrogels need to be improved. At this stage, through the structural design of hydrogels, basic mechanical properties tests such as bending, folding, and stretching can be realized. However, the reported hydrogels cannot meet the requirements for long-term use. Fatigue or deformation usually occurs after repeated stretching for about 1000 cycles, which seriously affects the stability of their electrochemical performance. This also cannot meet the requirements of wearable electronic products for flexible energy storage devices. The self-healing ability of hydrogels has aroused widespread interest among researchers. However, the current hydrogel self-healing process still requires external energy, such as irradiation or heating, and the whole process is still time-consuming. Therefore, how to realize the rapid self-healing process through the supramolecular interaction between self-molecules is a research direction worthy of attention in the future.

2. In order to meet different application requirements, the water content of the hydrogel electrolyte needs to be designed. It is worth noting that the water content in the hydrogel directly affects the ionic

conductivity and cycling stability of the battery. In order to avoid the volatilization of water in the hydrogel, the existing methods mainly deal with the rigid packaging of the hydrogel, which limits its application in wearable electronic devices. To stabilize the water content in hydrogels, it is necessary to use water molecules as the host polymer network or solvent, which is achieved by developing substances with strong interaction energy with water molecules. In addition, for different application fields, it is necessary to precisely control the water content inside the hydrogel. Therefore, how to realize long-cycle and stable flexible energy storage devices without being restricted by rigid packaging is a research direction that needs to be focused on in the future.

3. To expand potential applications in emerging smart applications, the multifunctional properties of hydrogels need to be exploited. In addition to the various functional properties of hydrogels mentioned in this chapter, such as extreme temperature tolerance, self-healing ability, electrochromic nature, and so on, some new functional properties still need to be actively explored. For example, to further expand the potential applications of hydrogels in robotics or artificial muscles, it is necessary to combine the ability to respond to stimuli through hydrogels with flexible properties. In addition, hydrogel-based flexible energy storage devices can also be systematically integrated with other functional devices to realize novel devices with collective functions. Therefore, how to realize the multifunctional integrated design of flexible energy storage devices is a research direction that needs to be focused on in the future.

References

1. E. Pomerantseva, F. Bonaccorso, X. Feng, Y. Cui, Y. Gogotsi, Energy storage: The future enabled by nanomaterials, *Science*. 366 (2019) eaan8285.
2. S. Wang, J. Xu, W. Wang, G.J.N. Wang, R. Rastak, F. Molina-Lopez, J.W. Chung, S. Niu, V.R. Feig, J. Lopez, T. Lei, S.K. Kwon, Y. Kim, A.M. Foudeh, A. Ehrlich, A. Gasperini, Y. Yun, B. Murmann, J.B.H. Tok, Z. Bao, Skin electronics from scalable fabrication of an intrinsically stretchable transistor array, *Nature*. 555 (2018) 83–88.
3. W. Liu, M.-S. Song, B. Kong, Y. Cui, Flexible and stretchable energy storage: Recent advances and future perspectives, *Adv. Mater.* 29 (2017) 1603436.
4. Z. Chen, J.W.F. To, C. Wang, Z. Lu, N. Liu, A. Chortos, L. Pan, F. Wei, Y. Cui, Z. Bao, A three-dimensionally interconnected carbon nanotube–conducting polymer hydrogel network for high-performance flexible battery electrodes, *Adv. Energy Mater.* 4 (2014) 1400207.

5. L. Li, Z. Wu, S. Yuan, X.B. Zhang, Advances and challenges for flexible energy storage and conversion devices and systems, *Energy Environ. Sci.* 7 (2014) 2101–2122.
6. Z. Wang, H. Li, Z. Tang, Z. Liu, Z. Ruan, L. Ma, Q. Yang, D. Wang, C. Zhi, Z.F. Hydrogel electrolytes for flexible aqueous energy storage devices, *Adv. Funct. Mater.* 28 (2018) 1804560.
7. F. Zhao, J. Bae, X. Zhou, Y. Guo, G. Yu, F. Zhao, J. Bae, X. Zhou, Y. Guo, G. Yu, Nanostructured functional hydrogels as an emerging platform for advanced energy technologies, *Adv. Mater.* 30 (2018) 1801796.
8. Y. Li, K.G. Blank, F. Sun, Y. Cao, Synthesis of novel hydrogels with unique mechanical properties, *Front. Chem.* 8 (2020) 942.
9. Z. Li, Z. Lin, Recent advances in polysaccharide-based hydrogels for synthesis and applications, Aggregate. 2 (2021) 21.
10. C.H. Li, C. Wang, C. Keplinger, J.L. Zuo, L. Jin, Y. Sun, P. Zheng, Y. Cao, F. Lissel, C. Linder, X.Z. You, Z. Bao, A highly stretchable autonomous self-healing elastomer, *Nat. Chem.* 8 (2016) 618–624.
11. X. Sun, F. Yao, J. Li, Nanocomposite hydrogel-based strain and pressure sensors: A review, *J. Mater. Chem. A.* 8 (2020) 18605–18623.
12. T. Su, M. Zhang, Q. Zeng, W. Pan, Y. Huang, Y. Qian, W. Dong, X. Qi, J. Shen, Mussel-inspired agarose hydrogel scaffolds for skin tissue engineering, *Bioact. Mater.* 6 (2021) 579–588.
13. S.M. Magami, P.K.T. Oldring, L. Castle, J.T. Guthrie, The physical–chemical behaviour of amino cross-linkers and the effect of their chemistry on selected epoxy can coatings' hydrolysis to melamine and to formaldehyde into aqueous food simulants, *Prog. Org. Coatings.* 78 (2015) 325–333.
14. N. Reddy, Y. Tan, Y. Li, Y. Yang, Effect of glutaraldehyde crosslinking conditions on the strength and water stability of wheat gluten fibers, *Macromol. Mater. Eng.* 293 (2008) 614–620.
15. E.S. Costa-Júnior, E.F. Barbosa-Stancioli, A.A.P. Mansur, W.L. Vasconcelos, H.S. Mansur, Preparation and characterization of chitosan/poly(vinyl alcohol) chemically crosslinked blends for biomedical applications, *Carbohydr. Polym.* 76 (2009) 472–481.
16. D. Buenger, F. Topuz, J. Groll, Hydrogels in sensing applications, *Prog. Polym. Sci.* 37 (2012) 1678–1719.
17. J. Sun, G. Jiang, Y. Wang, F. Ding, Thermosensitive chitosan hydrogel for implantable drug delivery: Blending PVA to mitigate body response and promote bioavailability, *J. Appl. Polym. Sci.* 125 (2012) 2092–2101.
18. C.J. Wu, A.K. Gaharwar, P.J. Schexnailder, G. Schmidt, Development of biomedical polymer-silicate nanocomposites: A materials science perspective, *Mater.* 3 (2010) 2986–3005.
19. S.J. Kim, K.J. Lee, I.Y. Kim, S.I. Kim, Swelling kinetics of interpenetrating polymer hydrogels composed of poly(vinyl alcohol)/chitosan, *J. Macromol. Sci. Part A.* 40 (2003) 501–510.
20. K. Hu, D.D. Kulkarni, I. Choi, V.V. Tsukruk, Graphene-polymer nanocomposites for structural and functional applications, *Prog. Polym. Sci.* 39 (2014) 1934–1972.
21. Y. Xu, K. Sheng, C. Li, G. Shi, Self-assembled graphene hydrogel via a one-step hydrothermal process, *ACS Nano.* 4 (2010) 4324–4330.
22. V.H. Luan, H.N. Tien, L.T. Hoa, N.T.M. Hien, E.S. Oh, J. Chung, E.J. Kim, W.M. Choi, B.S. Kong, S.H. Hur, Synthesis of a highly conductive and large surface

area graphene oxide hydrogel and its use in a supercapacitor, *J. Mater. Chem. A*. 1 (2012) 208–211.
23. H.P. Cong, X.C. Ren, P. Wang, S.H. Yu, Macroscopic multifunctional graphene-based hydrogels and aerogels by a metal ion induced self-assembly process, *ACS Nano*. 6 (2012) 2693–2703.
24. Z. Zhang, F. Xiao, Y. Guo, S. Wang, Y. Liu, One-pot self-assembled three-dimensional TiO_2-graphene hydrogel with improved adsorption capacities and photocatalytic and electrochemical activities, *ACS Appl. Mater. Interfaces*. 5 (2013) 2227–2233.
25. B.W. Yao, H.Y. Wang, Q.Q. Zhou, M.M. Wu, M. Zhang, C. Li, G.Q. Shi, Ultrahigh-conductivity polymer hydrogels with arbitrary structures, *Adv. Mater*. 29 (2017) 1700974.
26. G.P. Kim, H.H. Sun, A. Manthiram, Design of a sectionalized MnO_2-Co_3O_4 electrode via selective electrodeposition of metal ions in hydrogel for enhanced electrocatalytic activity in metal-air batteries, *Nano Energy*. 30 (2016) 130–137.
27. Y. Meng, Y. Zhao, C. Hu, H. Cheng, Y. Hu, Z. Zhang, G. Shi, L. Qu, All-graphene core-sheath microfibers for all-solid-state, stretchable fibriform supercapacitors and wearable electronic textiles, *Adv. Mater*. 25 (2013) 2326–2331.
28. M.F.L. De Volder, S.H. Tawfick, R.H. Baughman, A.J. Hart, Carbon nanotubes: Present and future commercial applications, *Science*. 339 (2013) 535–539.
29. J. Shen, B. Yan, T. Li, Y. Long, N. Li, M. Ye, Mechanical, thermal and swelling properties of poly(acrylic acid)-graphene oxide composite hydrogels, *Soft Matter*. 8 (2012) 1831–1836.
30. H. Bai, C. Li, X. Wang, G. Shi, A pH-sensitive graphene oxide composite hydrogel, *Chem. Commun*. 46 (2010) 2376–2378.
31. V. Alzari, D. Nuvoli, S. Scognamillo, M. Piccinini, E. Gioffredi, G. Malucelli, S. Marceddu, M. Sechi, V. Sanna, A. Mariani, Graphene-containing thermoresponsive nanocomposite hydrogels of poly(N-isopropylacrylamide) prepared by frontal polymerization, *J. Mater. Chem*. 21 (2011) 8727–8733.
32. M. Liao, P. Wan, J. Wen, M. Gong, X. Wu, Y. Wang, R. Shi, L. Zhang, Wearable, healable, and adhesive epidermal sensors assembled from mussel-inspired conductive hybrid hydrogel framework, *Adv. Funct. Mater*. 27 (2017) 1703852.
33. L. Pan, G. Yu, D. Zhai, H.R. Lee, W. Zhao, N. Liu, H. Wang, B.C.K. Tee, Y. Shi, Y. Cui, Z. Bao, Hierarchical nanostructured conducting polymer hydrogel with high electrochemical activity, *Proc. Natl. Acad. Sci. USA* 109 (2012) 9287–9292.
34. A.G. MacDiarmid, Y. Min, J.M. Wiesinger, E.J. Oh, E.M. Scherr, A.J. Epstein, Towards optimization of electrical and mechanical properties of polyaniline: Is crosslinking between chains the key?, *Synth. Met*. 55 (1993) 753–760.
35. Y. Wang, Y. Shi, L. Pan, Y. Ding, Y. Zhao, Y. Li, Y. Shi, G. Yu, Dopant-enabled supramolecular approach for controlled synthesis of nanostructured conductive polymer hydrogels, *Nano Lett*. 15 (2015) 7736–7741.
36. H. Kim, J. Hong, K.Y. Park, H. Kim, S.W. Kim, K. Kang, Aqueous rechargeable Li and Na ion batteries, *Chem. Rev*. 114 (2014) 11788–11827.
37. Q. Yang, S. Cui, Y. Ge, Z. Tang, Z. Liu, H. Li, N. Li, H. Zhang, J. Liang, C. Zhi, Porous single-crystal $NaTi_2(PO_4)_3$ via liquid transformation of TiO_2 nanosheets for flexible aqueous Na-ion capacitor, *Nano Energy*. 50 (2018) 623–631.
38. Y. Shi, J. Zhang, A.M. Bruck, Y. Zhang, J. Li, E.A. Stach, K.J. Takeuchi, A.C. Marschilok, E.S. Takeuchi, G. Yu, A tunable 3D nanostructured conductive gel

framework electrode for high-performance lithium ion batteries, *Adv. Mater.* 29 (2017) 1603922.
39. Y. Shi, X. Zhou, J. Zhang, A.M. Bruck, A.C. Bond, A.C. Marschilok, K.J. Takeuchi, E.S. Takeuchi, G. Yu, Nanostructured conductive polymer gels as a general framework material to improve electrochemical performance of cathode materials in Li-ion batteries, *Nano Lett.* 17 (2017) 1906–1914.
40. L. Zhong, Y. Lu, H. Li, Z. Tao, J. Chen, High-performance aqueous sodium-ion batteries with hydrogel electrolyte and alloxazine/CMK-3 anode, *ACS Sustain. Chem. Eng.* 6 (2018) 7761–7768.
41. L. Cao, M. Yang, D. Wu, F. Lyu, Z. Sun, X. Zhong, H. Pan, H. Liu, Z. Lu, Biopolymer-chitosan based supramolecular hydrogels as solid state electrolytes for electrochemical energy storage, *Chem. Commun.* 53 (2017) 1615–1618.
42. Z. Liu, H. Li, M. Zhu, Y. Huang, Z. Tang, Z. Pei, Z. Wang, Z. Shi, J. Liu, Y. Huang, C. Zhi, Towards wearable electronic devices: A quasi-solid-state aqueous lithium-ion battery with outstanding stability, flexibility, safety and breathability, *Nano Energy.* 44 (2018) 164–173.
43. Z. Guo, Y. Zhao, Y. Ding, X. Dong, L. Chen, J. Cao, C. Wang, Y. Xia, H. Peng, Y. Wang, Multi-functional flexible aqueous sodium-ion batteries with high safety, *Chem.* 3 (2017) 348–362.
44. L. Dai, O. Arcelus, L. Sun, H. Wang, J. Carrasco, H. Zhang, W. Zhang, J. Tang, Embedded 3D Li+ channels in a water-in-salt electrolyte to develop flexible supercapacitors and lithium-ion batteries, *J. Mater. Chem. A.* 7 (2019) 24800–24806.
45. K. Liu, X. Pan, L. Chen, L. Huang, Y. Ni, J. Liu, S. Cao, H. Wang, Ultrasoft self-healing nanoparticle-hydrogel composites with conductive and magnetic properties, *ACS Sustain. Chem. Eng.* 6 (2018) 6395–6403.
46. D. Huang, J. Liu, J. Wang, M. Hu, F. Mo, G. Liang, C. Zhi, An intrinsically self-healing NiCo||Zn rechargeable battery with a self-healable ferric-ion-crosslinking sodium polyacrylate hydrogel electrolyte, *Angew. Chemie Int. Ed.* 57 (2018) 9810–9813.
47. S. Huang, F. Wan, S. Bi, J. Zhu, Z. Niu, J. Chen, A self-healing integrated all-in-one zinc-ion battery, *Angew. Chemie Int. Ed.* 58 (2019) 4313–4317.
48. Z. Wang, F. Mo, L. Ma, Q. Yang, G. Liang, Z. Liu, H. Li, N. Li, H. Zhang, C. Zhi, Highly compressible cross-linked polyacrylamide hydrogel-enabled compressible $Zn-MnO_2$ battery and a flexible battery-sensor system, *ACS Appl. Mater. Interfaces.* 10 (2018) 44527–44534.
49. J. Duan, W. Xie, P. Yang, J. Li, G. Xue, Q. Chen, B. Yu, R. Liu, J. Zhou, Tough hydrogel diodes with tunable interfacial adhesion for safe and durable wearable batteries, *Nano Energy.* 48 (2018) 569–574.

14

Hydrogels for Flexible/ Wearable Supercapacitors

Jintao Guo, Xiaodong Qin, Wenkun Jiang, and Yinghui Han

CONTENTS

14.1 Introduction 273
14.2 Iterative Evolution of Hydrogels and Classification in
 Supercapacitors 274
14.3 Hydrogels Applied to Electrolytes of Supercapacitors 277
    14.3.1 Polyvinyl Alcohol Hydrogel 277
    14.3.2 Polyacrylamide Hydrogels 277
    14.3.3 Polyacrylic Acid Hydrogels 278
    14.3.4 Polyethylene Glycol-based Hydrogels 278
    14.3.5 Biopolysaccharide Hydrogels 279
    14.3.6 Composite Hydrogels 279
    14.3.7 Gelatin-based Conductive Hydrogels 280
    14.3.8 Alginate-based Conductive Hydrogels 281
    14.3.9 Cellulose-based Conductive Hydrogels 282
14.4 Hydrogel Applied to Electrodes of Supercapacitors 282
    14.4.1 PANI Hydrogels 282
    14.4.2 PPy Hydrogels 283
    14.4.3 PEDOT and Its Derivative Hydrogels 283
14.5 Technical Advantages and Challenges of Hydrogels in the
 Application of Supercapacitors 284
14.6 Conclusion 287
Acknowledgments 287
References 287

14.1 Introduction

Supercapacitors have become the mainstay of energy storage devices due to their high power density, fast charging and discharging, wide operating temperature window, and high cycle stability. The traditional supercapacitor is composed of five parts: positive and negative electrodes, electrolyte, diaphragm, collector, and shell. The electrodes used in traditional supercapacitors are often not flexible, and the electrolyte used is liquid, which is easy to leak, unstable, not environmentally friendly, and has security risks. All-solid-state supercapacitors have excellent characteristics such as high and low-temperature stability, low impedance, environmental protection, high ripple resistance, and high reliability, which have become the first choice for the energy supply of wearable devices/sensors. Solid electrolytes and flexible electrodes are the two main components of an all-solid supercapacitor. The material selection of solid electrolytes and flexible electrodes plays a key role in the electrochemical and mechanical properties of flexible wearable supercapacitors [1]. Hydrogel is a three-dimensional (3D) network structure formed by the physical or chemical interaction of polymer and expands without dissolving by absorbing water through the interaction with water molecules, so it can effectively blend the solutions containing conductive ions into its polymer network while realizing ion conductance to avoid the risk of leakage. As a typical soft wet material containing waterlocking groups, the hydrogel has both mechanical flexibility and ion conduction characteristics, as well as good stability under an electric field, even at a high voltage such as 15V, so it is considered a promising candidate material for all-solid flexible supercapacitors with efficient energy storage [2–5]. Furthermore, hydrogels are easy to shape into a variety of configurations, which can change shape under the action of external forces. Hydrogels can be functionally modified by manipulating the polymer composition and solution composition [6–8]. For example, hydrogels can effectively dissipate the potential energy of external force by introducing functional polymers or nanomaterials into the polymer network, so that hydrogels have high tensile properties and high toughness. Some studies have used water/ethylene glycol and water/glycerol as solutions in crosslinked polymer networks to improve the cold resistance and freezing resistance of hydrogels. Therefore, hydrogels have unique endowments of various functional properties, such as high strength, high tensile toughness, high cold resistance, and self-healing conditions, which are suitable for use as substrates for flexible wearable supercapacitors [9, 10].

14.2 Iterative Evolution of Hydrogels and Classification in Supercapacitors

The hydrogel can be traced back to 1894 and was originally used to describe the colloid of some inorganic salts. Over time, hydrogel now means something completely different than it did at the beginning. The world's first mature hydrogel product Ivalon (crosslinking of formaldehyde tree ester and ethylene) came out in 1949, and the advent of polyhydroxyethyl methacrylate (PHEMA) in 1960 promoted the prosperity of the hydrogel market.

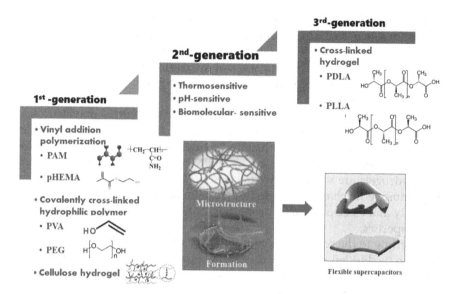

FIGURE 14.1
The history of hydrogel iteration.

Looking back on the development history of hydrogel, it can be generally divided into three generations, as shown in Figure 14.1.

The first generation of hydrogels is mainly divided into three categories. The first category is the polymer of the free radical-induced chain addition reaction of the one monomer, mainly represented by polyacrylamide (PAM) and PHEMA. These materials were invented more than 70 years ago and are still being used in a number of biologically relevant scenarios. The second category is covalently crosslinked hydrophilic polymers, mainly represented by polyvinyl alcohol (PVA) and polyethylene glycol (PEG), which are mainly used in tissue engineering. The third category is cellulosic-based hydrogels, which mainly serve as dispersion substrates to facilitate ion delivery.

Second-generation hydrogels are mainly PEG/polyester block copolymers. Compared with the first generation, the second-generation hydrogel can convert chemical energy into mechanical energy to achieve the specified function. Dating back to the 1970s, this class of stimulus-responsive hydrogels can respond to external environmental changes such as temperature or pH. The stimulus-response hydrogels can be generally divided into three categories. The first category is a temperature-sensitive hydrogel, which can show the phase transition between the gel state and the sol state from low temperature to high temperature. The main representatives are BASF Pluronics or poloxamers of the UK's Empire Chemical Industries. The second category is a pH-sensitive hydrogel, which is hydrolyzed at high or low pH, respectively.

The third category is biomolecule-sensitive hydrogels, which can respond to changes in the concentration of specific biomolecules through conformational changes. Such biomolecular sensitivity allows it to play an important role in wearable human-machine device materials. The third generation is "crosslinking" hydrogels, which mainly adjust the mechanical properties and degradation properties of hydrogels through stereo complex, clathrates, metal-ligand coordination, and synthesis of peptide chains. For example, one of the main applications of the stereocomplex method is to prepare injectable hydrogels from two amphiphilic copolymers polylactic acid (PLLA) and polylactic acid enantiomer (PDLA) blocks. There are also studies on the construction of hydrogels with hydrophobic cavities that can accommodate different molecules using cyclodextrin complexes. The "crosslinking" hydrogels are valued in wearable supercapacitor applications because they improve mechanical properties while maintaining their original conductivity and electrochemical properties, rather than sacrificing these superior properties.

According to the configuration design of flexible all-solid supercapacitors, they can be divided into vertical and planar types, as shown in Figure 14.2. The vertical supercapacitor consists of two working electrodes and a solid-state electrolyte assembled in a sandwich configuration, and the direction of electron transport is perpendicular to the electrode. In planar supercapacitors, two electrode materials are designed on the same plane, so that ions in the electrolyte can travel freely parallel to the electrode.

The performance of flexible solid-state supercapacitors mainly depends on the choice of electrode and electrolyte materials. Hydrogels have applications in both electrodes and electrolytes of flexible supercapacitors, and the research of hydrogels as solid electrolytes of flexible supercapacitors is mainstream. The application and classification of hydrogel materials in flexible supercontainers are shown in Figure 14.3. Some hydrogel materials with high conductivity and thermal stability, including polyvinyl alcohol, polyacrylamide, polyacrylic acid, polyethylene glycol, biopolysaccharide, composite hydrogel, gelatin, alginate, cellulose, and so on, are often selected

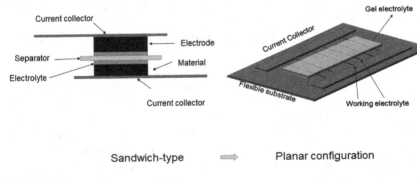

FIGURE 14.2
Two typical structures of a flexible supercapacitor.

Flexible/Wearable Supercapacitors

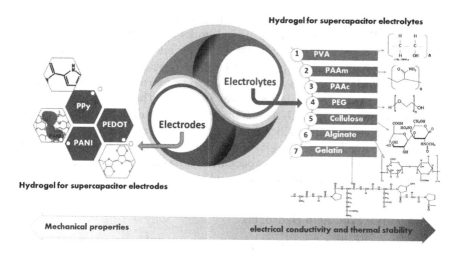

FIGURE 14.3
Classification of hydrogel for flexible supercapacitor electrodes and electrolyte.

as solid electrolytes of flexible supercapacitors, while stretchable hydrogels, such as PPy, PANI, PEDOT, and so on, are often used as flexible electrodes.

14.3 Hydrogels Applied to Electrolytes of Supercapacitors

14.3.1 Polyvinyl Alcohol Hydrogel

PVA is a linear polymer containing hydroxyl groups that help form intramolecular and intermolecular hydrogen bonds. PVA hydrogels have the advantages of excellent water solubility, biocompatibility, a non-toxic nature, and anti-corrosion properties, which can not only be used normally in acid/medium/alkaline conditions but also show higher mechanical strength compared with other polymer hydrogels. It is an ideal electrolyte for a solid flexible supercapacitor. In order to improve the electrochemical performance and achieve effective ion conduction, Kufian et al. [11] studied the conductivity of PVA electrolytes with H_3PO_4 as a proton donor. The addition of H_3PO_4 increased the conductivity of the PVA membrane, which may be caused by the increased number of H_3PO_4 charge carriers.

14.3.2 Polyacrylamide Hydrogels

Polyacrylamide (PAAm) hydrogels are a typical linear water-soluble polymer, which is widely used in artificial fats because of their excellent biocompatibility [12]. It also has good performances of high ionic conductivity, high water content, easy manufacturing, wide pH tolerance, and so on [13]. Dual networks [14], semi-interpenetrating networks [15], nanoparticle composite [15], and microgel enhancement [16] are often adopted to resist the shortcomings of PAAm such as poor temperature resistance and low hydrogel strength. In order to improve the mechanical properties of PAAm hydrogel, the strategy of "pre-stress-release" was invented. The supercapacitor uses CNT@PPy and CNT@MnO$_2$ as electrodes and PAAm composite hydrogel prepared by vinyl-grafted silica nanoparticles as a crosslinking agent as an electrolyte. After 500 stretch-relaxation cycles, the supercapacitor still maintains 96% of the initial specific capacitance. It has a high energy density (40 Wh/kg) and good cycle stability [17].

14.3.3 Polyacrylic Acid Hydrogels

Polyacrylic acid (PAAc) is a hydrophilic polymer with a large number of carboxyl groups in its molecular chain, which is easy to form intermolecular hydrogen bonds or ionic bonds. PAAc hydrogel has excellent stretchable and compressible properties and is a versatile non-electrolyte for flexible supercapacitors. As mentioned earlier, stretchable/compressible hydrogel electrolytes are often used in sandwich structure supercapacitors, but it is worth noting that they have technical bottlenecks that are prone to slip and delamination under large deformation, which destroy their electrochemical performance [18]. The PAAc@Co^{2+} hydrogel electrolyte can provide a solid 3D skeleton and construct a multistage pore structure through metal coordination and hydrogen bonding, which has good mechanical properties. The dynamic connection between the electrode and the interface enables the device to have the ability to self-repair to prevent slippage.

14.3.4 Polyethylene Glycol-based Hydrogels

PEG hydrogel has good biocompatibility and low toxicity, which can be used as wearable human-machine combination material [19]. Koh et al. [20] promoted the synthesis of PEDOT in PEG hydrogel by mixing PSS into PEG hydrogel and further improved the conductivity of hydrogel after H$_2$SO$_4$ treatment. This material can be used as a conductive tissue engineering scaffold material. Revzin et al. [21] prepared conductive hydrogel with polymer PEG as the skeleton and combined it with conductive polymer PEDOT, and the obtained hydrogel had good mechanical and electrical properties. By immersing hydrogel electrodes as biosensors into specific antibodies, the change information between antibodies and antigens can be converted into electrical signals, which can be used as a biosensor platform in the blood environment. Chung et al. [22] developed a PEG/AgNWs conductive hydrogel by a composite approach and designed the conductive hydrogel as a miniature image sensor for direct differentiation of neural stem cells and sheath-induced growth. In addition, conductive hydrogels based on PEG have also been applied in the hydrocondensing ion circuit. Kaplan et al. [23] prepared ion-conducting hydrogels by stabilizing highly conductive salt solution in a PEG hydrogel matrix based on phase separation technology between PEG and salt-soluble solution. Phase-separated ion solutions can provide polyethylene glycol hydrogels with highly ion-conducting circuits, which are stretchable, soft, and do not diffuse ions into the surrounding tissue medium, which has potential in bio-flexible energy storage systems.

14.3.5 Biopolysaccharide Hydrogels

Supercapacitors fabricated with dual electrolytes have high safety and reliability, which are especially suitable for implantable medical devices and wearable electronic devices [24]. Biological polysaccharide-based hydrogels containing a large number of sugar units, such as chitosan, sodium alginate, cellulose, and agarose, are often regarded as the favorite materials for biological solid-state supercapacitors because of the large number of hydroxyl groups on the sugar units. Zeng et al. [25] reported a biosolid supercapacitor made from edible seaweed. In seaweed, flat or leafy leaves have layered, porous, and interconnected structures that help improve ion storage and transport in electrodes, ensuring beneficial multiplication properties. The proposed supercapacitor has a maximum specific capacitance of 227 F/g and its rate performance is comparable to that of supercapacitors based on liquid electrolytes. After 10,000 charge and discharge cycles, the capacitor retention rate reaches 98.5%, and the maximum energy density and power density reach 8 Wh/kg and 5000 W/kg, respectively.

14.3.6 Composite Hydrogels

Due to the free radical polymerization of traditional hydrogels using monomers and chemical crosslinking agents, hydrogels have different polymer chain lengths and crosslinking points distribution, and their mechanical properties are not ideal. An effective solution is to combine materials capable of dispersing tensile stress with traditional hydrogel networks to form enhanced composite hydrogels such as double nets or 3D interpenetrating nets. Supramolecular forces such as hydrogen bonds, ionic bonds, and electrostatic action can also be exploited to improve the tensile strength of hydrogel by sacrificing supramolecular force. The conductivity of the electrolyte can be improved by blending two or more polymers because blending can provide more complex positions, promote ion transformation and exchange, and thus improve ion conductivity [26].

1. *Polyvinyl alcohol-biopolysaccharide hydrogel.* Although the hydroxyl content of PVA hydrogel electrolyte is high, its self-healing efficiency is generally poor, because the presence of a large number of supporting electrolyte inorganic ions severely hinders the formation

of hydrogen bonds between water molecules and PVA chains. The corresponding hydrogen bond interaction between PVA molecules may lead to the aggregation of PVA chains, which greatly reduces the mobility of PVA chains, thus affecting their self-healing performance [27]. Kim et al. [28] prepared a self-healing borax/PVA/agarose hydrogel electrolyte, in which the dynamic covalent crosslinking between the borax and PVA hydroxyl group enables the hydrogel to repair itself quickly and reversibly.

2. *Polyacrylamide biopolysaccharide hydrogel.* Although the dual-network hydrogel consisting of two interpenetrating networks can achieve energy dissipation by sacrificing covalent bond breaking, it causes permanent damage to the covalent bond. Therefore, the self-healing ability of traditional dual-network hydrogels is weak. Liu et al. [29] used Al^{3+} crosslinked alginate as a sacrificed but recoverable secondary network that dissipates energy through ionic bonds breaking during stress loading and recovers its bond during unloading.

3. *Poly(acrylic acid-coacrylamide) hydrogel.* Many structures of flexible supercapacitors have been designed, such as 3D structures [30], braiding [31], waveforms [32], and rubber elastic protective layers [33], to improve flexibility. However, these methods that rely solely on structural design to improve the performance of supercapacitors exhibit limitations due to the lack of suitable electrode and electrolyte materials, such as increasing capacitor weight/volume, decreasing specific capacitance, and manufacturing process complexity [34]. PAAm is the main matrix for constructing self-healing hydrogels [35]. Due to the difficulty in obtaining high tensile properties for pure PAAm hydrogels, Dai et al. [36] stated a P(AA-co-AAm)/Co^{2+} hydrogel electrolyte crosslinked by Co^{2+} ions. The hydrogel is formed by intramolecular and intermolecular hydrogen bonds and metal coordination between –coo– and Co^{2+} ions and has excellent mechanical properties (elongation strain >1200%, compressive stress >600 kPa). The 3D network structure of hydrogen bonds and metal coordination gives hydrogels good ionic conductivity. The capacitive performance is slightly improved under bending and twisting operations.

14.3.7 Gelatin-based Conductive Hydrogels

Gelatin is a kind of biopolymer that is produced by the hydrolytic degradation of collagen. Gelatin-based conductive hydrogel materials are competitive materials for wearable devices due to their high biocompatibility and biodegradability [37]. Wang et al. [38] prepared gelatin hydrogel via physical crosslinking, and then formed conductive gelatin hydrogel in $(NH_4)_2SO_4$ solution by simple immersion process, with a tensile strength of 4.23MPa

Flexible/Wearable Supercapacitors 281

and an electrical conductivity of 5×10^{-2} S/cm. Ma et al. [39] grafted polyaniline with gelatin and polymerized in situ with genipin, a natural biological crosslinking agent, to form a degradable conductive hydrogel with conductivity up to 4.54×10^{-4} S/cm, which can be used as a wearable bionic scaffold material. In addition, gelatin-based conductive hydrogels can also be applied as flexible sensors. Chen et al. [40] introduced multi-walled carbon nanotubes into the gelatin solution to obtain gelatin-based conductive hydrogels with glutaraldehyde as the crosslinking agent. Electrical conductivity reaches 5×10^{-4} S/cm, which can be used as the electronic skin to sense the changes in fingers. Turng et al. [41] introduced AgNWs into the grafted sulfhydryl gelatin to obtain hydrogels and then immersed the obtained hydrogels in the solution of Na_2SO_4 to form conductivity complex hydrogels. The grafting of sulfhydryl groups enhanced the interaction between gelatin molecules and AgNWs. The introduction of AgNWs not only enhances the mechanical strength of the hydrogel but also endows the hydrogel with electrical conductivity.

14.3.8 Alginate-based Conductive Hydrogels

Alginate is an anionic polysugar isolated from seaweed. It can be formed by ionic bond interactions between the carboxyl group on the alginate skeleton and cationic crosslinking agents, such as Ca^{2+}, Zn^{2+}, and Al^{3+}. Due to its good gel stability and biocompatibility, it has been widely used in the field of flexible wearable electronic devices [42]. Kohane et al. [43] introduced gold nanowires (AuNWs) into alginate saline gel to obtain composite hydrogels for cardiac microelectronic devices. Considering the weak mechanical properties of pure alginate conductive hydrogels, Guo et al. [44] synthesized a double-network hydrogel with conductivity by using core-shell hybrid nanoparticles of crosslinked PAAm as the first network and Ca^{2+} crosslinked sodium alginate as the second network. The resulting hydrogel has 2422% elongation at break and can be cycled for 40 min when the stretch shape becomes 1000%. Xu et al. [45] added AA monomer to sodium alginate on the basis of introducing AM monomer. Under the action of $ZnSO_4$ and APS, a double-network hydrogel with ionic conductivity was formed. The results show that the hydrogel has a tensile strain of 4273.08% at 0.21 MPa, and it can be recovered within 20 min when the tensile strain is 4000%. Likewise, based on dual-network hydrogels, alginate-based conduction-based hydrogels can also be used in the field of supercapacitors. Wei et al. [46] designed a supercapacitor, which chose sodium alginate/polyacrylamide double-network hydrogels containing conductive substances CNT and PEDOT: PSS as electrodes. This kind of hydrogel contains the conducting ion Na_2SO_4 as the electrolyte. When the current density is 1 mA/cm^2, the specific capacitance is 128 mF/cm^2, and the energy density is up to 3.6 $\mu Wh/cm^2$. In addition, the hydrogel can be combined with wireless electronic devices to assemble wireless retractable sensors or wearable sensors to monitor the movements of fingers, wrists, elbows, neck, knees, and other joints, and even realize responses to tiny human movements such as breathing and speaking.

14.3.9 Cellulose-based Conductive Hydrogels

Cellulose is the most abundant natural polymer, which is not only found in plants but can also be synthesized by some specific microorganisms, such as rhizomes. The development of substrates for flexible wearable supercapacitors based on good mechanical properties, dielectric properties, and piezoelectric properties of cellulose has also been widely studied [47]. Yan et al. [48] synthesized the cellulose/polyaniline conductive hydrogel electrolyte by low-temperature in-situ polymerization in cellulose solution and then assembled flexible all-solid-state supercapacitors using nitrogen-doped GO hydrogel as an electrode. The maximum energy density and power density of the asymmetric supercapacitor are 45.3 Wh/kg and 742.0 W/kg, respectively. In addition, cellulose-based conductive hydrogels can be used for biomimetic electronic skin. Yu et al. [49] constructed an ion-conducting hydrogel material based on cellulose, ionic liquid, and water based on the design principle of the hydrogen bond topological network, which can realize the control of the structure and performance of cellulose ionic hydrogel in the dynamic hydrogen bond network driven by humidity. The obtained hydrogel ionic conductivity can reach 4×10^{-2} S/cm, and the bionic electronic skin designed with it shows good ductility and flexibility. It can be seamlessly attached to the human wrist and can be removed without tissue damage. He et al. [50] introduced allyl glycerol into a cellulose solution dissolved in NaOH/urea to form an ion-conducting hydrogel through free radical polymerization. The maximum tensile strain is 126% and the maximum compressive strain is 80%. Natural polymer hydrogels have higher mechanical strength compared with unmodified cellulose or chitosan. Zhi et al. [51] introduced AM monomer into cellulose suspension to synthesize an ion-conducting hydrogel. The hydrogel not only had good mechanical properties but also had an ionic conductivity of 2.28×10^{-2} S/cm at room temperature. The designed battery can be stitched once, and the anti-shear ability of the battery after stitching can reach 43 N.

14.4 Hydrogel Applied to Electrodes of Supercapacitors

14.4.1 PANI Hydrogels

PANI hydrogels have been proven to be electrochemical electrodes with high performance [52]. The biggest problem faced by flexible supercapacitors is the electrochemical instability during bending, drawing, and machining. Li et al.

[53] prepared a flexible electroconductive poly (vinyl alcohol) (PVA) $-H_2SO_4$ hydrogel film with vertical straight channels by freeze-melting method and directional cold freezing method. APH-PANI membrane was obtained by in-situ polymerization of PANI on both sides of the membrane, which can be understood as both electrolyte and electrode of the supercapacitor. The APH-PANI supercapacitor has excellent electrochemical performance, good flexibility, and excellent self-healing properties. Liu et al. [54] produced a reduced graphene oxide (rGO) hydrogel film by the hydrothermal method by combining graphene oxide with a conductive junction. rGO hydrogel film was used as the working electrode for the electropolymerization of phenylamine. In the absence of additives, the prepared rGO/PANI hydrogel film consists of a symmetric solid-state supercapacitor with a specific capacitance of up to 741.8 F/g. After 8000 cycles at the current density of 20 A/g, the capacitance retention rate is 92.6%, which shows excellent electrochemical performance and good stability.

14.4.2 PPy Hydrogels

PPy is similar to PANI in that it is electrically conductive. Zang et al. [55] used PVA-H_2SO_4 as the base, pyrrole gas phase polymerization to prepare conductive hydrocoagulant PVA/PPy, and assembled an all-solid polymeric supercapacitor. The supercapacitor has good mechanical properties with a tensile strength of 20.83 MPa. The elongation rate of fracture is 377%, with a volume-specific heat capacity of 13.06 F/cm^3 and an energy density of 1160.9 µWh/cm^3. To improve the operating stability of the device, Sun et al. [56] constructed a flexible supercapacitor with a bulk electrode-electrolyte-electrode by embedding conducting PPy into the boron crosslinked PVA/KCl electrolyte hydrogel film through in-situ growth. The integrated flexible supercapacitor has high toughness and strength as well as a large area face of 224 mF/cm^2 and a good energy density. It can still maintain excellent electrochemical stability after multiple bending.

14.4.3 PEDOT and Its Derivative Hydrogels

Among conductive polymers, PEDOT has the advantages of high conductivity, stable charge-discharge effect, and so on. It is a widely used material for energy storage devices. Its energy storage mechanism is shown in Figure 14.4. The synergistic effect of different conductive polymers can be reduced by added oxygenation.

Pattananuwat et al. [57] prepared PPy/PEDOT/reduced oxyfossil ink (PPy/PEDOT/rGO) hydrogel through electropolymerization. Supercapacitor electrodes are prepared by a flexible hybrid layer on a transparent electroconductive plastic adhesive thin film (ITO-PET) substrate. The hybrid layer has a highly microporous structure, which is an effective way for supporting ions to diffuse to the inner surface. At the current density of 0.5 A/g,

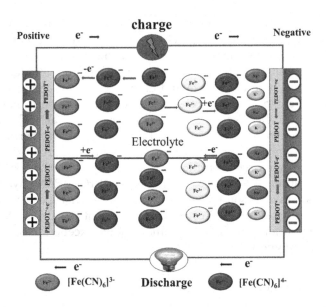

FIGURE 14.4
Charge storage mechanism of flexible supercapacitors using PEDOT/Fe(CN)$_6$ hydrogel as an example.

the specific capacitance of the hybrid material is up to 342 F/g. PEDOT can also be combined with biological components to improve its electrochemical properties. Saborío et al. [58] synthesized electrodes composed of biological substances poly-γ-glutamic acid (γ-PGA), PEDOT, and their derivatives by applying a two-step technique. PEDOT particles were dispersed into a γ-PGA hydrogel matrix, which was used as the polymerization nucleus, and the derivative poly(hydroxymethyl-3,4-ethyldioxythiophene) (PHMeDOT) was synthesized in an aqueous solution. The specific capacitance of the electrode is 45~47 mF/cm^2 and it has excellent cycle durability, which is promising for application in supercapacitors.

14.5 Technical Advantages and Challenges of Hydrogels in the Application of Supercapacitors

The advantages and disadvantages of hydrogels in wearable supercapacitor applications are shown in Figure 14.5.

At present, there are still many technical bottlenecks to overcome in the application of hydrogels, such as high interface impedance and poor structural stability. The presence of water-containing molecules limits its

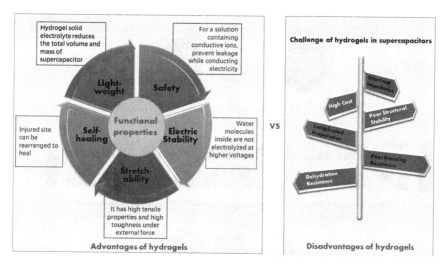

FIGURE 14.5
Functional advantages and challenges of hydrogel supercapacitors.

application in extreme environments. However, it must be admitted that compared with these unsolved shortcomings, the huge advantages of hydrogels, such as excellent stretchability, stability of high-voltage electric fields, self-healing, safety and leakage prevention, lightweight nature, and other characteristics, are attracting researchers to explore new depths to apply them to the practical industry as soon as possible.

The technological maturity of hydrogel materials and hydrogel supercapacitors is shown in Figure 14.6. From the perspective of technological maturity, some current hydrogel materials have commercial products and are used in the medical field. However, there are no commercial products for flexible electronics and flexible supercapacitors. The update and iteration of the material field are very fast, and hydrogel supercapacitors are expected to soon move from the laboratory to the shelf. The industrialization process of flexible supercapacitors depends not only on materials but also on the maturity of equipment and manufacturing methods. At present, the technical maturity of hydrogel supercapacitors is still in the initial stage of academic research. However, with the rapid increase in demand for flexible electronics, wearable devices, and human-machine devices, it is believed that the development of this process will also mature rapidly.

For the application research of hydrogels in flexible supercapacitors, researchers should not only consider the performance improvement of the hydrogel itself but also consider the manufacturing cost and process complexity, as shown in Figure 14.7. Only technology that has been tested by comprehensive performance evaluation can truly gain a foothold in the market.

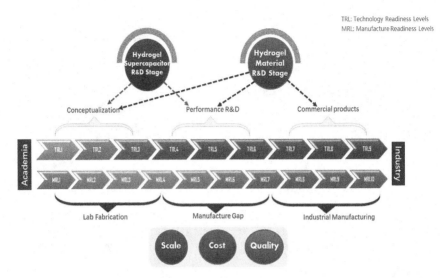

FIGURE 14.6
Technological maturity of hydrogel supercapacitors.

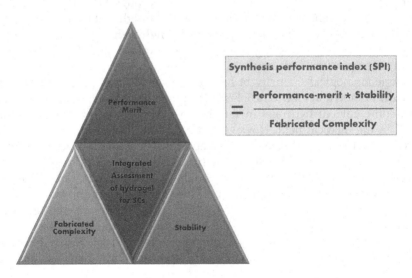

FIGURE 14.7
Integrated assessment of hydrogel supercapacitors.

14.6 Conclusion

In this chapter, the iterative history of hydrogels, their classification, and their application status in supercapacitors were reviewed, and the functional characteristics of hydrogels in two important components of electrolytes and electrodes were expounded, including mechanical flexibility, high cold resistance and freezing resistance, self-healing, and electric field stability. The technical maturity, influencing factors, and comprehensive evaluation indicators of hydrogel materials and supercapacitor devices were thoroughly analyzed. The challenges and technical bottlenecks of hydrogels in the development of flexible supercapacitors were summarized and countermeasures were analyzed. Finally, the development of flexible supercapacitors for hydrogel prospects was considered.

Acknowledgments

This research was funded by the Application Demonstration Project (Cultivation Project) of the Chinese Academy of Sciences (CAS-WX2021PY-0504)

References

1. P. Simon, Y. Gogotsi, Materials for electrochemical capacitors, *Nat. Mater.* 7 (2008) 845–854.
2. S. J. Kim, S. G. Yoon, Y. M. Lee, H. C. Kim, S. I. Kim, Electrical behavior of polymer hydrogel composed of poly (vinyl alcohol)–hyaluronic acid in solution, *Biosens. Bioelectron.* 19 (2004) 531–536.
3. J. Shang, Z. Shao, X. Chen, Electrical behavior of a natural polyelectrolyte hydrogel: Chitosan/carboxymethylcellulose hydrogel, *Biomacromolecules* 9 (2008) 1208–1213.
4. W. Zhang, P. Feng, J. Chen, Z. Sun, B. Zhao, Electrically conductive hydrogels for flexible energy storage systems, *Prog. Polym. Sci.* 88 (2019) 220–240.
5. F. Qiu, Y. Huang, G. He, C. Luo, X. Li, M. Wang, Y. Wu. A lignocellulose-based neutral hydrogel electrolyte for high-voltage supercapacitors with overlong cyclic stability, *Electrochim. Acta* 363 (2020) 137241.
6. E. M. Ahmed, Hydrogel: Preparation, characterization, and applications: A review, *J. Adv. Res.* 6 (2015) 105–121.
7. J. P. Gong, Y. Katsuyama, T. Kurokawa, Y. Osada, Double-network hydrogels with extremely high mechanical strength, *Adv. Mater.* 15 (2003) 1155–1158.

8. Y. S. Zhang, A. Khademhosseini, Advances in engineering hydrogels, *Science* 356 (2017) 6337.
9. C. Y. Chan, Z. Wang, H. Jia, P. F. Ng, L. Chow, B. Fei, Recent advances of hydrogel electrolytes in flexible energy storage devices, *J. Mater. Chem. A* 9 (2021) 2043–2069.
10. Y. Xu, Z. Lin, X. Huang, Y. Liu, Y. Huang, X. Duan, Flexible solid-state supercapacitors based on three-dimensional graphene hydrogel films, *ACS Nano* 7 (2013) 4042–4049.
11. F. Zhao, Y. Shi, L. Pan, G. Yu, Multifunctional nanostructured conductive polymer gels: Synthesis, properties, and applications, *Acc. Chem. Res.* 50 (2017) 1734–1743.
12. Y. Y. Lee, H. Y. Kang, S. H. Gwon, G. M. Choi, S. M. Lim, J. Y. Sun, Y. C. Joo, A strain-insensitive stretchable electronic conductor: PEDOT: PSS/acrylamide organogels, *Adv. Mater.* 28 (2016) 1636–1643.
13. Q. Ding, X. Xu, Y. Yue, C. Mei, C. Huang, S. Jiang, Q. Wu, J. Han, Nanocellulose-mediated electroconductive self-healing hydrogels with high strength, plasticity, viscoelasticity, stretchability, and biocompatibility toward multifunctional applications, *ACS Appl. Mater. Interfaces* 10 (2018) 27987–28002.
14. S. Choi, S. I. Han, D. Kim, T. Hyeon, D. H. Kim, High-performance stretchable conductive nanocomposites: Materials, processes, and device applications, *Chem. Soc. Rev.* 48 (2019) 1566–1595.
15. D. Gan, L. Han, M. Wang, W. Xing, T. Xu, H. Zhang, K. Wang, L. Fang, X. Lu, Conductive and tough hydrogels based on biopolymer molecular templates for controlling in situ formation of polypyrrole nanorods, *ACS Appl. Mater. Interfaces* 10 (2018) 36218–36228.
16. J. Lin, Q. Tang, D. Hu, X. Sun, Q. Li, J. Wu, Electric field sensitivity of conducting hydrogels with interpenetrating polymer network structure, *Colloids Surf. A* 346 (2009) 177–183.
17. A. K. Bajpai, J. Bajpai, S. N. Soni, Designing polyaniline (PANI) and polyvinyl alcohol (PVA) Bbased electrically conductive nanocomposites: Preparation, characterization and blood compatible study, *J. Macromol. Sci. Part A Pure Appl. Chem.* 46 (2009) 774–782.
18. P. Baei, S. Jalili-Firoozinezhad, S. Rajabi-Zeleti, M. Tafazzoli-Shadpour, H. Baharvand, N. Aghdami. Electrically conductive gold nanoparticle-chitosan thermosensitive hydrogels for cardiac tissue engineering, *Mater. Sci. Eng. C* 63 (2016) 131–141.
19. J. Zhu, Bioactive modification of poly (ethylene glycol) hydrogels for tissue engineering, *Biomaterials* 31 (2010) 4639–4656.
20. Y. S. Kim, K. Cho, H. J. Lee, S. Chang, H. Lee, J. H. Kim, W. G. Koh, Highly conductive and hydrated PEG-based hydrogels for the potential application of a tissue engineering scaffold, *React. Funct. Polym.* 109 (2016) 15–22.
21. D. S. Shin, Z. Matharu, J. You, C. Siltanen, T. Vu, V. K. Raghunathan, G.Stybayeva, A. E. Hill, A. Revzin, Sensing conductive hydrogels for rapid detection of cytokines in blood, *Adv. Healthcare Mater.* 5 (2016) 659–664.
22. J. M. Lee, J. Y. Moon, T. H. Kim, S. W. Lee, C. D. Ahrberg, B. G. Chung, Conductive hydrogel/nanowire micropattern-based sensor for neural stem cell differentiation, *Sens. Actuators B*, 258 (2018) 1042–1050.
23. S. Zhao, P. Tseng, J. Grasman, Y. Wang, W. Li, B. Napier, B. Yavuz, Y. Chen, L. Howell, J. Rincon, F. G.Omenetto, D. L. Kaplan, Programmable hydrogel ionic

circuits for biologically matched electronic interfaces, *Advanced Materials*, 30 (2018) 1800598.
24. Z. Wang, M. Zhu, Z. Pei, Q. Xue, H. Li, Y. Huang, C. Zhi, Polymers for supercapacitors: Boosting the development of the flexible and wearable energy storage, *Mater. Sci. Eng.: R: Rep.* 139 (2020) 100520.
25. J. Zeng, L. Wei, X. Guo, Bio-inspired high-performance solid-state supercapacitors with the electrolyte, separator, binder and electrodes entirely from kelp, *J. Mater. Chem. A* 5 (2017) 25282–25292.
26. M. H. Buraidah, A. K. Arof, Characterization of chitosan/PVA blended electrolyte doped with NH_4I, *J. Non-Cryst. Solids* 357 (2011) 3261–3266.
27. M. Teodorescu, S. Morariu, M. Bercea, L. Săcărescu, Viscoelastic and structural properties of poly (vinyl alcohol)/poly(vinylpyrrolidone) hydrogels, *RSC Advances*, 6 (2016) 39718–39727.
28. M. S. Kim, J. W. Kim, J. Yun, Y. R. Jeong, S. W. Jin, G. Lee, H. Lee, D. S. Kim, K. Keum, J. S. Ha, A rationally designed flexible self-healing system with a high performance supercapacitor for powering an integrated multifunctional sensor, *Appl. Surf. Sci* 515 (2020) 146018.
29. Z. Liu, G. Liang, Y. Zhan, H. Li, Z. Wang, L. Ma, Y. Wang, X. Niu, C. Zhi, A soft yet device-level dynamically super-tough supercapacitor enabled by an energy-dissipative dual-crosslinked hydrogel electrolyte, *Nano Energy* 58 (2019) 732–742.
30. G. Nyström, A. Marais, E. Karabulut, L. Wågberg, Y. Cui, M. M. Hamedi, Self-assembled three-dimensional and compressible interdigitated thin-film supercapacitors and batteries, *Nat. Commun.* 6 (2015) 7259.
31. Y. M. Chen, Z. Li, X. W. D. Lou, General formation of $M_x Co_{3-x} S_4$ (M=Ni, Mn, Zn) hollow tubular structures for hybrid supercapacitors, *Angew. Chem.* 127 (2015) 10667–10670.
32. K. S. Kim, Y. Zhao, H. Jang, S. Y. Lee, J. M. Kim, K. S. Kim, J. H. Ahn, P. Kim, J. Y. Choi, B. H. Hong, Large-scale pattern growth of graphene films for stretchable transparent electrodes, *Nature*, 457 (2009) 706–710.
33. Z. F. Liu, S. Fang, F. A. Moura, J. N. Ding, N. Jiang, J. Di, M. Zhang, X. Lepró, D. S. Galvão, C. S. Haines, N. Y. Yuan, S. G. Yin, D. W. Lee, R. Wang, H. Y. Wang, W. Lv, C. Dong, R. C. Zhang, M. J. Chen, Q. Yin, Y. T. Chong, R. Zhang, X. Wang, M. D. Lima, R. Ovalle-robles, D. Qian, H. Lu, R. H. Baughman, Hierarchically buckled sheath-core fibers for superelastic electronics, sensors, and muscles, *Science*, 349 (2015) 400–404.
34. D. K. Patel, A. H. Sakhaei, M. Layani, B. Zhang, Q. Ge, S. Magdassi, Highly stretchable and UV curable elastomers for digital light processing based 3D printing, *Adv. Mater.* 29 (2017) 1606000.
35. J. Liu, C. S. Y. Tan, Z. Yu, N. Li, C. Abell, O. A Scherman, Tough supramolecular polymer networks with extreme stretchability and fast room-temperature self-healing, *Adv. Mater.* 29 (2017) 1605325.
36. H. Wang, L. Dai, D. Chai, Y. Ding, H. Zhang, J. Tang, Recyclable and tear-resistant all-in-one supercapacitor with dynamic electrode/electrolyte interface, *J. Colloid Interface Sci.* 561 (2020) 629–637.
37. Z. Shi, X. Gao, M. W. Ullah, S. Li, Q. Wang, G. Yang, Electroconductive natural polymer-based hydrogels, *Biomaterials*, 111 (2016) 40–54.
38. C. Liu, H. J. Zhang, X. You, K. Cui, X. Wang, Electrically conductive tough gelatin hydrogel, *Adv. Electron. Mater.* 6 (2020) 2000040.

39. L. Li, J. Ge, B. Guo, P. X. Ma, In situ forming biodegradable electroactive hydrogels. *Polym. Chem.* 5 (2014) 2880–2890.
40. L.-Y. Hsiao, L. Jing, K. Li, H. Yang, Y. Li, P.-Y. Chen, Carbon nanotube-integrated conductive hydrogels as multifunctional robotic skin, *Carbon* 161 (2020) 784–793.
41. X. Jing, X.-Y. Wang, H.-Y. Mi, L.-S. Turng, Stretchable gelatin/silver nanowires composite hydrogels for detecting human motion, *Mater. Lett.* 237 (2019) 53–56.
42. K. Y. Lee, D. J. Mooney. Alginate: Properties and biomedical applications, *Prog. Polym. Sci.* 37 (2012) 106–126.
43. T. Dvir, B. P. Timko, M. D. Brigham, S. R. Naik, S. S. Karajanagi, O. Levy, H. Jin, K. K. Parker, R. Langer, D. S. Kohane, Nanowired three-dimensional cardiac patches, *Nat. Nanotechnol.* 6 (2011) 720–725.
44. S. Xia, S. Song, G. Gao, Robust and flexible strain sensors based on dual physically cross-linked double network hydrogels for monitoring human-motion, *Chem. Eng. J.*, 354 (2018)817–824.
45. H. Huang, L. Han, J. Li, X. Fu, Y. Wang, Z. Yang, X. Xu, L. Pan, M. Xu, Super-stretchable, elastic and recoverable ionic conductive hydrogel for wireless wearable, stretchable sensor, *J. Mater. Chem. A* 8 (2020) 10291–10300.
46. J. Zeng, L. Dong, W. Sha, L. Wei, X. Guo, Highly stretchable, compressible and arbitrarily deformable all-hydrogel soft supercapacitors, *Chem. Eng. J.* 383 (2020) 123098.
47. D. Zhao, Y. Zhu, W. Cheng, W. Chen, Y. Wu, H. Yu, Cellulose-based flexible functional materials for emerging intelligent electronics, *Adv. Mater.* 33 (2021) 2000619.
48. H. Wang, J. Wu, J. Qiu, K. Zhang, J. Shao, L. Yan, In situ formation of a renewable cellulose hydrogel electrolyte for high-performance flexible all-solid-state asymmetric supercapacitors, *Sustainable Energy Fuels* 3 (2019) 3109–3115.
49. D. Zhao, Y. Zhu, W. Cheng, G. Xu, Q. Wang, S. Liu, J. Li, C. Chen, H. Yu, L. Hu, A dynamic gel with reversible and tunable topological networks and performances, *Matter* 2 (2020) 390–403.
50. R. Tong, G. Chen, D. Pan, H. Qi, R. Li, J. Tian, F. Lu, M. He, Highly stretchable and compressible cellulose ionic hydrogels for flexible strain sensors, *Biomacromolecules* 20 (2019) 2096–2104.
51. D. Wang, H. Li, Z. Liu, Z. Tang, G. Liang, F. Mo, Q. Yang, L. Ma, C. Zhi, A nano fibrillated cellulose/polyacrylamide electrolyte-based flexible and sewable high-performance Zn-MnO$_2$ battery with superior shear resistance, *Small* 14 (2018) 1803978.
52. L. Pan, G. Yu, D. Zhai, H. R. Lee, W. Zhao, N. Liu, H. Wang, B. C.-K. Tee, Y. Shi, Y. Cui, Z. Bao, Hierarchical nanostructured conducting polymer hydrogel with high electrochemical activity, *PNAS* 109 (2012) 9287–9292.
53. W. Li, X. Li, X. Zhang, J. Wu, X. Tian, M.-J. Zeng, J. Qu, Z.-Z. Yu, Flexible poly (vinyl alcohol)–polyaniline hydrogel film with vertically aligned channels for an integrated and self-healable supercapacitor. *ACS Appl. Energy Mater.* 3 (2020) 9408–9416.
54. Z. Liu, Z. Zhao, A. Xu, W. Li, Y. Qin, Facile preparation of graphene/polyaniline composite hydrogel film by electrodeposition for binder-free all-solid-state supercapacitor. *J. Alloys Compd.* 875 (2021) 159931.
55. L. Zang, Q. Liu, J. Qiu, C. Yang, C. Wei, C. Liu, L. Lao, Design and fabrication of an all-solid-state polymer supercapacitor with highly mechanical flexibility based on polypyrrole hydrogel. *ACS Appl.Mater. Interfaces* 9 (2017) 33941–33947.

56. K. Sun, E. Feng, G. Zhao, H. Peng, G. Wei, Y. Lv, G. Ma, A single robust hydrogel film based integrated flexible supercapacitor. *ACS Sustainable Chem. Eng.* 7 (2019) 165–173.
57. P. Pattananuwat, D. Aht-Ong, One-step method to fabricate the highly porous layer of poly (pyrrole/(3, 4-ethylenedioxythiophene)/) wrapped graphene hydrogel composite electrode for the flexible supercapacitor. *Mater. Lett.* 184 (2016) 60–64.
58. M. C. G. Saborío, S. Lanzalaco, G. Fabregat, J. Puiggalí, F. Estrany, C. Alemán, Flexible electrodes for supercapacitors based on the supramolecular assembly of biohydrogel and conducting polymer. *J. Phys. Chem. C* 122 (2018) 1078–1090.

15
Hydrogels for Wearable Electronics

Zahra Karimzadeh, Mansour Mahmoudpour,
Abolghasem Jouyban, and Elaheh Rahimpour

CONTENTS

15.1 Introduction 293
　15.1.1 Flexible Wearable Electronics 293
15.2 Design Principles and Preparation Methods of Hydrogel-Based
　Sensors 295
15.3 Applications of Hydrogels for Wearable Electronics 298
　15.3.1 Hydrogels as a Sensor 298
　15.3.2 Hydrogels for Human Motion Monitoring 299
　15.3.3 Hydrogels for Prosthetics or Artificial Tissue as
　　　　　Implantable Electronics 300
　15.3.4 Hydrogels for Human-Machine Interfaces 301
　15.3.5 Hydrogels for Energy Storage Devices 302
　15.3.6 Hydrogel for Electromagnetic Shielding 304
15.4 Conclusion and Future Perspectives 305
Acknowledgments 306
References 307

15.1 Introduction

15.1.1 Flexible Wearable Electronics

Flexible electronics express electronic and circuit constituents that can preserve their performances under folding, rolling, bending, or stretching circumstances. This concept was invented in the 1960s as flexible and thin solar cells with greater power density were considered for satellites [1]. After that, innovations in materials with large processability and higher flexibility, such as amorphous silicon, organic semiconductors, and conductive polymers, have increasingly laid the basis for flexible devices. In the last few years, there has been an increase in the popularity of investigations for novel materials and construction methods that integrate high-efficiency electronic constituents directly onto flexible scaffolds. Compared with the more tough

DOI: 10.1201/9781003351566-18

and rigid conventional electronic tools, flexible devices have some distinctive benefits, for example, bendability, flexibility, portability, a lightweight nature, adaptability, and foldability, which are favorable for various applications [2]. Nowadays, technological progress in cloud computing, virtual reality, augmented reality, and artificial intelligence, together with the growing knowledge of individual health management, have accelerated the appearance of artificial electronic structures (bioelectronic wearable systems) and flexible wearable systems interfacing the living biological tissues [3].

Flexible wearable electronics benefiting from self-healing capability, human skin-inspired tactile sensing ability, stretchability, and flexibility have drawn considerable interest in next-generation electronic devices over the past few years [4]. For electronic devices, conventional elastomer substrates such as polyurethane and silicone rubbers offer higher elastic moduli than that of human tissues. This lack of mechanical properties coupled with biological function deficiency prompted several problems of foreign-body reaction, tissue trauma, and poor conformability. On account of biocompatibility, tissue-like softness, and excellent chemical and electrical properties of the efficient wearable electronics, hydrogels have been moved to the forefront to profit from their stimuli-responsiveness, tissue resemblance, high water content, high surface area, bioinspired structure, and adjustable conductive channels providing a unified interface among the biology and the electronics [5]. Variations in the molecular constituents, type, and degree of crosslinkers (chemical covalent/non-covalent), ionic charge, and functional groups significantly affect the stiffness, viscoelastic behavior, and transition temperatures of hydrogels. Despite substantial research evolution on hydrogel-based wearable electronics in recent years, numerous problems, such as narrow detection range, easy damage, instability, low sensitivity, poor adhesion, and weak mechanical strength, limit their applicability in flexible/wearable electronics. To feature hydrogel with additional sensing performances, interesting functionalities, and enhanced mechanical properties, the inclusion of conductive nanomaterials into hydrogels can be an effective, yet simple method [6]. Accordingly, the contribution of nanocomposite elastic thin hydrogel materials could offer a hopeful perspective for the design of wearable devices that can be reversibly compressed and elongated under external force [7]. The hydrogel matrix could offer a hopeful host network for nanoparticle integration and synthesis media. Generally, this class of heterogeneous materials combines the features of both nano-sized fillers and hydrogel components.

Regarding these perspectives, the recent progress in flexible hydrogel-based electronics is provided in this chapter. Also, a brief introduction of representative design strategies and approaches is presented, highlighting efficient hydrogel-based wearable electronics, as well as their anti-freezing capabilities, self-adhesion, and self-healing properties. In the conclusion, we describe future research directions and current challenges.

15.2 Design Principles and Preparation Methods of Hydrogel-Based Sensors

The outstanding features of hydrogels make them appropriate applicants to serve as wearable electronic devices. To broaden their wide application in sensory systems, some significant properties of hydrogels, for example, sensitivity, mechanical performance, and electrical conductivity, should be improved upon. In strain sensory applications, for instance, the electrical conductivity of pure hydrogels can be enhanced despite their inherent ionic conductivity. For that, integrating ionic polymers (zwitterion, anionic, cationic) and free or crosslinked mobile ions (Cl^-, SO_4^{2-}, Ca^{2+}, Li^+, and Fe^{3+}) can deliver additional electrical conductivity to hydrogels [8]. Otherwise, the migration of mobile ions across the hydrogel with a 3D porous structure results in the enhancement of ionic conductivity along with strain sensitivity. Furthermore, multivalent metal ions can also form coordination bonds with charged ligands, which increases the mechanical properties of hydrogels. Regardless of the profits of ionic conductive hydrogels, there is a tradeoff between significant strain sensitivity and excellent ionic conductivity. To address this matter, conductive fillers like MXenes, metal particles/wires (Cu, Au, Ag), poly(3,4-ethylenedioxythiophene), polystyrene sulfonate (PSS), polythiophene, polypyrrole, polyaniline, conductive polymers, graphene and its derivatives, carbon-based materials, ionic liquids, and liquid metals can be integrated into the matrix of the hydrogel. In this regard, the existence of these conductive agents equal/above to their electrical percolation threshold concentration can trigger conductive pathways inside the hydrogel network to transfer electrons by the tunneling effect and contact [9].

Conductive nanofillers can be configurated within the hydrogel matrix by several approaches. Among these, in-situ polymerization is the most common and simple incorporation method for the development of conductive hydrogels. At this state, Pan et al. [10] used various concentrations of carbon nanofibers (CNFs) in the agar/polyacrylamide (PAM) hydrogel to improve its electronic conductivity. It has been revealed that the addition of CNFs (0.75 wt%) decreased the electrical resistance of the hydrogel along with delivering remarkable strain sensitivity with a tunneling effect at large strains. Furthermore, they have the ability to directly attach to the skin as well as clothing to monitor human body motions in the wrist joint, neck, knee joint, and fingers.

In contrast, the employment of novel designs such as the impregnation of conductive polymers into the hydrogel media, utilizing nanomaterials with hollow sphere/fiber morphologies in the hydrogel matrix, infiltration of hydrogel into porous conductive filler foam, and aligning the conductive filler assisted with microchannel-shaped hydrogels can promote the performance of hydrogel-based strain sensors.

Despite the electrical conductivity features of hydrogels, the improvement of their mechanical properties including dynamic durability, toughness, and stretchability is essential for enduring frequent and dynamic loading through practical use. Indeed, introducing functional groups to conductive nanofillers encourages abundant chemical/physical interactions in the hydrogel network, enhancing mechanical properties. Until now, just a few NCHs have possessed moderate stretchability or low toughness resulting from the aggregation of nanofillers, which limits their toughening function [11]. In pursuit of promoting these reinforcing functions, numerous tactics like double or hydrophobic-associated crosslinking, employing an entire/semi-interpenetrating network, using ionic polymers within the hydrogel network, and hybridization/surface modification of hydrophobic nanofillers are proposed. For example, He et al. [12] employed carbon nanoclusters (CNC)-MXene in the $AlCl_3$/tamarind gum/PAM double-network hydrogel to develop a tough and conductive strain sensor. For the as-prepared hydrogel, the appropriate strain sensitivity and electrical conductivity of these NCHs were endowed with the existence of Cl^- and Al^{3+} ions together with MXene electronic conductors inside the hydrogel media. While the mechanical properties of the hydrogel were reinforced by hydrogen bonds between CNC-Mxenes/tamarind gum/PAM, coordination bonds between CNCs and Al^{3+}, and chemical bonds within the tamarind gum/PAM double network.

Besides the essential facilities discussed already, many demands have recently increased for improving and regulating the strain sensitivity of hydrogel-based wearable electronics. They demonstrate high electrical resistance for hydrogel under deformation. To demonstrate the strain sensitivity of hydrogels, various mechanisms containing crack propagation, disconnection, tunneling phenomenon, piezo-resistivity, and piezo-capacitance mechanism have been recommended. Apart from the excellent piezo-resistivity of several materials (such as metal (oxide) nanowires, thin films of carbon nanotube (CNT), and semiconductors), the wide-range performance of polymer-integrated materials was limited due to the lack of sufficient interfacial connection. However, hydrogels illustrate inherent piezo-resistivity while enduring large strains caused by the simple deformation of their porous structure under loading [13].

The sliding of rigid conductive fillers within the polymer network separates the overlapped fillers, prompting the disconnection mechanism. However, the tunneling effect reveals the strain sensitivity of the hydrogels, where the stiffed conductive fillers such as Ag nanowires, CNT, and nanographene with high cut-off distance are disconnected in the hydrogel matrix under distortion. For instance, Pan and coworkers [10] designed a PAM/CNF/agar hydrogel and investigated their sensing mechanisms during stretching. There was a partial separation of the CNFs and subsequent tunneling effect among them by control sensing mechanisms. When the diffused/mounted layer of electronic conductors on the hydrogel is reversibly distorted over stretching, crack propagation occurs. In this regard, Cai et al. [14] developed a new nanoprobe by inserting chitosan/calcium alginate/PAM hydrogel into the graphene foam with a porous structure. The designed hydrogel demonstrated the tunneling mechanism of graphene sheets, the crack propagation mechanism for graphene surfaces, and inherent piezo-resistivity for large, moderate, and small strains. This significant variation in gauge factor (GF) from 7–1800, 0–500% strain confirms the greater influence of the tunneling effect than crack propagation on the sensing. To this end, Shi et al. [15] studied the effect of deforming and restructuring the polypyrrole (PPy)-coated polyvinyl alcohol (PVA) hydrogel separately on its admirable sensitivity and excellent recoverability. To expand the functional range of sensitive hydrogel-based electronics, the sensing mechanisms can be combined. Based on these features, their ability to detect various tiny physiological signals, for example, distinguishing the pronunciation of words by vocal cord vibration, the bending of the knuckles at different frequencies, and detecting the number of pulse beats were evaluated (Figure 15.1).

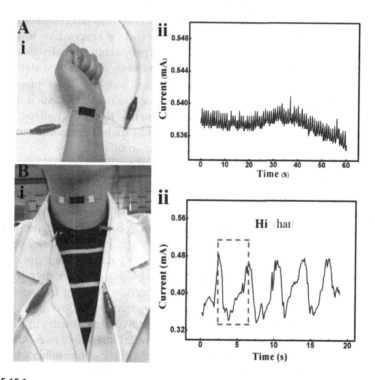

FIGURE 15.1
(a) (i) Photograph of a sensing device for pulse testing mounted on the wrist; (ii) wrist pulse signal. (b) (i) Photograph of a sensing device attached to the throat; (ii) the sound signals for "Hi" words. (Adapted with permission [15], Copyright [2020], American Chemical Society.)

15.3 Applications of Hydrogels for Wearable Electronics

Hydrogel-based wearable electronics have shown rapid demand in recent years regarding their probable applications in different fields. Although these hydrogel-based electronic devices can be introduced to the human skin for motion monitoring from subtle to large deformations, they can also be fixed into the human body to determine physiological signs as artificial tissue [16]. Some general applications for these hydrogels are discussed in the following sections.

15.3.1 Hydrogels as a Sensor

Sensors are machines or devices that diagnose external signals by mimicking the inherent functions of human sensory organs. The conversion of chemical or physical stimuli into another form of energy is the main function of sensors. In pursuit of providing long-lasting contact with human tissues or skin, these sensors have to be highly flexible and soft to be ergonomic, non-toxic, and highly biocompatible. Hydrogel-based wearable sensors are generally used for the detection of biological signals and medical diagnosis. To impart their functionality as sensors, hydrogels should be responsive to external stimuli to yield feedback signals. The unique properties of hydrogels including high permeability, stimulation responsiveness, and high water content make hydrogel sensors superior to traditional sensors. Therefore, diverse hydrogels could be approved to respond to various stimuli, such as strain and temperature, depending on the requirements. Recently, hydrogel-based biosensors have been employed in several fields, including monitoring, environmental monitoring, disease detection, and so on [17].

As an example of a wearable chemical sensor, Xu et al. [18] proposed a microfluidic-based electrochemical sensor based on conductive poly-3,4-ethylenedioxythiophene: polystyrene sulfonate (PEDOT: PSS) hydrogel for accurate determination of uric acid in sweat. The quantification of uric acid was primarily attained through chromogenic and enzyme-antigen-antibody substrates. The concentration of uric acid could be assessed by the way that uric acid concentration is directly related to the optical density recorded at 450 nm. This designed wearable sensor demonstrated a low detection limit and high sensitivity. For humidity sensing fields, Yu et al. [19] fabricated mechanically robust conductive hydrogels based on cellulose nanofibers. The $CaCl_2$/sorbitol addition to the hydrogel results in a relative humidity detection range of 23–97%, a water retention rate of over 90%, and a temperature resistance of −50°C. Furthermore, cellulose nanofibers' addition directed to the construction of a front-layered honeycomb-like structure and dynamic linking bridges in the hydrogel enhances mechanical properties. This wearable humidity sensor is a probable biocompatible electronic to be used in the human body as cellulose nanofibers are a bioderived material.

Also, a novel turn-on fluorescence nanoprobe structure for the determination of morphine in biological fluid using an in-situ synthesis approach was developed. For that, PVA hydrogel was formed in the presence of both gold ions and Zr-based metal-organic frameworks, UiO-66, and reduced to gold nanoparticles (AuNPs) without the addition of any reducing agent. The AuNPs@UiO-66 part of the NCH has been hired as a fluorescent sensing probe. The designed AuNPs@UiO-66/PVA NCH demonstrated a low detection limit of 0.016 µg/mL with a linearity range from 0.02 to 2.0 µg/mL. Its precise assay for morphine in human plasma resulted in acceptable outcomes and good recoveries. Based on the instrument-free characteristics, and robust, and portable features of the proposed platform, it displays excellent perspectives besides the wearable/stretchable characteristics for on-site clinical monitoring [20].

15.3.2 Hydrogels for Human Motion Monitoring

Inspired by previously discussed attractive features, hydrogels are qualified as appropriate strain sensors detecting human motion. They offer strains up to 4000%, which is vital for an extensive range of strain sensing [21]. Alternatively, the self-healing and anti-fatigue properties of tough hydrogels offer continuous monitoring of cyclic and frequent human motions. Additionally, their self-adhesiveness and sufficient biocompatibility enable them to contact the human body easily and directly. Hence, conductive and tough hydrogels are commonly formed to distinguish subtle and large (vital signs containing breathing and pulse beating) human activities. In an article conducted by Liu and coworkers [22], a wearable strain sensor was fabricated by attributing polydopamine/PVA hydrogel to the two copper terminals located on the stretchable acrylic tape substrate. The developed electronic device precisely monitored knee and finger bending by employing it on the volunteer's body. Endowed with sufficient softness and self-adhesiveness of the prepared hydrogel, vital signs such as breathing and the heart rate of volunteers were also monitored by attachment to the throat and wrist without using any adhesive tape. In another work, Han et al. [23] constructed PDA-PPy/PAM hydrogel capable of directly linking to the human skin. The electrocardiogram (ECG) and electromyography (EMG) signals obtained from the hydrogel-attached volunteer's hand were the same as the commercial devices' information.

In spite of the sufficient performance of the aforementioned sensors for distinctive motion monitoring, the lack of large GF at minuscule deformations limits their precise human biosignal detection such as diastolic and systolic blood pressure [24]. Regarding these, Wang et al. [25] assembled a wearable sensor for distinguishing systolic (P1) and diastolic blood pressure (P2) peaks. This device profited from exceptional strain sensitivity at tiny deformations in which GF was 11 under very low compression and 5.7 below 0.3% stretching. The as-prepared hydrogel as an epidermal diagnostic device was able to detect heart arterial stiffness (augmentation index (AIr)).

15.3.3 Hydrogels for Prosthetics or Artificial Tissue as Implantable Electronics

Hydrogels are capable of implantation into internal living organs to diagnose physiological signs from deep tissues and imitate the behavior of real tissue. Additionally, a long-lasting examination of hydrogels' biodegradability and biocompatibility is provided by their in vivo implantation. Han et al. [26] developed an electronic device comprising graphene oxide/PDA/PVA conductive NCHs. The as-prepared device was implanted into a rabbit's dorsal tissue to measure EMG signals. Compared with signals from surface electrodes, higher EMG detection signals using implanted hydrogel essentially imply the value of hydrogel implantation for precise signal diagnosis. Remarkably, the implantation of hydrogels into the head of the model can exhibit brain electrical activity by delivering electroencephalogram (EEG) signals, which could be appropriate for identifying brain disorders like brain death, anesthesia, and coma [27]. For instance, Gan and coworkers [28], fabricated conductive and self-adhesive NCHs using PEDOT-PDA/PAM-modified sulfonated graphene oxide. Integrated non-toxic materials like graphene oxide and PDA show satisfactory biocompatibility to be embedded into the rabbit's head. Using the as-prepared hydrogel-based device, the model's EEG signals were monitored as well as diverse movements such as walking and chewing. In an alternative investigation, NCHs based on nanohydroxyapatite/poly(ethylene glycol) diacrylate/PAM/LiCl were designed by Wang et al. [29] with the capability for monitoring the electrooculography and EEG with the same accuracy as the commercial electrode.

Besides the precise detection of signals from in vivo tissues through direct contact, implanted hydrogels can similarly control nerve/cell electrical function. To this end, Liu et al. [30] applied PEDOT conductive hydrogel into the tissue of a mouse to electrically stimulate its sciatic nerve using neuromodulation with low voltage. With advances in the design of different hydrogels, the development of hydrogels with twisted conductive channels could be impressive for signal intensity improvement in implantable electronic devices. Inspired by this, Liu and coworkers [31] formed microfluidic channels with different configurations inside the alginate/PAM hydrogel. To enhance the sensitivity of the NCHs, biocompatible indium and gallium liquid metals were perfused into the microchannels. The results demonstrated that the employment of a more complex conducive channel in strain sensors provides extra distinct signals by analyzing subtle human motions. The designed hydrogel was finally implanted into the heart of a rabbit model to evaluate its cardiac biosignals. In this study, cardiac electrophysiological signals were resembled by stretching of the conductive cardiac patch through heart tissue movement. To detect subtle human motions (vocal vibrations), designed NCHs were attached to human skin as well as fabricating hydrogel near-field communication (NFC) tags to permit wireless communications of hydrogel electronics in pig skin. The implanted NFC tag successfully recorded and captured the physiological parameters of biological tissues wirelessly from underneath the pig's skin.

15.3.4 Hydrogels for Human-Machine Interfaces

To date, despite the number of touch panels (e.g., organic light-emitting diodes) suffering from crack forming under deformation, hydrogels are considered admirable alternates endowed with their excellent deformability, anti-fatigue properties, tunable transparency, good strain sensitivity, and intrinsic ionic conductivity. Thanks to ionic conductive hydrogels displaying quick response against strain, using them in touch panel technology is in the spotlight [32]. In this context, efforts have focused on the NCH-based touch panel. For example, Kweon et al. [33] developed an ionic conductive and stretchable sodium polyacrylate/N-isopropyl acrylamide/borax/PVA hydrogel with self-healing, transparent, and thermo-responsive properties. To construct a smart touch sensor, two polyethylene terephthalate films comprising four copper electrodes captured the prepared hydrogel with the surface capacitive mechanism. The quick response of the designed sensor enabled its rapid detection of finger forces. Additionally, Kim et al. [34] employed transparent lithium chloride/PAM hydrogel in a touch panel system that endured more than 1000% stretching.

In the 2D electronic skin, a combination of a well-ordered array of hydrogels coated by silver/copper electrodes is essential to investigate pressure/force distribution. At this state, Wang et al. [25] comprised a multi-array (6×8) hydrogel with extraordinary electrical resistance variations to produce diverse letters under finger pressing. The electrical signals achieved from hydrogel-based strain sensors have unlocked the opportunity of actuating robotic systems. Therefore, integrated hydrogels with soft robots can act as artificial organs [35]. In this context, Liu and coworkers [36] fabricated a soft device using poly(ionic liquid) conductive hydrogel. The materials were identified by the prepared device based on their surface roughness, temperature, and triboelectric coefficient. Voltage signals produced by this device during the holding of diverse materials originated from the triboelectric effect. In another work, Zhou et al. [37] designed an artificial ligament by mounting NaCl/hydroxypropyl cellulose/PVA hydrogel onto the robotic hand. The electronic signal reached from the fabricated hydrogel identified the position of the ligament perfectly.

The development of ionic conductive hydrogel-based soft robots could make them appropriate devices for internal and external inducing. In this regard, a five-layer elastomer hydrogel/elastomer was prepared by Larson et al. [38] as a feedback sensory unit. The capacitance of the embedded hydrogels was changed by their deformation under applying pressure as an external stimulus. The change in capacitance of the hydrogel leads to actuation degree-dependent luminesce emitting light. Alternatively, Cheng et al. [39] mounted LiCl/PAM conductive hydrogel on the elastomer actuator modified with silane to donate new functionality. The self-adhesive and highly stretchable hydrogel was able to deform along with the pneumatic actuator during the inflation without detachment of the elastomer and hydrogel. The air pressure was also evaluated by the developed devise through relative calibration between bending angle and electrical resistance variations.

15.3.5 Hydrogels for Energy Storage Devices

With advances in flexible wearable electronics, the energy supply of these wearable devices has drawn attention. Hydrogels stand up as a potential candidate to create unprecedented energy storage devices from their tunable and highly abundant chemistry. The electrolytic ions can be effectively located and attracted in the polymeric network by means of the chain's charged functional groups, however, great quantities of solvent water can be trapped and absorbed in the frame. The high-water content characteristics of hydrogels allow them to show ionic conductivity equal to liquids while preserving the dimensional stability of solids, which is appreciated for flexible energy storage devices [40].

In principle, electrochemical capacitors store charge via electrolyte ions desorption/adsorption on the surface of the electrode. For example, Fu et al. [41] fabricated hybrid hydrogel capacitors based on anti-freezing zinc with great interfacial adhesion and high electrical conductivity. In this study, acrylic acid and [2(methacryloyloxy)ethyl]dimethyl-(3-sulfopropyl) are used as monomers of the hydrogel. To initiate monomer polymerization, cellulose nanocrystals coated with tannic acid reacted with ammonium persulfate followed by forming hydrogel through $ZnCl_2$ addition. The developed capacitor confirmed a capacity of 84.6% at low temperatures and an energy density of 80.5 Wh/kg. Furthermore, the capacitor outperformed advanced stretchable zinc-ion hybrid capacitors at room temperature. In another work, Huang et al. [42] described novel electrolytes comprising rechargeable Zn–air and Zn//NiCo batteries along with sodium polyacrylate hydrogel (PANa). Both electrolyte species demonstrated ultra-long cycle stability much higher than the greatest solid-state counterpart.

Supercapacitors offer a large storage capacity and higher capacitance, attaining Farad-level capacitance, compared with traditional capacitors. Electrochemical energy in supercapacitors is primarily stored through the surface adsorption of electrolytic ions, which are regularly categorized as fast surface REDOX reactions or double-layer capacitors. Qin and coworkers [43] developed a combined charging energy storage system made up of a cellulose-based supercapacitor and triboelectric nanogenerator. The designed hydrogel displayed an electrical conductivity of 1.92 S/m and light transmittance of 93% operating normally at −54.3°C. Profiting from their capability in storing and harvesting energy even in extremely harsh conditions, the system is exploited as a source of extreme power supply for applications in wearable electronic devices.

Batteries as alternative energy storage devices classically contain two active electrochemical electrodes and an ionic conductive barrier between them. In the context of rechargeable or galvanic batteries, hydrogels provide more flexible batteries compared with traditional electrodes for stretchable/flexible batteries in wearable electronics [44]. To this end, Ye et al. [45] fabricated a novel all-hydrogel battery comprised of glycerol/PAM/CNTs. When the dehydrated hydrogel electrode was placed in contact with the hydrogel electrolyte, the dehydrated hydrogel was able to be rehydrated. After reaching water equilibrium, a fusion interface was formed between the hydrogel electrode and hydrogel electrolyte forming a hydrogel-derived battery with high stability under encapsulation into isolation conditions. Regarding their entirely hydrogel-based construction, Young's modulus of the flexible battery preferably matched with human skin. Additionally, the designed battery exhibited a specific capacity of a psychological consultation battery (370 mAh/g) and a lithium-ion battery (83 mAh/g) at a density of 0.5 A/g.

Park et al. [46] fabricated a mixed energy storage battery composed of poly(acrylic acid) hydrogel electrolytes soaked in water-in-bisalt (WIBS). In this study, the devised hybrid energy storage full cells took advantage of both lithium manganese oxide ($LiMn_2O_4$) and nanosulfur multivalent redox chemistries as illustrated in Figure 15.2. Sulfur was employed as a capable electrode material thanks to its high theoretical capacity and low cost. Furthermore, its high ionic conductivity and nanoscale confinement were

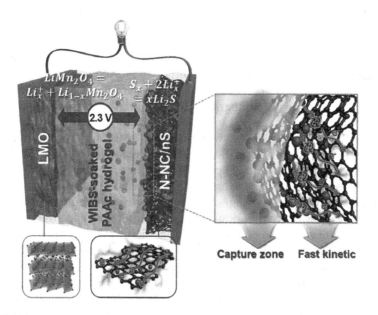

FIGURE 15.2
Schematic design of poly(acrylic acid) hydrogel electrolytes soaked in WIBS for hybrid energy storage full cells. (Adapted with permission [46], Copyright [2020], American Chemical Society.)

attained owing to the hierarchical porosity of nitrogen-incorporated nanoporous carbon/nano-sized sulfur. These NCHs demonstrated high energy and power density, an energy efficiency of 98% over 2000 cycles, and a retention rate of 78.7%.

In another work conducted by Yang et al. [47], a new strategy was proposed using a hygroscopic hydrogel electrolyte to avoid thermal runaway in water-based zinc-ion batteries. In this investigation, the hydrogel electrolyte was fabricated using zinc chloride solution and PAM. Moreover, MnO_2/carbon cloth fiber and porous zinc foam were used as the cathode and anode, respectively. At this state, the fall and rise of temperature in the hydrogel resulting in the reversible water evaporation and regeneration process can eventually tune the migration of ions in the hydrogel. It is proved that the water in the electrolyte evaporated after heating causing a stop or slowing down in ion transportation among the two electrodes.

15.3.6 Hydrogel for Electromagnetic Shielding

Electromagnetic interference, which is around us, affects human lives permanently. The negative effects of electromagnetic waves on scientific experiments or military missions are inevitable [48]. A magnetic or conductive material such as electromagnetic shielding (EMS) can decrease electromagnetic interference by isolating electronic substrates from the surrounding environment. Improving the electrical conductivity of the materials is the key approach to designing efficient EMS materials. The shielding capability of hydrogels arose from the repetitive absorption and reflection of the microwave porous network, which was supported by the liquid-infilled hydrogel. Generally, water can cause polarization losses and electromagnetic wave attenuation in the terahertz and gigahertz bands. An enhanced attenuation of the penetrating wave can be accomplished with no reflection of whether adequate water molecules integrate into a material with moderate electrical conductivity. Electromagnetic waves are typically absorbed by conductive hydrogels with more than 90% water content. EMS properties of hydrogels could be stronger by adding fillers with EMS features to water-rich hydrogels. MXene hydrogels with 3D porous structures are ideal electromagnetic interference shielding materials as they have more scattering centers for internally reflected electromagnetic waves than spacer nanosheets [49]. The existence of a porous network and water in the pores encourage repeated absorption and reflection of most incoming microwaves. At this state, Zhu et al. [50] established new hydrogel-based shielding materials using poly(acrylic acid) and MXene. The designed hydrogel with excellent recyclability and stretchability exhibited absorption-dominated EMS performance because of the accommodation of an internal water-rich environment and conductivity generated by porous MXene (Figure 15.3). The reliability, flexibility,

Hydrogels for Wearable Electronics

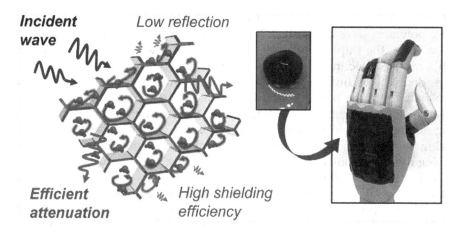

FIGURE 15.3
Scheme presentation of absorption or dominated EMS mechanism of the MXene-based NCHs and their shape adhesiveness and adaptability. (Adapted with permission [50], Copyright [2021], American Chemical Society.)

self-healing capability, and hydrogel type of this shielding material show promise for applications in artificial skin and wearable electronics.

Using cellulose nanofibers, Guand et al. [48] reported CNTs-based robust 3D dual NCHs using a low energy consumption, eco-friendly, fast, and large-scale strategy. The proposed CNTs/cellulose nanofiber construction exhibited the best EMS compared with other reported shielding materials, which allows it to be exploited in the design of EMS cars, buildings, and so on. Particularly, it could shield vehicles with sophisticated electrical equipment from several electromagnetic interferences equal to 100 dB where data inside EMS buildings and the security of electronic equipment can be protected.

15.4 Conclusion and Future Perspectives

In today's world, the advancement of hydrogel and NCH-based electronic devices as an effective substrate for an array of different utilizations has witnessed tremendous progress. Herein, we aim to provide a comprehensive review of debating concepts and present development in the improvement of NCH-based electronics, comprising design approaches, preparation tactics, and their applications. Particularly, we have described some imperative methods to overcome the restrictions of conventional hydrogels as flexible electronics. To increase the hydrogel's conductivity, methods like double-network, the inclusion of conductive fillers/dopants, and conductive polymers were established. Mechanical weakness is another important matter for the improvement of NCH-based flexible electronic devices. In this regard, selecting approaches like interpenetrating networks, nanofillers, double crosslinking, reversible physical bonds, and ionic/zwitterionic polymers into the hydrogel network is helpful to enhance the mechanical features of hydrogels. Self-healing is a vital factor for electronic skin configurations. Numerous techniques are explored for their relevance in realizing self-healing ability, which can be classified into two main tactics using dynamic non-covalent and covalent bonds. Environmental resistance and long-term stability are challenges for NCH-derived flexible electronic devices. To solve the adhesion problem, encapsulation with elastomers and presenting humectant are some accessible possible solutions. Although NCH-derived platforms hold excellent capacity as flexible electronic devices, several significant challenges still prevent hydrogel-based flexible electronics from becoming a harmonious part of our daily lives. The incorporation of hydrogel-derived flexible electronic devices is the first challenge. Nowadays, the singular and limited functions of hydrogel-derived flexible devices trigger the design of future models of ideally multifunctional hydrogel-based substrates. For example, an excellent bioelectronic tool must combine functions of sampling, data transmission and generation, and self-powered and timely interactions with humans on top of basic properties of comfortability, biocompatibility, conformability, and flexibility. Another critical problem is the easy adaptability and tolerability of hydrogel-derived flexible devices to realize diverse requirements of daily life under different conditions. Further approaches must be dedicated to how to incorporate hydrogel implantability or wearability tools with biological structures. In respect of engineering features, self-repairing capability, geometry, temperature correlation, and long-term stability of the hydrogels should be enhanced, as well as other practical demands of communications and data collection. Nevertheless, acceleration in technology advances and synergies, including the discovery of new functions and properties of traditional materials, the exploration of novel materials, developments in wireless communications, and revolutions in additive manufacturing, will bring forward effective flexible devices in the near future.

Acknowledgments

The authors would like to acknowledge the financial support from the Pharmaceutical Analysis Research Center, Tabriz University of Medical Science (Tabriz, Iran) (registration no: 68133).

References

1. R. Crabb, F. Treble, Thin silicon solar cells for large flexible arrays, *Nature* 213 (1967) 1223–1224.
2. J. Guo, Y. Yu, L. Cai, Y. Wang, K. Shi, L. Shang, J. Pan, Y. Zhao, Microfluidics for flexible electronics, *Mater. Today* 44 (2021) 105–135.
3. T.R. Ray, J. Choi, A.J. Bandodkar, S. Krishnan, P. Gutruf, L. Tian, R. Ghaffari, J.A. Rogers, Bio-integrated wearable systems: A comprehensive review, *Chem. Rev.* 119 (2019) 5461–5533.
4. Z. Bao, X. Chen, Flexible and stretchable devices, *Wiley Online Library* (2016) 4177–4179.
5. H. Yuk, B. Lu, X. Zhao, Hydrogel bioelectronics, *Chem. Soc. Rev.* 48 (2019) 1642–1667.
6. Z. Qin, X. Sun, Q. Yu, H. Zhang, X. Wu, M. Yao, W. Liu, F. Yao, J. Li, Carbon nanotubes/hydrophobically associated hydrogels as ultrastretchable, highly sensitive, stable strain, and pressure sensors, *ACS Appl. Mater. Interfaces* 12 (2020) 4944–4953.
7. C. Dannert, B.T. Stokke, R.S. Dias, Nanoparticle-hydrogel composites: from molecular interactions to macroscopic behavior, *Polymers* 11 (2019) 275.
8. C.-J. Lee, H. Wu, Y. Hu, M. Young, H. Wang, D. Lynch, F. Xu, H. Cong, G. Cheng, Ionic conductivity of polyelectrolyte hydrogels, *ACS Appl. Mater. Interfaces* 10 (2018) 5845–5852.
9. X. Sun, F. Yao, J. Li, Nanocomposite hydrogel-based strain and pressure sensors: A review, *J. Mater. Chem. A* 8 (2020) 18605–18623.
10. K. Pan, S. Peng, Y. Chu, K. Liang, C.H. Wang, S. Wu, J. Xu, Highly sensitive, stretchable and durable strain sensors based on conductive double-network polymer hydrogels, *J. Polym. Sci.* 58 (2020) 3069–3081.
11. G. Su, J. Cao, X. Zhang, Y. Zhang, S. Yin, L. Jia, Q. Guo, X. Zhang, J. Zhang, T. Zhou, Human-tissue-inspired anti-fatigue-fracture hydrogel for a sensitive wide-range human–machine interface, *J. Mater. Chem. A* 8 (2020) 2074–2082.
12. P. He, R. Guo, K. Hu, K. Liu, S. Lin, H. Wu, L. Huang, L. Chen, Y. Ni, Tough and super-stretchable conductive double network hydrogels with multiple sensations and moisture-electric generation, *J. Chem. Eng.* 414 (2021) 128726.
13. L. Shao, Y. Li, Z. Ma, Y. Bai, J. Wang, P. Zeng, P. Gong, F. Shi, Z. Ji, Y. Qiao, Highly sensitive strain sensor based on a stretchable and conductive poly (vinyl alcohol)/phytic acid/NH2-POSS hydrogel with a 3D microporous structure, *ACS Appl. Mater. Interfaces* 12 (2020) 26496–26508.
14. Y. Cai, J. Qin, W. Li, A. Tyagi, Z. Liu, M.D. Hossain, H. Chen, J.-K. Kim, H. Liu, M. Zhuang, A stretchable, conformable, and biocompatible graphene strain sensor based on a structured hydrogel for clinical application, *J. Mater. Chem. A* 7 (2019) 27099–27109.
15. W. Shi, G. Han, Y. Chang, H. Song, W. Hou, Q. Chen, Using stretchable PPy@PVA composites as a high-sensitivity strain sensor to monitor minute motion, *ACS Appl. Mater. Interfaces* 12 (2020) 45373–45382.
16. Z. Karimzadeh, M. Mahmoudpour, E. Rahimpour, A. Jouyban, Nanomaterial based PVA nanocomposite hydrogels for biomedical sensing: Advances toward designing the ideal flexible/wearable nanoprobes, *Adv. Colloid Interface Sci.* (2022) 102705.

17. M. Sepantafar, R. Maheronnaghsh, H. Mohammadi, F. Radmanesh, M.M. Hasani-Sadrabadi, M. Ebrahimi, H. Baharvand, Engineered hydrogels in cancer therapy and diagnosis, *Trends Biotechnol.* 35 (2017) 1074–1087.
18. Z. Xu, J. Song, B. Liu, S. Lv, F. Gao, X. Luo, P. Wang, A conducting polymer PEDOT: PSS hydrogel based wearable sensor for accurate uric acid detection in human sweat, *Sens. Actuators B Chem.* 348 (2021) 130674.
19. J. Yu, Y. Feng, D. Sun, W. Ren, C. Shao, R. Sun, Highly conductive and mechanically robust cellulose nanocomposite hydrogels with antifreezing and antidehydration performances for flexible humidity sensors, *ACS Appl. Mater. Interfaces* 14 (2022) 10886–10897.
20. Z. Karimzadeh, A. Jouyban, A. Ostadi, A. Gharakhani, E. Rahimpour, A sensitive determination of morphine in plasma using AuNPs@ UiO-66/PVA hydrogel as an advanced optical scaffold, *Anal. Chim. Acta* 1227 (2022) 340252.
21. F. Lin, Z. Wang, Y. Shen, L. Tang, P. Zhang, Y. Wang, Y. Chen, B. Huang, B. Lu, Natural skin-inspired versatile cellulose biomimetic hydrogels, *J. Mater. Chem. A* 7 (2019) 26442–26455.
22. S. Liu, R. Zheng, S. Chen, Y. Wu, H. Liu, P. Wang, Z. Deng, L. Liu, A compliant, self-adhesive and self-healing wearable hydrogel as epidermal strain sensor, *J. Mater. Chem. C* 6 (2018) 4183–4190.
23. L. Han, L. Yan, M. Wang, K. Wang, L. Fang, J. Zhou, J. Fang, F. Ren, X. Lu, Transparent, adhesive, and conductive hydrogel for soft bioelectronics based on light-transmitting polydopamine-doped polypyrrole nanofibrils, *Chem. Mater.* 30 (2018) 5561–5572.
24. X. Jing, H.-Y. Mi, Y.-J. Lin, E. Enriquez, X.-F. Peng, L.-S. Turng, Highly stretchable and biocompatible strain sensors based on mussel-inspired super-adhesive self-healing hydrogels for human motion monitoring, *ACS Appl. Mater. Interfaces* 10 (2018) 20897–20909.
25. Z. Wang, J. Chen, Y. Cong, H. Zhang, T. Xu, L. Nie, J. Fu, Ultrastretchable strain sensors and arrays with high sensitivity and linearity based on super tough conductive hydrogels, *Chem. Mater.* 30 (2018) 8062–8069.
26. L. Han, X. Lu, M. Wang, D. Gan, W. Deng, K. Wang, L. Fang, K. Liu, C.W. Chan, Y. Tang, A mussel-inspired conductive, self-adhesive, and self-healable tough hydrogel as cell stimulators and implantable bioelectronics, *Small* 13 (2017) 1601916.
27. N. Sharma, V. Sharma, Y. Jain, M. Kumari, R. Gupta, S. Sharma, K. Sachdev, Synthesis and characterization of graphene oxide (GO) and reduced graphene oxide (rGO) for gas sensing application, *Macromolecular Symposia*, Wiley Online Library (2017) 1700006.
28. D. Gan, Z. Huang, X. Wang, L. Jiang, C. Wang, M. Zhu, F. Ren, L. Fang, K. Wang, C. Xie, Graphene oxide-templated conductive and redox-active nanosheets incorporated hydrogels for adhesive bioelectronics, *Adv. Funct. Mater.* 30 (2020) 1907678.
29. Z. Wang, L. Chen, Y. Chen, P. Liu, H. Duan, P. Cheng, 3D printed ultrastretchable, hyper-antifreezing conductive hydrogel for sensitive motion and electrophysiological signal monitoring, *Research* 2020 (2020) 1426078.
30. Y. Liu, J. Liu, S. Chen, T. Lei, Y. Kim, S. Niu, H. Wang, X. Wang, A.M. Foudeh, J.B.-H. Tok, Soft and elastic hydrogel-based microelectronics for localized low-voltage neuromodulation, *Nat. Biomed. Eng.* 3 (2019) 58–68.

31. Y. Liu, T. Yang, Y. Zhang, G. Qu, S. Wei, Z. Liu, T. Kong, Ultrastretchable and wireless bioelectronics based on all-hydrogel microfluidics, *Adv. Mater.* 31 (2019) 1902783.
32. Q. Peng, J. Chen, T. Wang, X. Peng, J. Liu, X. Wang, J. Wang, H. Zeng, Recent advances in designing conductive hydrogels for flexible electronics, *InfoMat* 2 (2020) 843–865.
33. O.Y. Kweon, S.K. Samanta, Y. Won, J.H. Yoo, J.H. Oh, Stretchable and self-healable conductive hydrogels for wearable multimodal touch sensors with thermoresponsive behavior, *ACS Appl. Mater. Interfaces* 11 (2019) 26134–26143.
34. C.-C. Kim, H.-H. Lee, K.H. Oh, J.-Y. Sun, Highly stretchable, transparent ionic touch panel, *Science* 353 (2016) 682–687.
35. A.M. Hubbard, W. Cui, Y. Huang, R. Takahashi, M.D. Dickey, J. Genzer, D.R. King, J.P. Gong, Hydrogel/elastomer laminates bonded via fabric interphases for stimuli-responsive actuators, *Matter* 1 (2019) 674–689.
36. Z. Liu, Y. Wang, Y. Ren, G. Jin, C. Zhang, W. Chen, F. Yan, Poly (ionic liquid) hydrogel-based anti-freezing ionic skin for a soft robotic gripper, *Mater. Horiz.* 7 (2020) 919–927.
37. Y. Zhou, C. Wan, Y. Yang, H. Yang, S. Wang, Z. Dai, K. Ji, H. Jiang, X. Chen, Y. Long, Highly stretchable, elastic, and ionic conductive hydrogel for artificial soft electronics, *Adv. Funct. Mater.* 29 (2019) 1806220.
38. C. Larson, B. Peele, S. Li, S. Robinson, M. Totaro, L. Beccai, B. Mazzolai, R. Shepherd, Highly stretchable electroluminescent skin for optical signaling and tactile sensing, *Science* 351 (2016) 1071–1074.
39. S. Cheng, Y.S. Narang, C. Yang, Z. Suo, R.D. Howe, Stick-on large-strain sensors for soft robots, *Adv. Funct. Mater.* 6 (2019) 1900985.
40. Z. Wang, H. Li, Z. Tang, Z. Liu, Z. Ruan, L. Ma, Q. Yang, D. Wang, C. Zhi, Hydrogel electrolytes for flexible aqueous energy storage devices, *Adv. Funct. Mater.* 28 (2018) 1804560.
41. Q. Fu, S. Hao, L. Meng, F. Xu, J. Yang, Engineering self-adhesive polyzwitterionic hydrogel electrolytes for flexible zinc-ion hybrid capacitors with superior low-temperature adaptability, *ACS Nano* 15 (2021) 18469–18482.
42. Y. Huang, Z. Li, Z. Pei, Z. Liu, H. Li, M. Zhu, J. Fan, Q. Dai, M. Zhang, L. Dai, Solid-state rechargeable Zn//NiCo and Zn–air batteries with ultralong lifetime and high capacity: the role of a sodium polyacrylate hydrogel electrolyte, *Adv. Energy Mater.* 8 (2018) 1802288.
43. C. Qin, A. Lu, Flexible, anti-freezing self-charging power system composed of cellulose based supercapacitor and triboelectric nanogenerator, *Carbohydr. Polym.* 274 (2021) 118667.
44. Y.-Z. Zhang, J.K. El-Demellawi, Q. Jiang, G. Ge, H. Liang, K. Lee, X. Dong, H.N. Alshareef, MXene hydrogels: Fundamentals and applications, *Chem. Soc. Rev.* 49 (2020) 7229–7251.
45. T. Ye, J. Wang, Y. Jiao, L. Li, E. He, L. Wang, Y. Li, Y. Yun, D. Li, J. Lu, A Tissue-Like Soft All-hydrogel battery, *Adv. Mater.* 34 (2022) 2105120.
46. J.M. Park, M. Jana, P. Nakhanivej, B.-K. Kim, H.S. Park, Facile multivalent redox chemistries in water-in-bisalt hydrogel electrolytes for hybrid energy storage full cells, *ACS Energy Lett.* 5 (2020) 1054–1061.
47. P. Yang, C. Feng, Y. Liu, T. Cheng, X. Yang, H. Liu, K. Liu, H.J. Fan, Thermal self-protection of zinc-ion batteries enabled by smart hygroscopic hydrogel electrolytes, *Adv. Energy Mater.* 10 (2020) 2002898.

48. Q.-F. Guan, Z.-M. Han, K.-P. Yang, H.-B. Yang, Z.-C. Ling, C.-H. Yin, S.-H. Yu, Sustainable double-network structural materials for electromagnetic shielding, Nano Lett. 21 (2021) 2532–2537.
49. F. Zhao, J. Bae, X. Zhou, Y. Guo, G. Yu, Nanostructured functional hydrogels as an emerging platform for advanced energy technologies, Adv. Mater. 30 (2018) 1801796.
50. Y. Zhu, J. Liu, T. Guo, J.J. Wang, X. Tang, V. Nicolosi, Multifunctional Ti3C2T x MXene composite hydrogels with strain sensitivity toward absorption-dominated electromagnetic-interference shielding, ACS Nano 15 (2021) 1465–1474.

16

Hydrogels for the Removal of Water Content from Liquid Fuels

Letícia Arthus, Patrícia Bogalhos Lucente Fregolente, Maria Regina Wolf Maciel, and Leonardo Vasconcelos Fregolente

CONTENTS

16.1 Introduction 311
16.2 Liquid Fuels 312
    16.2.1 Forms of Water Contamination in Fuels 313
    16.2.2 Effect of Water on Fuel Processes and Engines 314
    16.2.3 Technologies for Removing Water from Fuels 315
16.3 Application of Hydrogels for Water Removal from Liquid Fuels 317
    16.3.1 Hydrogels for Removing Water from Petroleum-based Fuels 318
    16.3.2 Hydrogels for Removing Water from Biodiesel and Its Diesel Blends 322
    16.3.3 Hydrogel Design for Water Removal from Liquid Fuels 326
        16.3.3.1 Ionic-charged Hydrogels for High Water Removal Capacities 327
        16.3.3.2 Hydrogel Mass/Fuel Volume Ratio Effect on Water Removal Efficiency 328
        16.3.3.3 Temperature Effect on Water Removal Efficiency 328
        16.3.3.4 High Crosslinking Density as Means for Mechanically Stable Hydrogels 329
    16.3.4 Reuse of Hydrogels for Removing Water from Fuels 330
16.4 Conclusions 330
References 331

16.1 Introduction

Liquid fuels are the largest source of energy in the world, being used in transportation, commerce, industry, and domestic economies [1]. Due to its worldwide importance, specification standards are in place to regulate the quality of fuel available on the market. The maximum water content is among the main parameters of fuel specification. Water contamination is one of the major concerns in the production, handling, transporting, and storing

DOI: 10.1201/9781003351566-19   *311*

of fuels [2]. It occurs either by the direct contact of the fuel with water during the reaction and separation steps of its production process or by contact with air of high relative humidity, in which the fuel spontaneously dissolves water. At room temperature, for example, fuels composed of hydrocarbons present a solubility of water ranging from 50 to 200 ppm; in biofuels with structures containing hydrophilic polar groups, such as biodiesel, the solubility exceeds 1300 ppm [3]. Moreover, the addition of biodiesel to diesel, for example, further increases the water retention capacity of the blend [4].

Considering the global efforts to reduce CO_2 emissions and even to increase the performance of the final fuel, several countries have adopted public policies and legislative measures to increase the use of biofuels in their energy matrices. This significantly raises concerns in relation to water contamination; since the greater the polarity of the compounds in the mixture, the greater the hydrophilicity of the fuel [4]. The presence of water in fuels leads, most importantly, to microbiological activity and the generation of sludge, which saturates filter elements faster and impairs the functioning of engine elements [5, 6]. Technologies commercially available for water removal include gravitational equipment, centrifuges, and coalescing filters, which remove only free or suspended water. For the removal of dissolved water, energy-intensive processes are used such as heating and vacuum flash evaporators or packed beds of desiccant salts, which can lead to fuel contamination.

Considering the need for a new and complete technology that removes free and dissolved water and is simple to operate, at a competitive cost for industrial and domestic (integrated into vehicles) scales, hydrophilic polymeric hydrogels have been developed, showcasing their potential for removing water from liquids fuels, either in batches or in continuous processes. Generally, hydrogels can be used to dehydrate petroleum-based fuels, reaching lower water contents than that required by standard specifications. Specifically with diesel, good removal efficiency may be attained by employing even the most common type of hydrophilic hydrogels group, the neutral polyacrylamide. The use of ionic-charged synthetic polymers as the poly(acrylamide-co-sodium acrylate) hydrogel was essential to achieve better water removal efficiencies from hydrophilic fuels such as biodiesel and its blends.

16.2 Liquid Fuels

Crude oil and petroleum refining products – such as gasoline, diesel fuel, fuel oils, aviation kerosene, and marine fuels – as well as biofuels – such as ethanol and biodiesel – are categorized as liquid fuels. Standard specifications regulate fuel quality, guaranteeing their performance and compliance with environmental restrictions, which has been increasingly demanded by the market. Among the main parameters of fuel specification are viscosity, density, minimum flash point, maximum sulfur content, and maximum water content [7]. Water content may be a particularly challenging parameter to meet for some fuels since they can spontaneously dissolve water. Additionally, water is present everywhere in the fuel production and distribution chain; from being dissolved in the air to being used in industrial purification and cleaning processes. The solubility of water in fuel is a function of its chemical composition, temperature, and pressure to which it is subjected. Petroleum-derived fuels are hydrocarbons, consisting mainly of carbon and hydrogen; whereas biofuels are composed of hydrocarbon chains with oxygenated groups attached to them, such as acid, alcohol, and ester groups, a characteristic that gives biofuels a more hydrophilic behavior.

Currently, ethanol and biodiesel are the two most common types of biofuels in use. Although ethanol production is much higher, biodiesel production has grown faster since 2010, and more than quadrupled from 2010 to 2022 [8]. Biodiesel is a fuel consisting of fatty acid ester chains, generally produced via the transesterification reaction of vegetable oils and animal fats. Biofuels can offer considerable greenhouse gas emissions benefits, improve vehicle performance, and reduce demand for petroleum. For these reasons, several nations have implemented governmental policies and legislation to increase the use of biofuels in their energy matrices.

In the North American market – according to the US Energy Information Administration – the average percentage of biofuels added to gasoline, diesel, and jet fuel is expected to increase from 7.3% in 2019 to 9.0% in 2040 [9]. In the Brazilian market, the mandatory use of biodiesel in the mixture with diesel was set at 2% v/v in 2005, reaching 12% v/v in 2020 [10], with a predicted adjustment of up to 15% v/v in 2023 [11]. These measures lead to an increase in the hydrophilicity of commercialized fuels, raising further concerns about the quality of fuels regarding water content.

16.2.1 Forms of Water Contamination in Fuels

Water can be present in fuels in three states: dissolved, suspended, or free. Dissolved water is found when all the droplets are solubilized by the fuel and it is invisible to the naked eye, a condition in which the fuel is referred to as saturated. Suspended water appears as a dull, hazy, or cloudy appearance that takes time to coalesce or settle down [12]. When free water is present, there is a phase separation, and the water accumulates at the bottom of the tanks since it is denser than the fuel. Fuels are able to dissolve water spontaneously and have specific water solubilization capabilities, especially considering polar group concentration, carbon chain length, temperature, and pressure. Generally, the greater the polarity of the compounds in the mixture, the greater the hydrophilicity of the fuel [4]. Table 16.1 presents the solubility of water in a few fuels at 20, 40, and 50°C.

TABLE 16.1
Solubility of Water in Different Fuels as a Function of Temperature

Fuel	Solubility of water (ppm)			Reference
	20°C	40°C	50°C	
Biodiesel	1720.0	1873.0	1980.0	[4]
Diesel oil	116.0	158.0	189.0	[4]
Marine diesel oil	88.0	109.0	134.0	[13]
Jet fuel A	49.0	90.0	120.0	[12]

Biodiesel, for example, absorbs more moisture than petroleum diesel due to the hygroscopic nature of its fatty acid ester chains. The addition of biodiesel to fossil diesel increases the water-holding capacity of the mixture. Pure biodiesel is capable of absorbing up to 1980 ppm of water at 50°C, which is 10 to 15 times higher than pure diesel at the same temperature [4]. Although water can be considered a fuel contaminant, its use is inherent to many fuel production processes. During oil refining, steam is injected into the distillation columns for product rectification. In biodiesel production, wet washing is the most used technique to remove impurities, such as excess catalyst, glycerol, and residual alcohol, which are produced in the transesterification reaction. The water content of biodiesel that is purified through a wet washing method is typically higher than 1000 ppm, and further reduction of the water content is required to meet the standard specification [14].

Other means of water contamination of fuels include condensation of air humidity on the walls in storage tanks, precipitation of dissolved water from the fuel itself at low temperatures, water left behind after cleaning operations in tanks or transport vehicles, and rainwater that can enter through poorly sealed tanks [12]. In practice, the exposure of fuels to water is inevitable. Water contamination can occur during the production process, transportation, and in storage, either by direct contact of the fuel with water or by contact with air of high relative humidity.

16.2.2 Effect of Water on Fuel Processes and Engines

The presence of water in fuel oils can result in numerous operational problems from its production to its use in vehicles. The presence of water in the biodiesel production process reduces the activity of the catalyst and can lead to the hydrolysis of vegetable oil and glycerol to free fatty acids, which react with the catalyst to produce a saponified product [15]. Soap formation and reduced catalyst effectiveness result in low biodiesel conversion, in addition to hampering the product separation and purification stages [16].

Water Content Removal from Liquid Fuels 315

TABLE 16.2
Maximum Water Content for Some Fuels According to ASTM and ANP Resolution

Fuel	International specification	Limits[1] (ppm)	ANP Resolution	Limits[1] (ppm)
Biodiesel	ASTM D6751	500	No. 45 (2014/08/25)	250 (manufacturer) 350 (distributor)
Diesel fuel	ASTM D95	500	No. 50 (2013/12/23)	200
Marine diesel oil	ISO 3733	3000	No. 52 (2010/12/29)	3000

[1] *Limits converted to ppm.*

The consequences of this contaminant in fuel supply systems include foam formation in tanks, cavitation at pump suction, and corrosive action [17]. In the final product, the growth of microorganisms is suggested, which generate films and lead to the clogging of filters, reducing the stability and the specific heat of the product [18]. These effects compromise the functioning of motor vehicle components, resulting in loss of propulsion and corrosion of equipment. Regulatory agencies, such as the American Society for Testing and Materials (ASTM) and the Brazilian National Agency for Petroleum, Natural Gas, and Biofuels (ANP), establish fuel specifications to ensure product quality. Table 16.2 shows the maximum water content established by these agencies for different fuels. Notably, these regulatory agencies adopt test methods that differ in their accuracy in detecting the water content in fuels. For biodiesel, for example, the methods allowed by ASTM D6751 are the ASTM D2709 or D1796, which are tests based on centrifuging the sample and reading the meniscus of a conical test tube. The ANP Resolution No. 45 (08/25/2014) allows the analysis method ASTM D6304, a technique that employs automatic volumetric Karl Fischer titration to detect traces of water.

16.2.3 Technologies for Removing Water from Fuels

Depending on fluid characteristics – such as droplet size, water concentration, operating conditions such as the water removal requirement, and process feed flow rate – different water separation technologies are commercially available and have been industrially employed for removing water from fuels.

Gravitational separators are based on the different densities of fluids being applied to remove only free water. The size of conventional separators is indicated by the retention time required for gravitational separation to occur; this time is directly proportional to the volume of fluid. Thus, gravitational separators have low efficiency since high residence times, and therefore, large vessels, are required for the small droplets of free water to coalesce into drops large enough to migrate to the oil-water interface [19].

Centrifugal separators are used to remove free and lightly suspended water and provide efficient separation with short residence times for batches and continuous processes [20]. The limitations of centrifuges include high operating and capital costs. They are an energy-intensive process and do not remove dissolved water from the fuel. Coalescing separators, or simply coalescers, assist small water droplets to combine into larger droplets, separating from the oil more easily. Despite having operational advantages, when compared with centrifuges, such as lower energy consumption, more compact units, and high efficiency in separating low-viscosity fluids, they are also unable to remove dissolved water, and they are sensitive to fluids that have solid impurities, which can clog the coalescing membranes and reduce its efficiency [20].

Anhydrous salts packed beds are units containing solid dehydrating compounds capable of removing free and dissolved water. The salts combine with the water in the fluid to form an aqueous solution that can be separated from the fuel flowing through the unit. Sodium chloride is the cheapest option employed in industrial processes but it reduces the residual water level to about 75% of relative saturation; whereas calcium chloride, which is normally considered a highly cost-effective combination, is capable of removing about 50% of the saturation level [21]. Although drying salts are efficient for removing water from fuels, they require the replacement of the salt beds used in the process, introducing the need for maintenance interventions. Additionally, the generation of brine is inherent to this process, leading it to be sent to effluent treatment plants. Furthermore, the contamination of fuels with sodium, potentially originating from drying salts, was proposed as a relevant factor in incidents of filter clogging and fouling in nozzles [22, 23].

Vacuum flash evaporators can also effectively remove free, suspended, and dissolved water from fuels. The technique consists of heating the oil to approximately 60°C and spraying it through a nozzle into a vacuum chamber of 0.6–0.7 bar [24]. The combination of temperature and vacuum causes water to flash to steam at a temperature low enough to not cause thermal damage to the oil. Fuel dispersion makes the drying process more efficient – dry fuel droplets are collected at the bottom of the vacuum chamber. The main disadvantage of these dryers is the need for a vacuum system that can pump steam out of the chamber, meaning expensive investments in equipment and high energy operating costs [25].

Thus, commercially available technologies employed to remove free water from fuels include centrifugal separators, gravitational separators, and coalescing filters (coalescers). For dissolved water, heating/vacuum flash evaporators and anhydrous salt-packed beds are frequently employed. However, each of these technologies has limitations, such as high energy and installation costs, the possibility of fuel contamination, or low dissolved water removal efficiency [26]. Thus, hydrophilic polymeric hydrogels emerge

Water Content Removal from Liquid Fuels

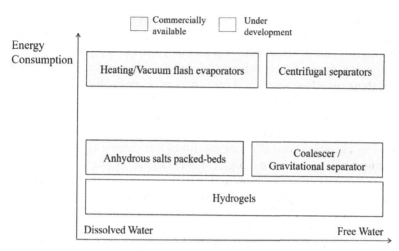

FIGURE 16.1
Technologies for water removal from fuels.

as a promising technology for removing water from fuels; they are able to remove both free and dissolved water, are easy to operate, have a competitive cost, and can be efficiently reused and applied on industrial and domestic scales. Figure 16.1 shows a comparison among commercially available technologies and hydrogels based on their energy consumption and water removal capacity.

16.3 Application of Hydrogels for Water Removal from Liquid Fuels

On account of their hydrophilic characteristics, polymeric hydrogels have been applied to remove water from liquid fuels [13, 26–37]. They are cross-linked three-dimensional materials capable of retaining a large amount of water in their structures [32]. Hydrophilic functional groups that are bonded to the main chain of the polymer confer water-retaining abilities to the hydrogels; while its elastic properties and mechanical resistance, as well as the insolubility of the polymer matrix in an aqueous medium, result from the magnitude of the crosslinking of the network [38]. Among the advantages of applying systems with hydrogels to remove water from fuels, it is possible to highlight their versatility regarding polymeric composition, which allows obtaining different polymeric matrices according to the need for the external stimulus to which the hydrogel will be subjected; thus, obtaining free and dissolved water removal [26, 32] and good chemical stability. Additionally, the hydrogels can be recovered, regenerated, and reused [39, 40].

Synthetic polymer systems have been applied to remove water from fuels due to their usual nontoxic and inert nature, long chain lengths, ability to preserve their shape and mechanical strength, and convenient adjustment of their mechanical, chemical, and biophysical properties [41]. Those materials are usually obtained by free radical polymerization – which is the most commonly employed crosslinking method for hydrogel synthesis – with the advantages of high reactivity, high conversion, and reaction under moderate conditions [42]. Synthetic homopolymer hydrogels as polyacrylamide, sodium polyacrylate, poly(acrylic acid), and copolymers hydrogels polyacrylamide-based – being poly(acrylamide-co-sodium acrylate) (poly(AAm-co-SA)), poly(acrylamide-co-acrylonitrile) (poly(AAm-co-AN)), and poly(acrylamide-co-acid acrylic) (poly(AAm-co-AA)) – have been employed to remove water from liquid fuels, such as diesel, marine diesel, and biodiesel and its diesel blends [13, 26–37, 43, 44, 45]. Hybrid sources of hydrogel, which contain a natural and synthetic polymer using *Moringa oleifera*-loaded keratin-co-sodium acrylate, have also been successfully applied for diesel purification [43]. Jet fuels present very low water affinity and, to remove traces of water, Li et. al (2021) used poly(N-isopropyl acrylamide) (PNIPAm) hydrogel containing hygroscopic chloride polypyrrole (PPy-Cl); they were able to obtain high removal efficiencies even after several dehydration cycles. Thus, the results reported in the literature have demonstrated that polymeric hydrogels are promising materials for removing both free and dissolved water from liquid fuels, either in batches or in continuous systems, as well as being reusable.

16.3.1 Hydrogels for Removing Water from Petroleum-based Fuels

Free, suspended, and dissolved water is satisfactorily removed from petroleum-based fuels employing hydrogels, reaching comfortable water levels – lower than those required by specifications. Due to their non-polar structure and lower water affinity in comparison with polar fuels, petroleum-based fuels, such as diesel, have been successfully purified even by applying the most common type of hydrophilic hydrogels group, the neutral polyacrylamide [31, 32]. In this case, 1–2 h of contact with diesel was enough for the hydrogel to absorb dissolved water and reduce it to the levels allowed by standard specification (<100 ppm).

Traces of dissolved water in petroleum-based jet fuel could be reduced from 47 to 7 ppm within 6 h of using poly(N-isopropyl acrylamide)/chloride polypyrrole hydrogel [27]. Table 16.3 and Table 16.4 show the results and experimental conditions of studies employing hydrogels to remove water from petroleum-based fuels, either in batches or in continuous processes, respectively. In the batch processes, 80.4% and 98.2% efficiencies were achieved in the removal of free, suspended, and dissolved water from

TABLE 16.3
Application of Hydrogels for Removing Water from Petroleum-based Fuels in Batch Processes

Reference	Fuel	Water form	Hydrogel composition	Ratio hydrogel/fuel (mg/ml)	Contact time (h)	Initial water content (ppm)	Removal efficiency
[31]	Diesel (B0)	Dissolved	Polyacrylamide	10.0	1	220	60%
[32]	Diesel	Dissolved	Polyacrylamide	50.0	2	240	65%
		Free and dissolved				350	80.4%
[26]	Diesel	Dissolved	Poly(acrylamide-co-sodium acrylate)	50.0	24	170	59.2%
			Poly(acrylamide-co-acid acrylic)				58.3%
[43]	Diesel	Free and dissolved	Moringa oleifera-loaded keratin-co-sodium acrylate	80.0	3	18,000	95%
[27]	Jet Fuel	Dissolved	Poly(N-isopropyl acrylamide)/chloride polypyrrole	10.0	6	47	85.1%
[13]	Marine diesel oil	Free and dissolved	Poly(acrylamide-co-sodium acrylate)	20.0	2	5,000	98.2%
			Poly(acrylamide-co-acrylonitrile)				98%

TABLE 16.4
Application of Hydrogels for Removing Water from Petroleum-based Fuels in Continuous Processes

Reference	Fuel	Water form	Hydrogel composition	Flow rate (ml/min)	Packed-bed characteristics	Initial water content (ppm)	Removal efficiency
[34]	Diesel	Dissolved	Poly(acrylamide-co-sodium acrylate)	1.0	$D = 30$ mm; $L_{packed\text{-}bed} = 3$ mm; porosity $= 0.4$	N/D	~50%
	Diesel	Dissolved	Poly(acrylamide-co-acid acrylic)				~50%
[35]	Diesel	Suspended and dissolved	Sodium polyacrylate	3.0 - 6.5	$Di = 17$ mm; $L_{packed\text{-}bed} = 45\text{-}60$ mm	630	84%

diesel oil (initial water content of 350 ppm) and marine diesel oil (initially at 5000 ppm), respectively, by using polyacrylamide [32] and poly(AAm-co-SA) hydrogels [13], in that order. Diesel oil containing only dissolved water (initial concentration ranging from 170 to 240 ppm) obtained water content levels lower than 100 ppm when treated with neutral polyacrylamide hydrogels [31, 32], poly(AAm-co-SA) and poly(AAm-co-AA) [26].

Studies focused on fuel water removal with hydrogels in continuous processes are particularly interesting since they address industrial operating conditions. Particles of poly(AAm-co-AA) and poly(AAm-co-SA) hydrogels filled in a random packing column showed similar performances, removing 50.0% water from saturated diesel oil, with a flow rate of 1 ml/min [34]. Furthermore, an unprecedented study demonstrated the ability of a packed bed filled with lyophilized sodium polyacrylate hydrogel to continuously treat cloudy diesel (Figure 16.2). The fixed bed containing 4 g of hydrogel was able to reduce the turbidity of more than 5 L of diesel under a 6.5 ml/min flow rate, reducing the water content to about 100 ppm [35], meeting the regulatory requirement regarding the permissible water content for this fuel.

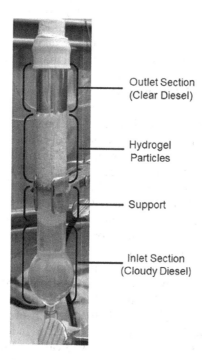

FIGURE 16.2
Fixed bed containing sodium polyacrylate hydrogel particles. (A colour version of this figure is available as part of the book's support materials at the following link: https://www.routledge.com/Hydrogels-Fundamentals-to-Advanced-Energy-Applications/Kumar-Gupta/p/book/9781032385129.) (Adapted with permission from [35], Copyright © [2022] Elsevier Ltd.)

16.3.2 Hydrogels for Removing Water from Biodiesel and Its Diesel Blends

The use of hydrogels to remove water from more hydrophilic fuels such as biodiesel and their diesel blends has been similarly investigated in batches (Table 16.5) and in continuous processes (Table 16.6).

In the batch process, water contents from 320 to 1100 ppm were obtained in saturated biodiesel (initial contents from 1300 to 2500 ppm) by applying homopolymer hydrogels such as polyacrylamide, sodium polyacrylate, poly(acrylic acid), and copolymers, such as poly(AAm-co-SA), poly(AAm-co-AN), and poly(AAm-co-AA). The efficiencies of removing water from saturated biodiesel varied according to hydrogel composition, hydrogel mass/biodiesel volume ratio, and temperature employed in this process. In the special case of saturated biodiesel, the use of ionic-charged synthetic polymers, such as anionic polyacrylamides hydrogels, was essential to achieve better efficiency. Anionic polyacrylamides comprise the class of polymers that contain acrylamide monomer at some level, copolymerized with acrylate salts derived from acrylic acid [47], such as the poly(AAm-co-SA). This material has greater water removal capacity due to the presence of dissociated ionic charges in the hydrogel structure, which results in an increase in the osmotic driving force due to the strong ion-dipole interaction. Water content below that specified by ASTM for biodiesel was reached by applying a moderate ratio of poly(AAm-co-SA) hydrogel mass/fuel volume, resulting in a reduction from 1514 to 453 ppm in a batch process [37]. Water content lower than the ASTM and ANP standards (for distributors) was also achieved by hydrogels in systems with temperatures around 40°C [28, 48]. Owing to its hydrophilic behavior, additional efforts have been put into hydrogel design in combination with process parameters setup to meet standard specifications for biodiesel, which will be discussed later in this chapter.

Regarding diesel-biodiesel blends, neutral polyacrylamide hydrogels reduced the water content of the saturated mixture containing 20% (v/v) and 40% (v/v) of biodiesel, B20 and B40, respectively, to levels below those allowed by ASTM [31]. However, under the same experimental conditions, the polyacrylamide hydrogel could not meet the specification requirements for diesel blends above 60% in biodiesel due to its hydrophilic structure.

Regarding the removal of water from biodiesel in a continuous process, poly(AAm-co-SA) and poly(AAm-co-AA) hydrogels continuously removed free water in a bed named by the authors as hydrogel discs on sieve trays (HDST) [36] (Figure 16.3). Hydrogels in the form of perforated discs were placed on polyamide 12 sieve trays produced by additive manufacturing. Efficiencies higher than 97.0% were observed varying the flow rate from 1 to 3 ml/min, obtaining turbidity-free biodiesel in the output stream. The system has demonstrated high potential for use as a guard bed [36]. Dissolved water was also removed from biodiesel in a packed bed containing poly(AAm-co-SA) and poly(AAm-co-AA) hydrogels, water content was reduced from

Water Content Removal from Liquid Fuels

TABLE 16.5
Application of Hydrogels for Removing Water from Polar Fuels in Batch Processes

Reference	Fuel	Water form	Hydrogel composition	Ratio hydrogel/fuel (mg/ml)	Contact time (h)	Initial water content (ppm)	Removal efficiency
[30]	Biodiesel	Dissolved	Poly(N-isopropylacrylamide) (pNIPAM)	5.0, 10.0 e 12.0	24	1800	47–58%
			Poly(NIPAM-co-vinyl propionate)				38–58%
			Poly(NIPAM-co-vinyl butylacrylate)				23–27%
			Poly(NIPAM-co-vinyl biphenyl)				36–58%
			Poly(NIPAM-co-vinyl laurate)				36–58%
			Poly(NIPAM-co-vinylpyridine)				25–34%
			Poly(NIPAM-co-acrylic acid)				42–58%
			Poly(NIPAM-co-vinyl laurate-coacrylic acid)				38–55%
[31]	Biodiesel	Dissolved	Polyacrylamide	10	4	2200	63.6%
	Diesel-biodiesel (B60)					~1270	58%
	Diesel-biodiesel (B40)					~750	57%
	Diesel-biodiesel (B20)					~550.0	64%
[32]	Biodiesel	Suspended and dissolved	Polyacrylamide	50	13	2100	63.3%
[33]	Biodiesel	Dissolved	Poly(acrylamide-co-sodium acrylate)	125	18.3	2100	39.5%
[28]	Biodiesel	Dissolved	Poly(acrylamide-co-acrylonitrile)	4	72	1219.0–1303.0	62–74%

(Continued)

TABLE 16.5 (CONTINUED)
Application of Hydrogels for Removing Water from Polar Fuels in Batch Processes

Reference	Fuel	Water form	Hydrogel composition	Ratio hydrogel/fuel (mg/ml)	Contact time (h)	Initial water content (ppm)	Removal efficiency
[26]	Biodiesel	Dissolved	Poly(acrylamide-co-sodium acrylate)	50	24	2583	53.4%
			Poly(acrylamide-co-acid acrylic)				35%
[29]	Biodiesel	Suspended and dissolved	Polyacrylamide	4	24	6977	82.1%
			Sodium polyacrylate				84.3%
			Polyacrylic acid				40.8%
[37]	Biodiesel	Dissolved	Polyacrylamide	9.3–500.0	48	1514	29–42%
			Poly(acrylamide-co-sodium acrylate)				42–70%
			Poly(acrylamide-co-acid acrylic)				33–52%
[46]	Biodiesel	Dissolved	Poly(acrylamide-co-acrylonitrile)	4	72	1333	62%–74%

TABLE 16.6
Application of Hydrogels for Removing Water from Polar Fuels in Continuous Processes

Reference	Fuel	Water form	Hydrogel composition	Flow rate (ml/min)	Packed-bed characteristics	Initial water content (ppm)	Removal efficiency
[34]	Biodiesel	Dissolved	Poly(acrylamide-co-sodium acrylate)	1	$D = 30$ mm; $L_{packed-bed} = 3$ mm; porosity $= 0.4$	1500	~54%
		Dissolved	Poly(acrylamide-co-acid acrylic)				~54%
[36]	Biodiesel	Dissolved	Poly(acrylamide-co-sodium acrylate)	1–3	Hydrogel discs (D = 26 mm with five holes of 2.0 mm) placed on sieve trays of polyamide 12 of D = 40.0 m with holes of diameter 2.0 mm.	2042	31.6%
		Free				50,000 and 100,000	98%
		Dissolved	Poly(acrylamide-co-acid acrylic)			2042	26%
		Free				50,000 and 100,000	98%

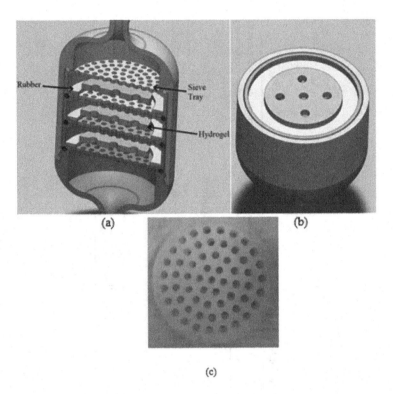

FIGURE 16.3
Bed scheme of hydrogel discs supported on sieve trays. Internal view (a), top view (b), and sieve tray (c). (Adapted with permission from [36], Copyright © [2021] AIDIC/CET Servizi S.r.l.)

1500 to around 750 ppm [34]. Notably, while US legislation accepts up to 500 ppm of water in biodiesel, measured through visual meniscus reading techniques, Brazilian legislation has much stricter criteria regarding the maximum admissible water content (250 ppm for producers and 350 ppm for distributors). A biodiesel quality control study conducted in Brazil observed water contents higher than 800 ppm along the biodiesel logistics: from the producer in the state of Mato Grosso (where water content was initially observed at 160 ppm) to the distributor located in the city of Manaus, in the state of Amazonas (a distance of 2514 km) [44]. The humid climate of some regions added to the hydrophilic character of biodiesel makes the maintenance of this content below 350 ppm a current challenge to be faced by distributors of biodiesel in Brazil.

16.3.3 Hydrogel Design for Water Removal from Liquid Fuels

The main characteristics required of hydrogels to remove water from fuels in industrial processes are (1) high water removal capacities and (2) good mechanical strength in their swollen state. The first feature is essential for higher water removal efficiency from hydrophilic fuels such as biodiesel.

Thus, considering the physical and chemical characteristics of fuels and their specific dewatering requirements in industrial processes, hydrogel attributes have been designed by tuning their chemical composition and structural network parameters, as well as combining them with process conditions, such as hydrogel mass/volume fuel ratio and temperature.

16.3.3.1 Ionic-charged Hydrogels for High Water Removal Capacities

Most published studies involving hydrogels for fuel dehydration applications are based on polyacrylamide uses in its pure form or copolymerized with other monomers. Polyacrylamide has stability characteristics, is inert, and presents viscoelastic and hydrophilic properties in addition to a competitive price, which makes it commercially relevant due to its versatility. However, as we have previously emphasized, hydrogels with higher removal capacity have been shown to be indispensable in removing water from biodiesel and its blends with diesel due to their greater affinities with water.

When a polymeric network is immersed in a fluid, the network begins to absorb water due to the thermodynamic compatibility of the polymeric chains and water. Opposing this osmotic force, there is an elastic retractive force, induced by the crosslinking of the network [45]. At equilibrium, there is no additional water absorption due to the balance between elastic and osmotic forces. In the case of ionic hydrogels, the ionization degree of the polymer chains and the ionic strength of the external solution should be considered. The water absorption phenomenon can be described in terms of the change in the chemical potential of water at constant temperature and pressure as:

$$\mu_1 - \mu_{1,0} = \Delta\mu_{mix} + \Delta\mu_{elastic} + \Delta\mu_{ionic} \tag{16.1}$$

In which, μ_1 is the chemical potential of the solvent in the polymeric gel, $\mu_{1,0}$ is the chemical potential of the pure solvent, and $\Delta\mu_{mix}$ and $\Delta\mu_{elastic}$ are the mixing and elastic contributions to the total change in chemical potential, respectively. $\Delta\mu_{ionic}$ is the contribution of the chemical potential change due to the ionic character of the hydrogel.

By inspection of the total chemical potential equation (Equation 16.1), if we are able to maintain or minimize changes to the structural crosslinking parameters of the network so that $\Delta\mu_{elastic}$ is not significantly affected, as well as maintain temperature and solvent type, we expect an increase in the total osmotic driving force due to the presence of dissociated ionic charges in the hydrogel structure, thus increasing its water removal capacity due to strong ion-dipole interaction. Thus, due to its improved water absorption characteristics and dominant share in the polyacrylamide market [47], anionic polyacrylamide, such as poly(AAm-co-SA), has shown to be a good candidate for hydrogel synthesis with improved water retention performance at a reasonable cost.

Studies that applied sodium acrylate hydrogels in their pure form (sodium polyacrylate hydrogels), or copolymerized with acrylamide (poly(AAm-co-SA) hydrogels), and compared them with other synthetic polymers have demonstrated that the efficiency of the process of removing water from fuels is in fact considerably increased, as predicted by thermodynamics. Under the same process conditions, Gonçalves et al., 2021 [37] showed that poly(AAm-co-SA) hydrogel was able to remove 45% more water from saturated biodiesel than neutral polyacrylamide hydrogel and 27% more than poly(AAm-co-AA) hydrogel. In the same study, water contents of about 860 ppm were observed in biodiesel (initially at 1,514 ppm) treated with poly(AAm-co-SA) hydrogel, applying a ratio of 8 mg hydrogel/mL of fuel; whereas, to achieve similar performances using neutral polyacrylamide and poly(AAm-co-AA) hydrogels, around 370 mg/ml and 304 mg/ml were needed, respectively.

In addition to increasing the water removal capacity, the poly(AAm-co-SA) hydrogel showed water removal kinetics about twice as fast as the poly(AAm-co-AA) hydrogel when immersed in distilled water for 4 h [26].

16.3.3.2 Hydrogel Mass/Fuel Volume Ratio Effect on Water Removal Efficiency

Gonçalves et al. (2021b) [37] demonstrated that by increasing the poly(AAm-co-SA) hydrogel/biodiesel ratio from 8 to 130 mg/ml, the water content in biodiesel is reduced from 860 ppm to 453 ppm at room temperature (~20°C), achieving a lower content than that required by ASTM for biodiesel. This result is comparable with the results achieved using more mature processes such as vacuum flash distillation [49]. However, the hydrogel-based process consumes less energy [37]. High mass/fuel ratios could be verified when a fluid moves through a packed-bed column. Polyacrylamide and poly(AAm-co-AA) hydrogels were also evaluated by the authors, and they were not able to reach the specification even at very high hydrogel/biodiesel ratios, showing once more the better performance of sodium acrylate-based hydrogels for removing water from biodiesel.

16.3.3.3 Temperature Effect on Water Removal Efficiency

Studies that applied hydrogels to remove dissolved water from biodiesel have shown that rising temperature results in greater removal kinetics and water holding capacity at equilibrium. An increase of about 20% in water removal capacity of poly(AAm-co-AA) and poly(AAm-co-SA) hydrogels at 40°C was obtained when compared with the same test performed at 25°C [48]. Efficiencies of up to 80% were achieved at the highest temperature, reaching a concentration of approximately 330 ppm, which is below the limits established by ASTM and ANP Resolution (for distributors). Similar behavior was observed using poly(AAm-co-AN) hydrogels, reaching a water content in biodiesel of 320 ppm at 40°C against 502 ppm at 25°C [28]. With the increase in temperature, the viscosity of the fluid is reduced, which provides greater wettability of the hydrogel by the fuel and, therefore, a better contact is established. Higher temperatures also contribute to the relaxation of polymer chains, making them softer and more flexible, increasing the elasticity of the hydrogel network structure. As a consequence of the superior expansion capacity of polymeric chains, more water can be held in their polymer network. All these factors allow higher incorporation of water into the hydrogel network at higher temperatures.

16.3.3.4 High Crosslinking Density as Means for Mechanically Stable Hydrogels

Synthesized hydrogels are generally dry, glassy in appearance, and very rigid. However, when the hydrogels are immersed in an aqueous medium and they swell, a loss of mechanical strength is observed as they become soft and flexible. Hydrogels with good mechanical strength are required for industrial purification, where they must withstand the pressure exerted by the column of liquid and the high shear rates resulting from high operating flow rates. Reducing the distance between crosslinking points, or equivalently increasing the density of the crosslinking, will increase the strength of the material, while simultaneously offering greater resistive force to the deformation of the chain for water retention. Consequently, highly crosslinked polymer networks typically show lower equilibrium water absorption capacities. Thus, although high water absorption capacities are necessary for good performance in fuel purification, these materials must still present compatible mechanical strength for this application.

Fregolente and Maciel [31] demonstrated that with increasing acrylamide monomer (AAm) content (2.0 to 8.0% by mass) combined with higher amounts of the crosslinking agent N,N'-methylenebisacrylamide (MBBAm) (1.0 to 4.0% by mass, in relation to the mass of acrylamide), there are more points of contact between the main chain and the crosslinking agent and, consequently, an increase in the crosslink density, resulting in a strong but fragile hydrogel – with high modulus of elasticity, albeit brittle – with a low water absorption capacity of W = 9.16 g water/g hydrogel. Hydrogels synthesized with the same ratio of monomer/crosslinking agent 2:1, with half the concentration of these species in the polymerization reaction (2.0% mass of AAm | 1.0% mass of MBBAm), did not exhibit sufficient handling resistance characteristics due to lower amounts of crosslinking points in their structures, despite having a water absorption capacity of 25.5 g water/g. Hydrogels with 4.0 wt% of Aam, combined with 3.25 wt% of MBBAm, showed sufficient water absorption capacity and crosslink density characteristics for its application, and they were able to reduce the biodiesel water content by 63.6%, from 2200 ppm to 800 ppm.

16.3.4 Reuse of Hydrogels for Removing Water from Fuels

The reusability of hydrogels is an important parameter for the economic viability of their application on an industrial scale due to their competitiveness against existing dewatering technologies. These polymeric gels have proven capable of maintaining the same performance of water removal from fuels after several reuse cycles, whether they are regenerated or not. Usually, the hydrogel regeneration process consists of immersing the material in a desiccant fluid such as ethanol followed by drying.

Li et al. [27] reused poly(N-isopropylacrylamide) (PNIPAm) and polycationic (N,N'-dimethylaminopropyl acrylamide) (PDMAPAA) hydrogels, without regeneration, to remove dissolved water from jet fuel for several cycles in batch runs. After each cycle, the hydrogels were removed from the fuel and immediately placed in another flask containing untreated fuel. The performance of these reused hydrogels remained stable, and they were able to reduce the water content from 45 ppm to ~7 ppm, even after seven cycles. Also in non-continuous tests, regenerated poly(AAm-co-SA) hydrogels applied to remove water from marine diesel oil (MDO) statistically maintained the same removal efficiency after four reuse cycles [13]. In continuous processes, similar performances were obtained when employing never-used and regenerated sodium polyacrylate hydrogels filled in a fixed bed to treat cloudy diesel [35]. The two hydrogels efficiently diminished the turbidity of diesel oil from 16.4 NTU at the inlet to <3 NTU at the outlet, appearing completely clean at the end of the columns as shown in Figure 16.2.

16.4 Conclusions

Synthetic hydrogels have been shown to be highly promising for the dehydration of liquid fuels. The selection of hydrogels for this application mainly depends on the chemical characteristics of the fuel, especially its polarity, and the required level of removal. Petroleum-based fuels such as diesel have a lower affinity for water and it can be efficiently removed by hydrogels in both batch and continuous processes to a level below that required by international standard specifications. For the purification of hydrophilic fuels, such as biodiesel and its diesel blends, ionic-charged hydrogels have been shown to be essential to increase the osmotic strength of the system and provide greater water removal. Hydrogels with ionic charges combined with moderate ratios of hydrogel mass/fuel volume and temperature conditions provide a route to obtain water content in biodiesel within the levels set by the regulatory agencies. New hydrogel technology to remove water from fuels has several advantages when compared with other commercially available technologies. They offer a complete solution since they remove free and dissolved water, are simple to operate, and can be reused. Moreover, hydrogel technology has a range of applications: industrial processes, distribution centers, gas stations, and in vehicle filters.

References

1. A.E. Atabani, A.S. Silitonga, I.A. Badruddin, T.M.I. Mahlia, H.H. Masjuki and S. Mekhilef, A comprehensive review on biodiesel as an alternative energy resource and its characteristics, *Renew. Sustain. Energy Rev.* 16 (2012), pp. 2070–2093.
2. Petrobras Technical Assistance Database, Manual Técnico: Óleo Diesel: Informações Técnicas (Versão Mar/2021), *Petróleo Brasileiro S.A.*, (2021), pp. 1–26.
3. B.B. He, J.C. Thompson, D.W. Routt and J.H. Van Gerpen, Moisture absorption in biodiesel and its petro-diesel blends, *Appl. Eng. Agric.* 23 (2007), pp. 71–76.
4. P.B.L. Fregolente, L.V. Fregolente and M.R. Wolf Maciel, Water content in biodiesel, diesel, and biodiesel-diesel blends, *J. Chem. Eng. Data* 57 (2012), pp. 1817–1821.
5. V.B. Veljković, I.B. Banković-Ilić and O.S. Stamenković, Purification of crude biodiesel obtained by heterogeneously-catalyzed transesterification, *Renew. Sustain. Energy Rev.* 49 (2015), pp. 500–516.
6. I.J. Stojković, O.S. Stamenković, D.S. Povrenović and V.B. Veljković, Purification technologies for crude biodiesel obtained by alkali-catalyzed transesterification, *Renew. Sustain. Energy Rev.* 32 (2014), pp. 1–15.
7. M. Latarche, Oil fuels chemistry and treatment, in *Pounder's Marine Diesel Engines and Gas Turbines*, M. Latarche, ed., Elsevier Ltd., Oxford, 2021, pp. 117–149.
8. *Monthly Energy Review database, October 2022 Monthly Energy Review (Report No DOE/EIA-0035(2022/10))*, Washington, DC 20585, 2022.
9. E. Shi and S. Hanson, EIA projects U.S. biofuel production to slowly increase through 2050. Available at https://www.eia.gov/todayinenergy/detail.php?id =43096 (accessed October 10, 2022).
10. *Brazilian Energy Research Company database, Análise de Conjuntura dos Biocombustíveis 2019 (Relatório EPE-DPG-SDB-Bios-NT-01-2020-r0)*, Brasília, DF, Brazil, 2020.
11. *Brazilian Energy Research Company database, Combustíveis renováveis para uso em motores do ciclo Diesel (Nota Técnica DPG-SDB No. 01/2020)*, Rio de Janeiro, RJ, Brazil, 2020.
12. S. Baena-Zambrana, S.L. Repetto, C.P. Lawson and J.K.W. Lam, Behaviour of water in jet fuel - A literature review, *Prog. Aerosp. Sci.* 60 (2013), pp. 35–44.
13. I.D. Perez, F.B. dos Santos, N.T. Miranda, M.G.A. Vieira and L.V. Fregolente, Polymer hydrogel for water removal from naphthenic insulating oil and marine diesel, *Fuel* 324 (2022), pp. 1–13.
14. H. Bateni, A. Saraeian and C. Able, A comprehensive review on biodiesel purification and upgrading, *Biofuel Res. J.* 4 (2017), pp. 668–690.

15. M. Balat and H. Balat, Progress in biodiesel processing, *Appl. Energy* 87 (2010), pp. 1815–1835.
16. I.M. Atadashi, M.K. Aroua, A.R.A. Aziz and N.M.N. Sulaiman, Refining technologies for the purification of crude biodiesel, *Appl. Energy* 88 (2011), pp. 4239–4251.
17. MarineProHelp Company, Which fuel quality parameters are most important?. Available at https://marineprohelp.com/questions/8323/which-fuel-quality-parameters-are-most-important (accessed October 24, 2022).
18. ASTM D975, *Standard Specification for Diesel Fuel* (2003), pp. 1–29.
19. S. Khurana, J.M. Torres and P.L. Ferguson, Alternatives to conventional gravity separation methods, *Offshore Mag.* 12 (1999), pp. 1–33.
20. M. Anez-Lingerfelt, Centrifugal vs. coalescing separation technologies. Available at https://biodieselmagazine.com/articles/7795/centrifugal-vs-coalescing-separation-technologies (accessed October 24, 2022).
21. M. Utriainen, F. Application, P. Data and E. Oy, (12) *Patent Application Publication (10) Pub . No .: US 2009 / 0312954 A1 (D Patent Application Publication*, 1 (2009), pp. 1–6.
22. J. Barker, S. Cook and P. Richards., Sodium Contamination of Diesel Fuel, its Interaction with Fuel Additives and the Resultant Effects on Filter Plugging and Injector Fouling, *SAE Int. J. Fuels Lubr.* 6 (2013), pp. 826–838.
23. S. Schwab, J.J. Bennett, S.S. Dell, J. Galante-Fox, A.M. Kulinowski and K.T. Miller, Internal Injector Deposits in High-Pressure Common Rail Diesel Engines, *SAE Int. J. Fuels Lubr.* 3 (2010), pp. 865–878.
24. A. Morynets, Methods for removing water and solid impurities from diesel fuel. Available at https://globecore.com/air-drying/methods-for-removing-water-and-solid-impurities-from-diesel-fuel/ (accessed October 23, 2022).
25. A. Alliance, Vacuum dryer advantages and disadvantages and vacuum thermal drying method. Available at https://www.antsalliance.com/news/Vacuum-dryer-advantages-and-disadvantages-and-vacuum-thermal-drying-method.html (accessed October 23, 2022).
26. H.L. Gonçalves, P. Bogalhos Lucente Fregolente, M.R. Wolf Maciel and L.V. Fregolente, Formulation of hydrogels for water removal from diesel and biodiesel, *Sep. Sci. Technol.* 56 (2021), pp. 374–388.
27. H. Li, H. Yang, W. Lv, X. Liu, S. Bai, L. Li et al., Hygroscopic Hydrogels for Removal of Trace Water from Liquid Fuels, *Ind. Eng. Chem. Res.* 60 (2021), pp. 17065–17071.
28. F.B. Santos, I.D. Perez, G.T. Gomes, M.G.A. Vieira, L.V. Fregolente and M.R.W. Maclel, Study of the kinetics swelling of poly(acrylamide-co-acrylonitrile) hydrogel for removal of water content from biodiesel, *Chem. Eng. Trans.* 80 (2020), pp. 265–270.
29. M.D.L. Dourado, P.B.L. Fregolente, M.R.W. Maciel and L.V. Fregolente, Screening of hydrogels for water adsorption in biodiesel using crosslinked homopolymers, *Chem. Eng. Trans.* 86 (2021), pp. 1129–1134.
30. H. Nur, M.J. Snowden, V.J. Cornelius, J.C. Mitchell, P.J. Harvey and L.S. Benée, Colloidal microgel in removal of water from biodiesel, *Colloids Surfaces A Physicochem. Eng. Asp.* 335 (2009), pp. 133–137.
31. P.B.L. Fregolente and M.R. Wolf Maciel, Water absorbing material to removal water from biodiesel and diesel, *Procedia Eng.* 42 (2012), pp. 1983–1988.

32. P.B.L. Fregolente, M.R. Wolf Maciel and L.S. Oliveira, Removal of water content from biodiesel and diesel fuel using hydrogel adsorbents, *Brazilian J. Chem. Eng.* 32 (2015), pp. 895–901.
33. P.B.L. Fregolente, H.L. Gonçalves, M.R.W. Maciel and L.V. Fregolente, Swelling degree and diffusion parameters of Poly(Sodium acrylate-Co-Acrylamide) hydrogel for removal of water content from biodiesel, *Chem. Eng. Trans.* 65 (2018), pp. 445–450.
34. C.D. Paula, G.F. Ferreira, L.F.R. Pinto, P.B.L. Fregolente, R.M. Filho and L.V. Fregolente, Evaluation of different types of hydrogels for water removal from fuels, *Chem. Eng. Trans.* 74 (2019), pp. 889–894.
35. L.V. Fregolente, H.L. Gonçalves, P.B.L. Fregolente, M.R.W. Maciel and J.B.P. Soares, Sodium polyacrylate hydrogel fixed bed to treat Water-Contaminated cloudy diesel, *Fuel* 332 (2023), pp. 1–10.
36. H.L. Gonçalves, P.B.L. Fregolente, G.M. De Andrade, M.R.W. Maciel and L.V. Fregolente, Development of a hydrogel column for water removal from fuels, *Chem. Eng. Trans.* 86 (2021), pp. 1117–1122.
37. H.L. Gonçalves, P.B. Lucente Fregolente, L.V. Fregolente and J.P.B. Soares, Application of polyacrylamide-based hydrogels for water removal from biodiesel in isothermal experiments, *Fuel* 306 (2021), pp. 1–4.
38. J. Maitra and V.K. Shukla, Cross-linking in Hydrogels - A Review, *Am. J. Polym. Sci.* 4 (2014), pp. 25–31.
39. M. Kaplan and H. Kasgoz, Hydrogel nanocomposite sorbents for removal of basic dyes, *Polym. Bull.* 67 (2011), pp. 1153–1168.
40. P. Mohammadzadeh Pakdel and S.J. Peighambardoust, A review on acrylic based hydrogels and their applications in wastewater treatment, *J. Environ. Manage.* 217 (2018), pp. 123–143.
41. C. Zhou, Q. Wu, Y. Yue and Q. Zhang, Application of rod-shaped cellulose nanocrystals in polyacrylamide hydrogels, *J. Colloid Interface Sci.* 353 (2011), pp. 116–123.
42. W. Wang, R. Narain and H. Zeng, Chapter 10 - Hydrogels, in *Polymer Science and Nanotechnology - Fundamentals and Applications*, Elsevier Inc., 2020, pp. 203–244.
43. P. Palanisamy, P. Pavadai, S. Arunachalam, S.R.K. Pandian, V. Ravishankar, S. Govindaraj et al., Removal of water and their soluble materials from fuels using Moringa oleifera loaded keratin-co-sodium acrylate hydrogel, *J. Porous Mater.* 28 (2021), pp. 515–527.
44. E.P. De Oliveira, T.C.S. De Oliveira, E.L. Da Silva, E. de Q. Lima, M. dos S. Souza, F.P. De Macedo et al., Investigação do teor de água no Biodiesel utilizado na composição do Diesel B comercializado por uma distribuidora de combustíveis em Manaus/AM / Investigation of water content in Biodiesel used in the composition of Diesel B marketed by a fuel distributo, *Brazilian J. Dev.* 7 (2021), pp. 89663–89680.
45. N.A. Peppas, Y. Huang, M. Torres-Lugo, J.H. Ward and J. Zhang, Physicochemical foundations and structural design of hydrogels in medicine and biology, *Annu. Rev. Biomed. Eng.* 2 (2000), pp. 9–29.
46. F.B. Santos, I.D. Perez, L.V. Fregolente and M.R.W. Maciel, Application of poly(acrylamide-co-acrylonitrile) hydrogel to remove soluble water from biodiesel and evaluation in the control mechanism of the mass transfer process in an adsorption process, *Chem. Eng. Trans.* 92 (2022), pp. 487–492.

47. L.A. Jackson, Polyelectrolyte polymers: Types, forms, and function, in *Water-Formed Deposits: Fundamentals and Mitigation Strategies*, Z. Amjad and K.D. Demadis, eds., Elsevier Inc., Amsterdam, Netherlands, 2022, pp. 747–764.
48. F.B. Santos, I.D. Perez, P.B.L. Fregolente, M.G.A. Vieira, L.V. Fregolente and M.R. Wolf Maciel, Estudo da remoção de água em biodiesel comercial utilizando hidrogel de acrilamida (AAm) co-polimerizado com acrilato de sódio e ácido acrílico, in XXXIX Congresso Brasileiro de Sistemas Particulados, 1 (2019), pp. 1–7.
49. P. Felizardo, M.J. Neiva Correia, I. Raposo, J.F. Mendes, R. Berkemeier and J.M. Bordado, Production of biodiesel from waste frying oils, *Waste Manag.* 26 (2006), pp. 487–494.

17
Current Challenges and Perspectives of Hydrogels

Dimpee Sarmah, Ashok Bora, and Niranjan Karak

CONTENTS

17.1 Introduction 335
17.2 Emerging Applications of Hydrogels 337
    17.2.1 Energy Sector 338
        17.2.1.1 Bioelectronics 338
        17.2.1.2 Electrocatalysis 339
        17.2.1.3 Metal-Ion Batteries 340
        17.2.1.4 Fuel Cells 341
        17.2.1.5 Flexible Batteries 342
        17.2.1.6 Supercapacitors 343
    17.2.2 Agricultural Applications 344
    17.2.3 Biomedical Fields 345
    17.2.4 Pollution Control 346
    17.2.5 Sensors 347
    17.2.6 Miscellaneous Applications 348
17.3 Current Challenges and Prospects 349
17.4 Conclusion and Recommendations 350
References 351

17.1 Introduction

In this recent era, hydrogel research is evolving at a rapid rate due to its most relevant real-life applications across the globe. With tailorable physiochemical properties, hydrogel-based materials have been immersed as a multifunctional material platform for energy and related applications along with traditional biomedical, agricultural, and personal hygiene applications, pollution control sectors, and so on. To manage the growing worldwide requirement for energy, remarkable attempts have been adopted to obtain a clean and renewable source of energy [1, 2]. In this vein, electrochemical energy sources have become an attractive alternative with the capability of storing electrical energy without climate threat [3]. The desirable swelling behavior, mechanical strength, stimuli-responsiveness, ionic/electronic conductivities, and so on. are the key properties that allow hydrogels to be employed as a template of electrochemically active materials with their three-dimensional (3D) architecture [4]. The tunable synthesis of hydrogels can be achieved by varying the starting element such as monomer/polymer building blocks, various crosslinking agents, and other additives with multifunctional properties. Moreover, the highly abundant starting materials for the preparation of hydrogels allow the introduction of novel functionality in the existing hydrogel, which leads to the desired property to fabricate astonishing energy storage devices with additional functionalities [5]. Additionally, electronic, and ionic conductive properties, required mechanical toughness, flexible structure, electrolyte permeability, and so on are the key parameters to improve the long-term performance of various energy storage devices including supercapacitors (SCs), flexible batteries, and wearable devices. In addition to this, the 3D scaffolds of hydrogels possess large surface areas with huge numbers of active sites for electrocatalysis reactions in electrochemical energy conversion systems such as fuel cells and metal-air batteries. In addition, hydrogel-based energy storage devices are endowed with various smart properties, such as self-healing abilities, stretchability, and shape memory, to enrich the applicability of hydrogels in these fields. Along with these smart properties, ionic conductivity is another property that qualifies these materials for various electrochemical devices such as SCs, flexible batteries, and wearable devices [4].

Apart from these emerging applications in energy-related fields, the biocompatible nature and similarity with the biological tissues of hydrogels provide great potential for a variety of biomedical applications such as drug delivery, wound recovery, and tissue engineering [6]. Moreover, the sensing ability of a specific biochemical compound provides another opportunity to detect both infectious and non-infectious diseases. Hydrogel-based sensors also found substantial applicability in the detection of various human movements [7]. Hydrogels also occupy an important role in the agricultural sector and have huge applicability as a water-holding agent in soil and encapsulation of various agrochemicals with the ability to control their release [8]. The pollution control ability of hydrogels and hydrogel-derived materials is also very pronounced, and they can strongly bind both charged and uncharged species in their 3D networks. Thus, hydrogels have a remarkable role in wastewater treatment by absorbing toxic dyes, metal ions, pesticides, and so on [9]. The important applications of hydrogels are summarized in Figure 17.1. Although hydrogels receive a high status in all these applications, there are still drawbacks associated with them and great opportunities to meet unmet demands. The scope of this chapter covers the recent challenges that exist in current hydrogel research. The chapter also provides concluding remarks on recent advancements associated with each application along with our perception to overcome these challenges. Moreover, the future perspectives of these applications are also briefly highlighted.

Current Challenges and Perspectives 337

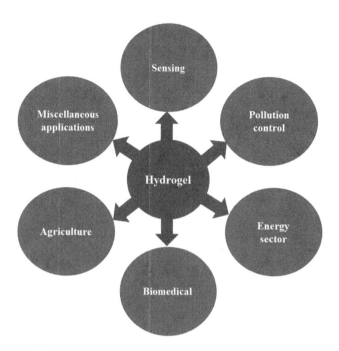

FIGURE 17.1
A few important applications of hydrogels.

17.2 Emerging Applications of Hydrogels

Hydrogels are 3D crosslinked, hydrated, elastic polymeric networks with interstitial spaces filled with water, which are several times higher than the dry weight, and thus the material remains "soft and wet". Along with water, the charged functional groups could effectively attract and retain electrolyte solutions within the framework. This provides liquid-like ionic conductivity within the network along with the retention of dimensional stability for the fabrication of flexible energy storage devices [5]. Moreover, the plentiful starting materials and the revolutionizing polymer technology endow with distinctive properties with tunable diverse design capabilities. Thus, hydrogels have found substantial use in various energy sectors such as SCs, fuel cells, metal-air batteries, electrocatalysis, CO_2 reduction, bioelectronics, and flexible/ wearable batteries. Figure 17.2 shows the diagrammatic representations of energy-related applications of the hydrogel.

However, along with energy-related applications, hydrogels occupy all sorts of utilization related to real-life applications including agriculture, biomedical, environment protection, personal hygiene care materials, and so on. Apart from these massive achievements, there is always scope for future modifications. In this chapter, the drawbacks in associated fields are shown to boost

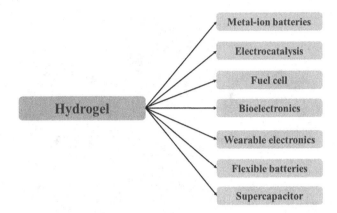

FIGURE 17.2
Various applications of hydrogels in the energy sector.

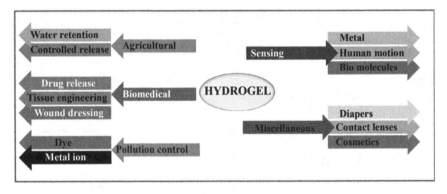

FIGURE 17.3
Flow chart of miscellaneous applications of hydrogels, excluding energy.

future research for rapid development in all these areas. Figure 17.3 shows the diagrammatic representation of various applications of hydrogel excluding energy sectors. The recent advancements in the production of multifaceted hydrogels reflect a new and booming achievement in materials science and technology. Although encouraging achievements have been obtained, there are still lots of drawbacks associated with them. The current challenges and prospects related to each application are discussed in this section.

17.2.1 Energy Sector

17.2.1.1 Bioelectronics

Bioelectronics is the interface between electronics engineering and biological systems. The biological tissues are soft and wet, while electronic systems are mostly hard and dry. To carry information, an electronic system uses electrons and/or holes. Similarly, in various biochemical transformations of our body, biological systems use various ions and molecules. Thus, a genuine connection between electronics and biological systems is essential. In this vein, hydrogels having a "soft and wet" porous structure with an ion-conducting nature is the ideal choice for the interface between biological systems and electronics [10]. Hydrogels with conductive properties are widely used in biomedical engineering and electronics along with other applications such as soft actuators and flexible wearable devices. Along with the regular ones, hydrogels with tissue adhesive properties have also wide applicability in wearable sensors and various electronic devices for real-time monitoring and repairing organs in the body [11].

Despite skyrocketing utilization of hydrogels, there are still many drawbacks and unmet demands. The first disadvantage of hydrogels for bioelectronics is the instability of their structures. Most of the hydrogels used for this application become hard and brittle after water loss. Thus, there is an urgent requirement to reduce the water loss rate during the performance. Moreover, weak strength, toughness, flexibility, and so on are other serious drawbacks that should be focused on in future studies. In addition to this, there are some other desired qualities for the utilization of hydrogels in this field, which include the sensitivity of the hydrogel system on both small and large scales, linearity of sense, and accuracy of a sensing device for long-term service. Moreover, various problems associated with the adhesive behavior also provide room for further research. For example, the reversibility of the adhesion mechanism, tunable adhesive property, and more emphasis should be given to improving the wet adhesion property of the hydrogel. Most importantly, regarding the utilization of hydrogels at the tissue/cell interfaces, more emphasis should be given to enhancing biodegradability and biocompatibility. Moreover, the incorporation of conductive hydrogels with biological tissues is one of the most challenging tasks because of the neuroinflammatory responses induced due to the chemical and mechanical mismatch that occurs between the inorganic bioelectronics and the biological tissues [12]. Thus, all these problems should be addressed and evaluated in further research to achieve more advanced materials.

17.2.1.2 Electrocatalysis

Electrocatalysis is one of the electricity-driven processes that lower the activation energy of the chemical process [13]. It is an alternative to fossil fuels and thus addresses a solution to global warming and energy scarcity. Hydrogels have received an increasing research interest in various electrocatalysis reactions, such as the hydrogen evolution reaction, oxygen reduction reaction, nitrogen reduction reaction, and oxygen evolution reaction, owing to their physicochemical and structural properties [13]. However, despite the continuous process and noteworthy success of hydrogel-derived electrocatalysts, many challenges should be overcome to achieve more efficient catalysts for electrochemical reactions. First, more attempts are needed to enhance the active sites of hydrogel-based catalysts to fully utilize the large surface present in the gel framework. One potential way is to provide numerous anchoring sites such as –COOH, –OH, and –NH$_2$ within the hydrogel network with optimized size and homogeneity to interact with the active species [14].

The electronic arrangement of active sites should be tuned by controlling the dopants to obtain an optimized intrinsic electrocatalyst. Heteroatom should be doped with the hydrogel nanocomposites containing metal oxides or carbon-based nanomaterials to result in structural defects. Although several research reports are available in this context, it is quite challenging to locate the position of the doped atom within the gel network. To overcome such drawbacks, theoretical studies should also be conducted to control the rational selection of dopants to obtain nanocomposite-based catalysts with increased intrinsic activity [14]. In addition to this, structural optimization of advanced electrocatalysts is of utmost importance, as the surface area and morphology have a great influence on catalytic activity. Optimization of the pore size with a larger surface area, and fast mass transfer are essential to acquire more active sites for electrons, reactants, and electrolytes to achieve enhanced utilization of the catalyst. Moreover, it is essential to broaden the applications of hydrogel-based electrocatalyst systems by including novel electrocatalysts in addition to photo-catalytic and other pronounced activities. It would be expected that the future research direction in hydrogel-based electrocatalysis systems would achieve a more fascinating platform with mechanical stability, required size and shape without a complicated electrode fabrication process, and minimal or no loss in electrocatalytic activity.

17.2.1.3 Metal-Ion Batteries

The replacement of expensive and flammable organic electrolytes with hydrogel electrolytes is an attractive field due to the reduction of the cost of battery fabrication. Li- and Na-ion batteries with high voltage and energy density have received noteworthy achievements recently. In addition to the high energy density with a longer lifespan, flexible energy storage devices are another attractive property for the fast development of Li- or Na-ion battery storage devices. Apart from these Li- or Na-ions, energy storage devices with multivalent cations including Al-ion, Zn-ion, and Mg-ion batteries have also gained attention. Despite the advancements of these metal-ions batteries, there are still numerous challenges for improvement [5].

The optimization of ionic conductivity for different ions is one of the most essential requirements for the fabrication of metal-ion batteries. The interaction between the functional groups present in the polymer chains and the electrolytic salts is the key factor for the variation of ionic conductivity in various metal-ion batteries. For example, poly(acrylic acid) can conduct H$^+$

ions, whereas it is not suitable for OH⁻ ions. Similarly, poly(vinyl alcohol) can also be used for H^+ ions widely, while without modification it cannot provide high conductivity with Zn^{2+} ions. Various factors such as valence state and the radius of ions influence polymer electrolyte interactions and thus the conductivity of the resulting device. Moreover, the degree of cross-linking and molecular weight of polymers, the side chain length of the polymer, and the solvent water, along with the ions present in the electrolytes, influence the interactions. Thus, to optimize all these factors, more advanced work with theoretical elucidation is needed to enhance the productivity of this field. Moreover, apart from Zn^{2+}, other multivalent ions such as Al^{3+} and Mg^{2+} should be investigated properly. Different hydrogel networks have different interactions with all these metal ions; hence their interactions should also be optimized accordingly [5].

The development of hydrogel electrolytes with various novel functionalities may enhance the adaptability of the system in numerous working environments. Although shape memory, self-healing, and highly stretchable hydrogel-based devices have been demonstrated, the repeatability is still limited several times. To integrate novel functionalities, the recent system requires further exploration and modification of polymer chemistry to render hydrogels with the desired properties.

Moreover, the development of a biodegradable battery manufacturing process is essential for the sustainable development of society. Although more reports are available on biopolymer-based devices, more research should be conducted in this sector. The development of novel hydrogel electrolytes that can suppress the electrochemically stable potential window is important to enhance the output voltage and energy density of devices. Thus, more research interest should be devoted to enriching this field for further development. Because of the limitations that exist in the two decomposition reactions of water, hydrogel electrolytes that can facilitate the improvement in power density and energy density are important. In recent times, the open circuit voltage of Zn-ion batteries is still much lower than the Li-ion batteries used with organic electrolytes. Thus, more attention is required to overcome these drawbacks [15, 16].

17.2.1.4 Fuel Cells

Fuel cells are electricity-producing devices where the chemical energy of fuels such as H_2, methanol, and ethanol, are converted into electricity, generally in the presence of the oxidizing agent. The oxidation process occurs at the anode and the reduction process occurs at the cathode separated by a cation exchange membrane [4]. In a fuel cell, the Nafion membrane is applied commercially with many pronounced properties, but the cost and loss of ionic conductivity after degradation are its main shortcomings. To overcome these challenges, Wang et al. investigated the ethanol fuel cell by using sodium polyacrylate hydrogel as an alkaline membrane and cathode by using a porous N, S-co-doped carbon catalyst. The polyacrylate hydrogel with super-absorbing attributes and water retention ability exhibits numerous extraordinary properties [17]. However, researchers have attempted to synthesize conductive, cheap materials to replace Nafion in fuel cells. Moreover, rational design and improved fabrication techniques should be implemented to achieve the required qualifications such as fuel permeability, high conductivity, swelling, and mechanical and thermal stability [18]. In addition, improved strategies can be utilized to achieve highly performed hydrogel-based fuel cells in combination with Nafion or a replacement to this material to improve properties.

17.2.1.5 Flexible Batteries

In today's world, flexible batteries have received huge research interest due to the devolvement of wearable intelligent devices with various smart properties such as self-healing, extreme deformations tolerance, anti-freezing and anti-dehydration abilities, environmental adaptability, and external stimuli responsiveness [19]. Remarkable efforts have been devoted to upgrading various mechanical properties of hydrogel electrolytes, and thus recently synthesized hydrogels can withstand various mechanical forces such as stretching, bending, and folding, as well as cutting without deformation. Despite these achievements, the mechanical strength of hydrogels is still far away from the desired range. Moreover, there is a trade-off between the mechanical property and ionic conductivity of hydrogels, and the mechanical properties should be retained even after multiple stretching or bending. Moreover, for most self-healable hydrogels, the long duration of healing restricts the applications. Thus, novel methodologies should be adopted to obtain ultrafast healing without any external stimuli. The bendable and stretchable electrodes well-matched with hydrogel-based electrolytes should be developed to achieve assembled batteries with the desired flexibility.

Although flexibility is the most important criterion for wearable electronic devices, there is not any accurate method for the precise measurement of it. Although bending and stretching tests are performed, more advanced experimental setups with accurate measurements of electrochemical performance with variation of bending behaviors, such as bending angles, radius, or times, should be established. The development of flexible energy storage devices with outstanding mechanical and electrochemical properties within a single system and assembled into the final stage is of utmost importance. However, at present, the newly developed energy storage devices are limited mostly to laboratory practice. In this context, highly flexible energy storage devices with higher power/energy densities with admirable cyclic stability are mostly desired [20].

Future technological evolution should be led in the direction of minimization of electronics with simultaneous increments in performance. Thus, there is an increasing demand to expand the scope of recent technologies to construct lightweight and thin flexible micro-batteries or SCs [21]. Most importantly, miniaturized electronic systems with diverse functionalities such as self-healing, biocompatibility with biodegradability, and optical transparency are highly recommended to overcome the current challenges of flexible energy storage systems. Moreover, 3D printing is the most promising way to fill the research gap for the development of flexible electronic devices. More emphasis should be given to simplifying the complicated fabrication process of hydrogel electrolytes and electrodes.

As the goal of any research output is commercialization, the performance of flexible energy storage devices should be investigated under extreme conditions. The extreme conditions include high humidity condition, high CO_2 concentration, high pressure, and high salinity in the case of underwater applications. The reaction mechanism between the hydrogel electrolyte and electrode should be thoroughly investigated. There may be both the possibilities of a reaction of polymeric chain with the metal-ion electrode such as Zn-electrode or the solvent-like DMSO, which can interact with the water molecules. Thus, more efforts are required to gain comprehensive knowledge about the electrode/electrolyte interface. The large-scale utilization of various electrochemical devices creates huge market demand and thus leads to problems for battery recycling and secondary use. Thus, from the environmental safety concern, biodegradable and recyclable hydrogels are required. Moreover, the electrodes used in the hydrogel-based batteries are inorganic substances, which is quite complicated in the separation process from the organic hydrogels during the recycling process. In the future, biomaterial-based electrodes should be developed to generate more eco-friendly batteries [20].

17.2.1.6 Supercapacitors

In the last few decades, hydrogel-based supercapacitors (SCs) and flexible wearable devices have received increased attention. The combination of hydrogel-matrix and conductive electrolyte introduces quite good capacitance and high conductivity into hydrogel-based electrolytes, which make them advantageous to be used as a component of SCs. Besides, due to the hydrophilic properties of the hydrogel, they can absorb water and accommodate abundant ions, which, together with higher mechanical properties, provide them with the potential to be used as electrodes in SCs [22]. Despite remarkable growth over decades, their high cost and weak mechanical and non-biocompatible properties restrain the extensive utilization of these materials.

In addition to this, conductive hydrogels carry water as a solvent, which automatically gets solidified at sub-zero temperatures and becomes dry under hot circumstances, which in turn restricts the operational temperature of hydrogel-based SCs. Therefore, the fabrication of hydrogels with improved mechanical, conductivity, and high-temperature tolerance properties are required to assist the development of flexible SCs. Further, hydrogel exhibits several excellent properties such as self-healing, stretchability, conductivity, and responsiveness. However, it is difficult for SCs to have these properties simultaneously. For instance, SCs with self-healing and stretchability face the issue of poor electrochemical performance. Therefore, to bring them into practical applications, it is important to improve self-healing and stretchable properties without compromising their electrochemical properties. Moreover, there is still a lot of effort needed to improve the overall energy storage parameters of hydrogel-based SCs to meet the electrochemical performance of conventional SCs. Although flexible SCs possess self-healing properties, their recoverability is relatively low due to which they face a long healing time as well as fewer self-healing cycles. Furthermore, hydrogel-based electrolytes still fail to possess ionic conductivities equivalent to that of liquid electrolytes and they are also not able to satisfy a longer cycle life for practical applications [4, 23].

Recently, wearable electronics have been developed rapidly to explore several applications such as health monitoring, skin sensor, and soft robotics. Hydrogel as a soft material has been extensively used for wearable electronics due to its flexibility, biocompatibility, and stretchability. Due to these properties, it is possible to fill the gap between hard artificial machinery and soft biological units. Despite that, the growth of hydrogel-based wearable electronics still faces many challenges including safety, limited stability of temperature, and hydrated and mechanical properties. Besides, the introduction of biodegradability into conducting hydrogels for stretchable devices is also difficult. Further, designing conducting hydrogel with numerous properties (like anti-freezing, self-adhesiveness, biocompatible, self-healing, etc.) is a challenge for the growth of hydrogel-based wearable electronics. The manufacturing of conducting hydrogel-based flexible wearable devices at a large scale and relatively low cost is a long way from the prerequisites of commercialization [24, 25].

17.2.2 Agricultural Applications

To overcome the water scarcity problem faced by the farmers of arid and semi-arid areas, hydrogels provide an alternative way as a water holding agent and soil conditioner. In addition to this, with the binding ability of various agrochemicals and fertilizers, hydrogels find an alternative way to control their release to reduce the cost and pollution caused by the leach-out materials [8]. Thus, several remarkable developments have been made to utilize hydrogels as a soil conditioner and a control release vehicle system in the last few decades. Although lots of hydrogels have been designed, the application part is confined to laboratory scale and only limited study was done on the actual agricultural field. Thus, more emphasis should be given to scaling up the product to ensure practical utilization in the real field. Scale-up is expected to face several challenges, including scale-up feasibility, consistency, and reproducibility. Therefore, future research should give more attention to overcoming these challenges [26].

Moreover, the unique physicochemical properties of hydrogels pave the way for soilless cultivation in the agricultural field. Due to the global population explosion every year around the world, soilless cultivation is an interesting method to meet the demand for food and other items. However, the high investment cost, the possibility of infectious disease in a closed system, and the need for high-skill labor are limitations of this process [27]. In this vein, 3D printing artificial soil proposes a solution to this problem. But a very small amount of literature exists in the current scenario for the design and development of 3D printable hydrogel for the agricultural sector. Current developments should be focused to develop a multifunctional hydrogel that can host plants and micro-organisms with the required pore size suitable for plant cultivation [27]. After overcoming the above-mentioned drawbacks, the utilization of hydrogel in the agricultural field will become more prominent but further improvements are expected through future research.

17.2.3 Biomedical Fields

The high-water retention capacity, similarities with biological tissues, biocompatibility, and so on make hydrogels suitable for diverse biomedical applications such as controlled drug delivery, wound dressing, and tissue engineering. Although several hydrogel-based materials are in clinical use, there is still scope for improvements and modifications to enrich their applications.

As an immerging biomaterial, hydrogels have drawn attention in controlled drug systems. The stimuli-responsive properties, biocompatibility, and non-toxicity with similarities with the biological tissues are the key parameters for the success of hydrogel as an efficient drug-carrying agent. However, there are some obstacles during the transition from the experimental to the clinical phase. For example, in most of the research studies, pharmacokinetics studies are conducted in vitro. But the release mechanisms in vivo remain ambiguous due to the presence of various therapeutic proteins, nucleic acids, and other components present in the human body. Moreover, the short-term cell viability test conducted in most of the research work could not give a guarantee that the material is non-toxic for long-term implementation.

Similarly, in most of the academic research in the tissue engineering field, small-duration investigations have been done using animal models on animal cells. However, it is essential that ultimate success is accessed by long-term studies on the real investigation of cartilage defects in humans. Thus, experimental design should be conducted including simultaneous pathologies, patient conditions, period of injury, and joint movements. Moreover, the development of multifunctional, smart biomaterials with stem cell/gene therapies could overcome the major challenges relying on the tissue engineering field. Further, ongoing research should provide particular attention to biocompatible polymers and suitable nanomaterials to understand the required advancement in the cartilage repair process [28]. Thus, extensive research should be conducted through the collaboration of scientists from various research fields such as medicine, chemistry, biology, engineering, and material science to devolve actual tissue regenerative processes in the market range to treat human cartilage defects [28].

However, in recent years, 3D printing hydrogels provide an improved platform for the utilization of hydrogels in tissue engineering applications. It is becoming a more powerful tool for the development of biocompatible natural polymer-based hydrogels. It can be expected that future research on 3D printing would provide the most important platform owing to its fine regulation of object structure [29]. Along with other biomedical applications, wound dressing using hydrogels has become the most competitive research area year by year. Although huge numbers of hydrogels with diverse properties including antibacterial, injectable, hemostatic, antioxidant, and sustained-release properties have been prepared for dressing purposes, still some unmet demands should be focused on. As wound repair is a complex process, several challenges such as providing on-demand functions on the dressing material are a future research direction. For example, it is vital to control inflammation during wound repair, but usually not needed in another phase of wound repair [30]. Second, as the pH of a chronic wound is alkaline, the development of alkaline pH-friendly dressing materials should be given more focus in the near future. Third, in the present research, mice or rats are usually used as models for investigation, but the dissimilarities with human skin may cause deviations throughout the experiment. Thus, a more suitable animal model should be explored in future results. The duration of the wound repair process is affected by numerous factors, hence more in-depth mechanism investigation is required in future research.

17.2.4 Pollution Control

Contamination of water with uncontrolled industrialization and urbanization and thus the scarcity of fresh water is one of the most threatening global concerns. Removal of harmful organic/inorganic contaminants from water resources is the most urgent demand and utmost importance has been given to obtaining pure water. In this regard, hydrogel-based adsorbents have considerable utilization in wastewater treatment with low toxicity, high penetration ability of charged or uncharged species during water absorption, and biodegradable nature. Moreover, the preparation of hydrogels with suitable properties such as highly porous 3D structure, large surface area, and multifunctionality with the binding ability for charged and uncharged groups is a potential way of overcoming water pollution caused by toxic pollutants [9]. However, there are some drawbacks associated with the current wastewater treatment process which should be further modified to obtain the best results. For example, although adsorption by using hydrogel is a recyclable process and the adsorbents could be used for several cycles, in most cases, the adsorption capacity decreased, and after some cycles, the adsorbents lost their active sites. Thus, more focused research should be conducted to overcome this problem. One possible solution is to utilize the adsorbents in further possible uses including catalysis and energy application for waste management. Moreover, highly biodegradable adsorbents should be synthesized without the inclusion of toxic chemicals to reduce the pollution caused by synthetic polymeric materials. Sometimes, the adsorbents are mechanically very weak and are not able to retain their integrity during performance, leading to secondary water pollution. One possible remedy for this problem is the generation of mechanically tough hydrogels with self-healing abilities both in a dry state and in a swelling state. Moreover, multipurpose adsorbents should be prepared to remove ionic, anionic, or neutral contaminates within a short time duration. Most importantly, industrial-scale research should be done in actual wastewater to examine the real efficiency of the prepared adsorbent.

17.2.5 Sensors

Hydrogel-based sensors have received remarkable attention over the past decade. Their chemical diversity and tunable physicochemical properties allow them to be applied as a sensing tool in various applications including energy, pollution control, biomedical field, humidity sensing, health monitoring, and soft robotics [7]. Although hydrogel-based biosensors are drawing huge attention due to similarities with human tissue, some vital modifications need to be obtained for a biocompatible with high accuracy, flexibility, and stability. Moreover, from a future perspective, much more effort is used to achieve multipurpose hydrogel-based sensors having sensing and actuation properties and optical/electrical responsiveness in a single system.

Selectivity and working range are two important parameters for sensing applications. The higher the selectivity, the higher the capturing power of the physiological movements of the body, while huge working capacity ensures prominent detection variety thus more human movements can be detected. Although hydrogels with different fillers have achieved high sensitivity, the equilibrium between sensitivity and other parameters should be investigated. Moreover, under harsh environments, hydrogel sensors are unable to work properly. Long-term stability and environmental stability are two important factors for wearable sensors. But, under dry conditions, hydrogels lose water and freeze under low temperatures. Although some strategies have been adopted, still more work should be conducted in this field. Thus, hydrogel-based sensors demand the integration of qualities such as higher conductivity and sensitivity, large working range, longer life span, mechanical stability, and so on to overcome the recent shortcomings present in this area [31].

To certify as an advanced sensing device, the device should have several characteristics such as high stretchability with superior biocompatibility, strong adhesiveness, anti-freezing property, fast self-healing, and self-recovery. However, most of the reported hydrogel-based sensors are not able to occupy all these features simultaneously, which limits their applications [32]. For the hydrogels used in metal sensing applications, more emphasis should be given to sensing in very minute concentrations with high accuracy. Moreover, the durability of the hydrogel for long-term use along with selectivity for a particular metal ion should be given more importance. Moreover, recyclability without deteriorating the performance of the hydrogel is another room for development in future research.

Hydrogels also have great potential in biomedical fields for sensing various molecules present in the human body. In this context, a hydrogel-based sensor is widely used for glucose sensing to detect diabetic patients [33]. The inherent properties such as flexibility, higher liquid content, biocompatibility, and biodegradability of hydrogel-based sensors offer accurate detection with improved comfortability, sensitivity, and speed. However, there must be some improved properties that should be included to enhance the performance of hydrogels in this field. For example, there is great potential to combine drug delivery systems such as sensing devices to improve the health and quality of life of patients. Moreover, much attention must be given to enhancing the service life and durability of hydrogel-based sensing devices with improved biodegradability.

17.2.6 Miscellaneous Applications

Apart from all the previously discussed emerging applications, hydrogels are also used for health hygiene products such as baby diapers, napkins, and sanitary pads. Generally, acrylic acid (AA)-based hydrogels are used for hygienic applications as superabsorbents due to their high-water absorption capacity. However, despite having excellent water absorption capacity, AA-based hydrogels raise some health issues such as skin burning, itching, and irritation, along with some serious environmental pollution due to their non-biodegradability nature [34]. Moreover, hydrogels are used for the preparation of several cosmetic formulations as they contain active cosmetic ingredients. Although bio-based materials are used as hydrogels in cosmetics applications, they may cause an allergic reaction, irritation, and inflammation in humans [35]. Further, hydrogels are used for ophthalmic drug delivery through contact lenses due to their biocompatibility and transparency [36]. However, high burst release of the drugs, low mechanical strength, and lower transparency are major challenges in hydrogel-based contact lenses [37]. Thus, future research should be conducted to overcome these drawbacks.

17.3 Current Challenges and Prospects

This chapter compiles recent challenges and drawbacks of hydrogel-based materials for various potential applications in numerous fields. In the ever-growing scientific achievements, there are always substantial scientific and technological challenges in conversions from laboratory-scale research to practical industrial applications. The main drawbacks of hydrogels are their low mechanical strength, non-biodegradability, time-consuming self-healing ability, the potential loss in activity on long-term performance, and so on. Considering these challenges, future directions for further research are highlighted and discussed briefly to achieve more insight. Hopefully, this chapter motivates current researchers to overcome these drawbacks and obtain more material in the future. The important drawbacks and prospects of hydrogels are tabulated in Table 17.1.

TABLE 17.1
Drawbacks and Future Prospective of Hydrogels in Various Applications

Application	Sub-class	Drawbacks	Future prospective
Energy	Bioelectronics	Instability of the hydrogel structure, weak strength, lower flexibility, incorporation of conductive hydrogel with biological tissue, etc.	Reduce the water loss rate, mechanically stable hydrogel, highly sensitive sensing device, etc.
	Electrocatalysis	Insufficient active sites	Enhance the active sites of gel-derived catalyst, improve structural optimization, pore size, and surface area
	Metal-ion batteries	Productivity is not up to the mark	Except Zn other cations should be investigated
	Fuel cell	Lower performance and high cost	Highly conductive with cheap technology
	Flexible batteries	Unstable mechanical properties after multiple stretching and laboratory-scale research	Ultrafast healing with mechanical stability, development of accurate flexibility method with practical field research
	SCs	Non-biocompatible, high cost, and weak mechanical properties	More advanced work to overcome present shortcomings

(Continued)

TABLE 17.1 (CONTINUED)
Drawbacks and Future Prospective of Hydrogels in Various Applications

Application	Sub-class	Drawbacks	Future prospective
Agriculture	Water held in soil, controlled release of agrochemicals	Confined with laboratory-scale research	Practical research, 3D printable hydrogels, etc.
Biomedical	Drug delivery	In vitro pharmacokinetics studies	Real-life research
	Tissue engineering	Short-term performance with a rat model	
	Wound dressing	On-demand function	
Pollution control	Metal ions and dye	Decrease in re-adsorption capacity, low mechanical strength, etc.	Enhanced mechanical strength with self-healing ability
Sensor	Metal sensing	Loss of recyclability after performance	A recyclable device with sensitivity in minute concentrations
	Biological molecules sensing	Only mono-purpose device	A multipurpose device with drug delivery along with a high-sensing ability
	Human motion sensing	Loss of activity under harsh conditions	High sensitivity with mechanical stability and a long lifespan
Miscellaneous applications	Contact lenses	Burst release, weak mechanical strength, and low transparency	More focus to overcome all these problems
	Diapers	Skin burning, itching, irritation, environmental pollution	More biocompatible material
	Cosmetics	Allergic action	Biocompatible hydrogels

17.4 Conclusion and Recommendations

In this chapter, we demonstrated the recent prospects and drawbacks of hydrogels in various fields from energy to biomedicine, agriculture, and other multidisciplinary areas. The numerous attractive properties with tunability according to the required application are the key factors for the success of these soft materials in every aspect of material science. However, a careful review of the literature reveals that despite the broad spectrum of applicability, there are still some challenges and scope for development in future studies. Moreover, in the ever-growing scientific world, a huge amount of research is conducted for the development of more advanced materials for the benefit of humankind. In this vein, the drawbacks of hydrogels can be overcome by more in-depth multidisciplinary research. However, in interdisciplinary research, the involvement of experts from different areas including chemistry, materials, physics, and biomedical engineering is necessary for the development of these materials. Thus, this chapter helps in improving hydrogel research by considering recent drawbacks.

References

1. N. Kittner, F. Lill, D. M. Kammen, Energy storage deployment and innovation for the clean energy transition, *Nat. Energy* 2 (2017) 1–6.
2. Q. Zhang, E. Uchaker, S. L. Candelaria, G. Cao, Nanomaterials for energy conversion and storage, *Chem. Soc. Rev.* 42 (2013) 3127–3171.
3. A. V. M. A. Manthiram, A. V. Murugan, A. Sarkar, T. Muraliganth, Nanostructured electrode materials for electrochemical energy storage and conversion, *Energy Environ. Sci.* 1 (2008) 621–638.
4. Y. Guo, J. Bae, Z. Fang, P. Li, F. Zhao, G. Yu, Hydrogels and hydrogel-derived materials for energy and water sustainability, *Chem. Rev.* 120 (2020) 7642–7707.
5. Z. Wang, H. Li, Z. Tang, Z. Liu, Z. Ruan, L. Ma, C. Zhi, Hydrogel electrolytes for flexible aqueous energy storage devices, *Adv. Funct. Mater.* 28 (2018) 1804560.
6. S. H. Aswathy, U. Narendrakumar, I. Manjubala, Commercial hydrogels for biomedical applications, *Heliyon*, 6 (2020) e03719.
7. X. Sun, S. Agate, K. S. Salem, L. Lucia, L. Pal, Hydrogel-based sensor networks: Compositions, properties, and applications-A review, *ACS Appl. Bio Mater.* 4 (2020) 140–162.
8. D. Sarmah, N. Karak, Biodegradable superabsorbent hydrogel for water holding in soil and controlled-release fertilizer, *J. Appl. Polym. Sci.* 137 (2020) 48495.
9. C. B. Godiya, L. A. M. Ruotolo, W. Cai, Functional biobased hydrogels for the removal of aqueous hazardous pollutants: Current status, challenges, and future perspectives, *J. Mater. Chem. A.* 8 (2020) 21585–21612.
10. H. Dechiraju, M. Jia, L. Luo, M. Rolandi, Ion-conducting hydrogels and their applications in bioelectronics, *Adv. Sust. Sys.* 6 (2020) 2100173.
11. S. Li, Y. Cong, J. Fu, Tissue adhesive hydrogel bioelectronics, *J. Mater. Chem. B.* 9 (2021) 4423–4443.
12. F. Fu, J. Wang, H. Zeng, J. Yu, Functional conductive hydrogels for bioelectronics, *ACS Mater. Lett.* 2 (2020) 1287–1301.
13. Y. Guo, Z. Fang, G. Yu, Multifunctional hydrogels for sustainable energy and environment, *Polym. Int.* 70 (2021) 1425–1432.
14. Z. Fang, P. Li, G. Yu, Gel electrocatalysts: An emerging material platform for electrochemical energy conversion, *Adv. Mater.* 32 (2020) 2003191.
15. H. Li, Z. Liu, G. Liang, Y. Huang, Y. Huang, M. Zhu, C. Zhi, Water proof and tailorable elastic rechargeable yarn zinc ion batteries by a cross-linked polyacrylamide electrolyte, *ACS Nano* 12 (2018) 3140–3148.
16. Q. Zhao, W. Huang, Z. Luo, L. Liu, Y. Lu, Y. Li, J. Chen, High-capacity aqueous zinc batteries using sustainable quinone electrodes, *Sci. Adv.* 4 (2018) eaao1761.
17. J. Wang, Z. Pei, J. Liu, M. Hu, Y. Feng, P. Wang, Y. Huang, A high-performance flexible direct ethanol fuel cell with drop-and-play function, *Nano Energy* 65 (2019) 104052.

18. Y. Sahai, J. Ma, V. Mittal, High Performance Polymer Hydrogel, In *Polymers for Energy Storage and Conversion*, 2013, Hoboken, John Wiley & Sons.
19. F. Mo, M. Cui, N. He, L. Chen, J. Fei, Z. Ma, Y. Huang, Recent progress and perspectives on advanced flexible Zn-based batteries with hydrogel electrolytes, *Mater. Res. Lett.* 10 (2022) 501–520.
20. S. Zhao, Y. Zuo, T. Liu, S. Zhai, Y. Dai, Z. Guo, M. Ni, Multi-functional hydrogels for flexible zinc-based batteries working under extreme conditions, *Adv. Energy Mater.* 11 (2021) 2101749.
21. W. Zhang, P. Feng, J. Chen, B. Zhao, Flexible energy storage systems based on electrically conductive hydrogels, *Prog. Polym. Sci.* 88(2019) 220–240.
22. X. Cao, C. Jiang, N. Sun, D. Tan, Q. Li, S. Bi, J. Song, Recent progress in multifunctional hydrogel-based supercapacitors, *J. Sci.: Adv. Mater. Dev.* 6 (2021) 338–350.
23. H. Dai, G. Zhang, D. Rawach, C. Fu, C. Wang, X. Liu, S. Sun, Polymer gel electrolytes for flexible supercapacitors: Recent progress, challenges, and perspectives, *Energy Storage Mater.* 34 (2021) 320–355.
24. B. Ying, X. Liu, Skin-like hydrogel devices for wearable sensing, soft robotics and beyond, *Iscience*, 24 (2021) 103174.
25. L. Wang, T. Xu, X. Zhang, Multifunctional conductive hydrogel-based flexible wearable sensors, *TrAC-Trends Anal. Chem.* 134 (2021) 116130.
26. R. A. Ramli, Slow release fertilizer hydrogels: A review, *Polym. Chem.* 10 (2019) 6073–6090.
27. L. M. Kalossaka, G. Sena, L. M. Barter, C. Myant, 3D printing hydrogels for the fabrication of soilless cultivation substrates, *Appl. Mater. Today* 24 (2021) 101088.
28. N. Eslahi, M. Abdorahim, A. Simchi, Smart polymeric hydrogels for cartilage tissue engineering: A review on the chemistry and biological functions, *Biomacromolecules* 17 (2016) 3441–3463.
29. Z. Bao, C. Xian, Q. Yuan, G. Liu, J. Wu, Natural polymer-based hydrogels with enhanced mechanical performances: Preparation, structure, and property, *Adv. Healthc. Mater.* 8 (2019) 1900670.
30. Y. Liang, J. He, B. Guo, Functional hydrogels as wound dressing to enhance wound healing, *ACS Nano* 15 (2021) 12687–12722.
31. X. Sun, F. Yao, J. Li, Nanocomposite hydrogel-based strain and pressure sensors: A review, *J. Mater. Chem. A.* 8 (2021) 18605–18623.
32. L. Tang, S. Wu, J. Qu, L. Gong, J. Tang, A review of conductive hydrogel used in flexible strain sensor, *Materials* 13 (2020) 3947.
33. J. Song, Y. Zhang, S. Y. Chan, Z. Du, Y. Yan, T. Wang, W. Huang, Hydrogel-based flexible materials for diabetes diagnosis, treatment, and management, *npj Flex. Electron.* 5 (2021) 1–17.
34. S. J. Hong, Y. R. Kwon, S. H. Lim, J. S. Kim, J. Choi, Y. W. Chang, D. H. Kim, Improved absorption performance of itaconic acid based superabsorbent hydrogel using vinyl sulfonic acid, *Polym.-Plast. Tech. Mater.* 60 (2021) 1166–1175.
35. S. Mitura, A. Sionkowska, A. Jaiswal, Biopolymers for hydrogels in cosmetics, *J. Mater. Sci. Mater. Med.* 31(2020) 1–14.
36. I. S. Kurniawansyah, S. R. Mita, K. Ravi, Drug delivery system by hydrogel soft contact lens materials: A review, *J. Pharm. Sci. Res.* 10 (2018) 254–256.
37. F. A. Maulvi, T. G. Soni, D. O. Shah, A review on therapeutic contact lenses for ocular drug delivery, *Drug Deliv.* 23 (2016) 3017–3026.

Index

3D Printing 73, 148, 149, 151, 152, 154, 156–160, 162, 164, 165, 343, 345, 346

AEM 36–39
Aerogels 60, 61, 170, 177, 179–184, 192
Aluminum-ion batteries (AIBs) 222, 223
Anion exchange membrane 21, 36, 37

Bioelectronics 121, 337–339

Calcium-Ion Batteries (Ca-ion batteries) 221
Cancer treatment 104, 115, 121, 160
Carbon nanofibers 28
Carbon nanotubes (CNT) 10, 28, 179–184, 205, 262, 278, 281, 296, 303
Catalysts 4, 11–13, 171, 173, 175, 184, 190, 191, 197–205, 208, 235, 237, 242, 245–247, 251, 339, 340
Chitosan 13, 21, 33, 60, 69, 86, 104, 107, 108, 110, 113, 114, 116, 117, 119, 138, 142, 154, 170, 176, 198, 205, 206, 208, 240, 241, 259, 264, 279, 282, 297
Carbon nanofiber (CNF) 28, 244, 295, 296
CO_2 reduction 170, 171, 173, 175–177, 179, 183, 184, 197, 203–205, 337

Devices 4, 6, 7, 13, 14, 20, 21, 23, 28, 31, 39, 40, 42–44, 48, 52, 116, 140, 162, 164, 190, 192, 194, 195, 213, 214, 217, 225, 236, 239, 242, 246, 247, 252, 257, 258, 263, 264, 266–269, 273, 274, 279, 280, 281, 283, 285, 287, 293–295, 298–303, 305, 306, 336, 337, 339–344, 348
Direct ink writing (DIW) 148, 151, 162

Drug delivery 48, 52, 83–87, 104, 108, 110, 111, 113, 114, 121, 138, 160, 164, 213, 345, 348

Electric double-layer capacitors (EDLC) 10, 25–29
Electrocatalyst 12, 184, 190, 191, 197–208, 237, 242, 245–247, 251, 330, 340
Electromagnetic shielding 304
Energy conversion 4, 11, 13, 237, 240, 336
Energy storage 6, 9, 10, 13, 14, 20, 25, 40, 43, 44, 48, 52, 118, 197, 202, 213, 214, 221, 223, 225, 227, 237–239, 243, 245, 247, 251, 252, 257, 258, 253, 264, 266–269, 273, 274, 279, 283, 302, 303, 336, 337, 340, 342, 343

Fused deposition method (FDM) 148, 149, 151, 157
Flexible 10, 12, 21, 23, 28, 39, 43, 48, 57, 104, 108, 118, 190, 192, 194, 195, 213, 214, 217, 218, 223, 225, 227, 228, 242, 246, 252, 257, 258, 259, 264–269, 274, 276–283, 285, 287, 293, 294, 298, 302, 303, 305, 306, 329, 336, 337, 339, 340, 342–344
Fuel cells 11, 21, 36, 121, 190, 197, 206, 208, 236, 237, 245, 246, 336, 341, 342

Gel 5, 9, 10, 12, 14, 20–24, 26, 32–34, 36, 38–40, 50–55, 58, 68, 74, 78, 84–89, 91–95, 97, 110, 130–132, 137, 152, 154–156, 175, 181, 183, 192–195, 198, 199, 201, 203–206, 208, 219, 222–224, 238, 275, 281, 327, 330, 340
Graphene 10, 26, 28–30, 49, 57–62, 71, 113, 118, 121, 179, 180, 183, 184, 198, 199, 202, 205, 214, 219, 223, 238, 240, 245, 246, 259, 261, 262, 283, 295–297, 300

353

Hydrogen evolution reaction (HER) 11,
  12, 200, 202, 203–206, 215, 246,
  247, 339
Hybrid supercapacitors 28
Hydrogels 4–10, 13, 20–30, 39, 42–62,
  67–78, 83, 84–90, 93, 95–97,
  104–121, 128, 130–133,
  135–143, 148, 152–165, 169–173,
  175–177, 179, 182–185, 191–205,
  208, 213–218, 225, 228, 237,
  240–247, 250–252, 258–265,
  267–269, 274–282, 284, 285,
  287, 294–296, 298, 299,
  300–306, 312, 216–318,
  321, 322, 326–330, 336–339,
  341–350

Implantable electronics 300

Li-ion batteries (LIBs) 4, 7–9, 31, 32, 34,
  36, 217, 236, 242, 246, 341
Lithium-sulfur batteries (LSBs) 34

Magnesium-ion batteries 224
Metal-air batteries 32, 33, 236–246,
  250–252, 336, 337
Metal-ion batteries 31, 213, 214, 217,
  221, 224, 227, 228, 235,
  265, 340
Metal-organic frameworks (MOFs)
  35, 36, 184, 191, 193, 198,
  237, 299

Na-ion batteries 263–265, 340

Oxygen evolution reaction (OER)
  11–13, 200, 201, 205–207,
  121, 215, 200, 220, 237,
  245–247, 339
Oxygen reduction reaction (ORR)
  11–13, 197, 198, 206, 207, 237,
  245–247, 339

Proton exchange membrane (PEM)
  36, 37
Photocatalyst 202, 205
Pollution control 335, 336, 346, 347

Poly(vinyl alcohol) PVA 12, 20, 37, 39, 54,
  56, 59, 69, 86, 107–109, 112–116,
  118, 138, 156, 169, 171, 204, 216,
  221–225, 238, 242, 243, 247, 262,
  264, 265, 277, 279, 280, 283, 297,
  299–301, 341
Polyaniline (PANI) 7, 10, 12, 27, 28, 35, 59,
  60, 117, 165, 191, 200, 204, 218,
  225, 238, 239, 242, 244, 245, 263,
  281–283, 295
Polymerization 5, 6, 10, 23, 28, 55, 56,
  59, 68, 70, 71, 75, 83, 85, 106,
  132–135, 156, 180, 193, 218, 223,
  225, 239, 244, 245, 247, 251, 279,
  282–284, 295, 302, 318, 329
Polypyrrole (PPy) 7, 9, 10, 27, 28, 57, 60,
  198, 221, 238–240, 244, 247, 263,
  297, 318
Polystyrene 12, 179, 295, 298
Potassium-ion batteries 220
Pseudocapacitors 25–27, 190

Redox-active 25–27, 196, 206
Reduced graphene oxide (RGO) 8, 29, 30,
  57, 60, 202, 205, 219, 283

Selective laser sintering (SLS) 148, 151
Self-healing 13, 21, 23, 39, 40, 43, 51, 106,
  192, 194, 208, 238, 240, 243, 265,
  268, 269, 274, 279, 280, 283, 285,
  287, 294, 299, 301, 305, 306, 336,
  341, 342–344, 347, 348
Sensors 48, 90, 116, 119, 160, 162, 184, 266,
  274, 278, 281, 295, 298–301, 339,
  347, 348
Stereolithography (SLA) 148, 149, 151
Sodium-ion batteries (SIBs) 8, 218, 219,
  264, 267
Starch 86, 117, 170, 246
Supercapacitors 7, 20, 25, 28, 29, 31, 43, 61,
  62, 118, 121, 214, 225, 238, 245,
  257, 273–287, 302, 336, 343

Tissue engineering 75, 84, 87, 104, 111,
  114, 121, 128, 131, 132, 137, 152,
  156, 157, 160, 213, 238, 275, 278,
  336, 345, 346

Index

Wastewater 119, 184, 208, 336, 346, 347
Wearable 11, 40, 194, 217, 223, 228, 236, 240, 242, 243, 246, 252, 258, 263–266, 268, 269, 274, 276, 278–282, 284, 285, 294–296, 298, 299, 302, 303, 305, 336, 337, 339, 342–344, 347

Wearable electronics 40, 240, 257, 268, 293–296, 298, 302, 303, 305, 344
Wound healing 84, 107, 110, 111, 114, 121

Zinc-ion batteries (ZIBs) 32, 215–217, 265, 304

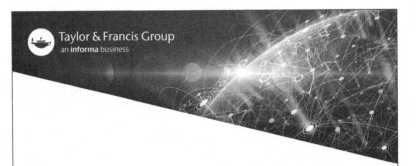

Printed in the United States
by Baker & Taylor Publisher Services